Medicinal Chemistry Case Study Workbook

Medicinal Chemistry Case Study Workbook

Bruce L. Currie, Ph.D.
Professor of Pharmaceutical Sciences, Chicago College of Pharmacy, Midwestern University, Downers Grove, Illinois

Victoria F. Roche, Ph.D.
Professor of Pharmaceutical Sciences, School of Pharmacy and Allied Health Professions, Creighton University, Omaha, Nebraska

S. William Zito, Ph.D.
Professor of Pharmaceutical Sciences, College of Pharmacy and Allied Health Professions, St. John's University, Jamaica, New York

A Lea & Febiger Book

Williams & Wilkins
A WAVERLY COMPANY

BALTIMORE • PHILADELPHIA • LONDON • PARIS • BANGKOK
BUENOS AIRES • HONG KONG • MUNICH • SYDNEY • TOKYO • WROCLAW

1996

Editor: Donna M. Balado
Managing Editor: Victoria Rybicki Vaughn
Production Coordinator: Danielle Santucci
Book Project Editor: Arlene C. Sheir-Allen
Designer: Nancy Hagan Abbott
Typesetter: Port City Press
Printer: Port City Press
Binder: Port City Press

351 West Camden Street
Baltimore, Maryland 21201-2436 USA

Rose Tree Corporate Center
1400 North Providence Road
Building II, Suite 5025
Media, Pennsylvania 19063-2043 USA

Accurate indications, adverse reactions, and dosage schedules for drugs are provided in this book, but it is possible they may change. The reader is urged to review the package information data of the manufacturers of the medications mentioned.

Printed in the United States of America

ISBN 0-683-02260-1

The Publishers have made every effort to trace the copyright holders for borrowed material. If they have inadvertently overlooked any, they will be pleased to make the necessary arrangements at the first opportunity

96 97 98 99
1 2 3 4 5 6 7 8 9 10

Reprints of chapters may be purchased from Williams & Wilkins in quantities of 100 or more. Call Isabella Wise in the Special Sales Department, (800) 358-3583.

Dedication

To our students and colleagues who challenge and stimulate us, and to our spouses for their patience, support, and encouragement.

Preface

The case study approach to instruction has recently received significant attention from academic pharmacy. There is now great enthusiasm for using cases to integrate important pharmaceutical and clinical science principles, and hone critical thinking and problem-solving skills. This Workbook was designed to assist medicinal chemistry instructors in illustrating the relevance of medicinal chemistry to the practice of pharmacy and the provision of pharmaceutical care. We trust it will become an invaluable educational tool for faculty and students alike.

The Workbook is organized into 10 sections that relate either to the organ system(s) affected (eg, Gastrointestinal Agents) or to the pharmacological action provided (eg, Antimicrobials) by the drugs discussed within the section. Two appendices have been provided to facilitate effective use of the Workbook. Appendix A is a drug list that provides a reference to each case where a particular pharmaceutical agent appears. All compounds discussed in the Workbook are listed alphabetically by generic name in Appendix A. Appendix B is a drug-interaction index that identifies only those cases where a chemical, pharmacological, or pharmacokinetic drug interaction is presented.

To enhance utility, this Workbook is presented in a perforated ("tear out") format. Students can use the Workbook for homework assignments and workshop exercises, as well as for self-study and in-class discussions. Most of the Workbook sections include one to three "starter" cases that are published with solutions. These sample cases are designed to help students develop the thinking processes necessary to solve medicinal chemistry case studies. The cases in the Instructor's volume are all solved. Some of the Workbook cases (supplemental) appear only in the Instructor's volume, and faculty may wish to include these cases on examinations.

We hope that all students and faculty find this Workbook to be an effective and enjoyable learning and teaching tool. We gratefully acknowledge all Workbook case contributors, as well as colleagues who reviewed the cases selected for publication for accuracy and relevance to contemporary pharmacy practice. Special thanks are due to two students for their help with various phases of the preparation of the final copy of the manuscript, Lisa Sikorski and Amy Elliott. We welcome comments and suggestions from all users of this Workbook.

Bruce L. Currie
Victoria F. Roche
S. William Zito

Case 1

Submitted by Victoria F. Roche

A dentist calls your pharmacy and asks you to prepare an antiseptic mouthwash for a patient with a painful bacterial infection of the oral cavity. You ask your second-year pharmacy intern to research the project, and she suggests combining bupivacaine hydrochloride (local anesthetic) with cetylpyridinium bromide (anti-infective).

1. Identify the acidic, basic, or neutral character of the two drug molecules in the case based on salt name.

Answer Bupivacaine hydrochloride is an ionized acid, whereas cetylpyridinium bromide is a neutral quaternary ammonium salt. All hydrochlorides are ionized acids and are made by reacting hydrochloric acid with an unionized basic amino group of a drug structure. The "ium" ending on cetylpyridinium, followed by an anionic halide ion, pegs this molecule as quaternary. Quaternary amines started as tertiary amines that were reacted with an alkyl halide. Because they give their lone pair of electrons to the alkyl halide when forming the salt, they are not basic. Because they have no hydrogens attached, they are not acidic. Lacking acidic or basic character, they are neutral.

2. Are these species ionized or unionized in salt form?

Answer As salts, both molecules are ionized and water soluble, so they would be appropriate for incorporation into an aqueous mouthwash preparation. The water solubility of the acid salt is dependent on the pH of the medium and would be most soluble at pHs at or below its pKa. The neutral quaternary salt is water soluble regardless of medium pH.

3. From an acid-base standpoint, is the mouthwash suggestion sound?

Answer From an acid-base standpoint, this suggestion is sound. Because the cetylpyridinium bromide has no basic character, it will not abstract the acidic proton from the cationic bupivacaine. If it had done so, the bupivacaine would have been left in unionized basic form. Because unionized organic molecules are, most commonly, water insoluble, precipitation would have resulted. As it is, the protonated bupivacaine is allowed to exist in the formulation and will remain water soluble.

Contributors

Robert L. Chapman, Ph.D.	Assistant Professor of Pharmaceutical Sciences, Chicago College of Pharmacy, Midwestern University, Downers Grove, Illinois
Mary L. Chavez, Pharm.D.	Associate Professor of Pharmacy Practice, Chicago College of Pharmacy, Midwestern University, Downers Grove, Illinois
Jeffrey J. Christoff, Ph.D.	Assistant Professor of Pharmaceutical Sciences, Chicago College of Pharmacy, Midwestern University, Downers Grove, Illinois
Bruce L. Currie, Ph.D.	Professor of Pharmaceutical Sciences, Chicago College of Pharmacy, Midwestern University, Downers Grove, Illinois
Marc W. Harrold, Ph.D.	Associate Professor of Pharmaceutical Sciences, School of Pharmacy, Duquesne University, Pittsburgh, Pennsylvania
Richard N. Herrier, Pharm.D.	Assistant Professor of Pharmacy Practice, College of Pharmacy, University of Arizona Health Sciences Center, Tucson, Arizona
Denise L. Howrie, Pharm.D.	Associate Professor of Pharmacy and Therapeutics, School of Pharmacy, University of Pittsburgh, Pittsburgh, Pennsylvania
Eric J. Lien, Ph.D.	Professor, Biomedicinal Chemistry & Pharmaceutics, School of Pharmacy, John Stauffer Pharmaceutical Sciences Center, University of Southern California, Los Angeles, California
Kimberly A. Linowiecki, Pharm.D.	Assistant Professor of Pharmacy Practice, Chicago College of Pharmacy, Midwestern University, Downers Grove, Illinois
Matthias C. Lu, Ph.D.	Associate Professor of Medicinal Chemistry, College of Pharmacy, University of Illinois at Chicago, Chicago, Illinois
Arnold R. Martin, Ph.D.	Professor of Medicinal Chemistry, College of Pharmacy, University of Arizona Health Sciences Center, Tucson, Arizona

Victoria F. Roche, Ph.D. Professor of Pharmaceutical Sciences, School of Pharmacy and Allied Health Professions, Creighton University, Omaha, Nebraska

Paul L. Schiff, Jr., Ph.D. Professor of Pharmaceutical Sciences, School of Pharmacy, University of Pittsburgh, Pittsburgh, Pennsylvania

Kathleen A. Snella, Pharm.D. Assistant Professor of Pharmacy Practice, Chicago College of Pharmacy, Midwestern University, Downers Grove, Illinois

George O. Udeani, Pharm.D. Clinical Assistant Professor of Pharmacy Practice, College of Pharmacy, University of Illinois at Chicago, Chicago, Illinois

Robert A. Wiley, Ph.D. Professor of Medicinal Chemistry, College of Pharmacy, University of Iowa, Iowa City, Iowa

Robin M. Zavod, Ph.D. Assistant Professor of Pharmaceutical Sciences, Chicago College of Pharmacy, Midwestern University, Downers Grove, Illinois

S. William Zito, Ph.D. Professor of Pharmaceutical Sciences, College of Pharmacy and Allied Health Professions, St. John's University, Jamaica, New York

Contents

Case 2

Submitted by Victoria F. Roche

BB, an elderly arthritic patient who has been taking aluminum aspirin (pKa = 3.5), calls you at your community pharmacy complaining of severe tinnitis. She had been having moderate arthritis pain, but is confused as to how many tablets she has taken. The patient's urinary pH is 4.5.

1. What is the acid-base character of aluminum aspirin?

Answer Aluminum aspirin is an ionized base. Aspirin, in unionized form, is acidic. Aluminum aspirin is made by reacting aspirin in free acid form with aluminum hydroxide. Aspirin gives up its acidic hydrogen to hydroxide ion, forming water and leaving behind the anionic conjugate species of aspirin and aluminum ion. Because the resulting drug species is anionic, it is capable of accepting a proton (i.e., it is basic). Salts are ionic crystals made up of multiple ions (most commonly two) with balancing charge. Aluminum serves as the counterion in this drug salt to keep the charges balanced. Basic salts can always be recognized by the metal ion in their name.

2. What is the ratio of ionized to unionized drug forms in this patient's urine?

Answer All functional groups that are ionized in basic form (which means unionized in acidic form) follow the same Henderson-Hasselbalch expression: $pH - pKa = \log [:B^-]/[H:B]$. For the aluminum aspirin in BB's urine, the ratio of ionized base to unionized acid conjugate would be 10:1. That means there is 10 times more ionized basic form than unionized acidic form in her urine. Ionized species are water soluble and excretable, whereas unionized conjugates are usually lipophilic and able to be reabsorbed across renal tubular membranes. In this situation, a 10:1 ratio of ionized to unionized forms is not sufficient to avoid toxicity (tinnitis is a sign of acetylsalicylic acid overdose), so excretion of the aspirin must be enhanced.

3. Would you advise BB to visit her neighborhood outpatient emergency clinic for administration of IV $NaHCO_3$ in D_5W, or send your pharmacy intern right over with NH_4Cl tablets?

Answer To enhance the excretion of a drug that is ionized in basic form, the pH of the urine must be increased. The base added to the media will take a proton from any molecules of unionized acid, causing a shift in the equilibrium to form more ionized, excretable conjugate base. Because $NaHCO_3$ basifies aqueous media by forming sodium hydroxide on dissociation, it should be able to speed up the elimination of the aspirin. Ammonium chloride would compound the problem by acidifying the urine and allowing more of the unionized, absorbable acidic conjugate to form.

Case 3

Submitted by Victoria F. Roche

A distraught mother calls the poison control center where you work. Her 6-year-old son just ate half a bottle of NH_4Cl tablets meant for the family cat. As a complication of morbid obesity, he is hypertensive (the son, not the cat), and has been on timolol maleate (Blockadren) for 6 months. A significant fraction of timolol is excreted unchanged in the urine, and an overdose can result in bradycardia and hypotension.

1. **From an acid-base standpoint, is this a potential medical emergency? Why, or why not?**

Case 4

Submitted by Eric J. Lien

A patient has frothy urine, plus other indications of a urinary tract infection. Culture has disclosed *E. coli*. The physician suspects bladder-intestinal fissure that has allowed intestinal bacteria to get into the bladder. To confirm the diagnosis, the physician wants to give an unabsorbable, nontoxic dye to the patient. This dye will appear in the urine if a fissure exists.

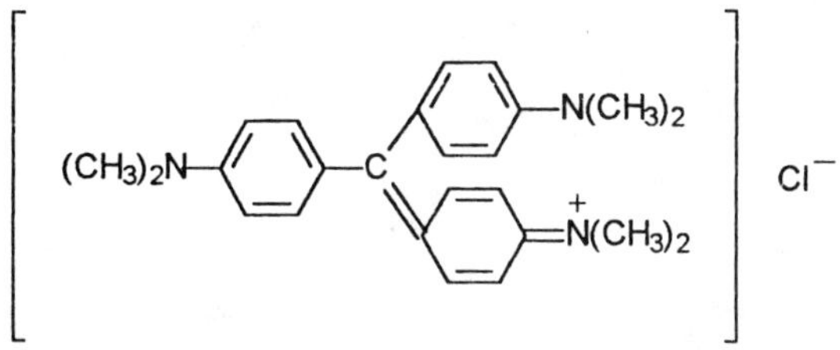

4.1

1. **The physician asks the pharmacist to recommend a dye; explain chemically how it would work.**

Case 5

Submitted by Victoria F. Roche

ZT, a 26-year-old single mother of four with a history of depression, is rushed unconscious to the emergency room of the hospital where you work. The attendant hands you an empty bottle that had contained 10 tablets of amobarbital sodium (pKa = 8.0). After stabilizing respiration and performing gastric lavage, you decide to enhance the renal elimination of the drug. ZT has a urinary pH of 6.0.

1. What is the acid-base character of amobarbital sodium?

2. What would be the ratio of ionized to unionized drug forms in this patient's urine?

3. Would you recommend the IV administration of NH_4Cl or $NaHCO_3$?

4. The ER nurse is preparing to add phentermine hydrochloride to the IV bag containing the pH modifier you have just selected. Would you stop him? Why, or why not?

Case 6

Submitted by Victoria F. Roche

You are refilling a prescription for fenoprofen calcium (an anti-inflammatory agent) in your community pharmacy for BP, an athletic patient with an inflamed knee who has been training for a local marathon run. He complains that, even though he has drastically reduced his running, he experiences mild pain between doses. Upon questioning the patient further, you find that he consumes approximately 1 gallon of a popular sports drink containing a high concentration of sodium citrate (which basifies the urine) per day. BP has a urinary pH of 7.5, and the pKa of this anti-inflammatory agent is 4.5.

1. What is the acid-base character of fenoprofen calcium?

2. What would be the ratio of ionized to unionized drug forms in BP's urine?

3. How can you explain the lack of adequate anti-inflammatory therapy?

4. **Would you counsel this patient to increase his sports drink intake (if that's possible), or eliminate this drink from his diet (or at least cut back)?**

Case 7

Submitted by Victoria F. Roche

A child is rushed to the hospital emergency room where you work as a clinical pharmacist; the child has swallowed 20 tablets of chlorprothixene hydrochloride, an antipsychotic agent (pKa = 8.4). As one therapeutic measure, you seek to enhance the elimination of the drug via the urine. The pH of the child's urine is 7.4.

1. **Is chlorprothixene an acid, a base, or a neutral species in hydrochloride salt form?**

2. **Is the hydrochloride salt ionized or unionized?**

3. **What is the ratio of ionized to unionized drug forms in this child's urine?**

4. **Which conjugate species must be promoted to enhance the renal excretion of this drug?**

5. **Would you suggest the administration of NH_4Cl or $NaHCO_3$?**

Case 8

Submitted by Denise L. Howrie and Paul L. Schiff, Jr.

LF, a 55-year-old male friend, has been battling his dependence upon cigarettes for many years. He has been a "pack a day" smoker for over 35 years and admits "It's bad for me." He has tried many methods to quit smoking, from "cold turkey" to counseling sessions. He has chronic bronchitis every winter and is now turning to you for help. He has recently changed physicians and has a new prescription for nicotine polacrilex gum (Nicorette—SmithKline Beecham). Because of the "horrible taste" of the gum, LF likes to rinse his mouth with various juices (particularly orange and apple) immediately after chewing. He complains that he really doesn't believe that the gum is working well, and is quite discouraged. Finally, LF comments that he has heard from other reformed smokers that on cessation of smoking there is "a change in the effect of other drugs that one is taking."

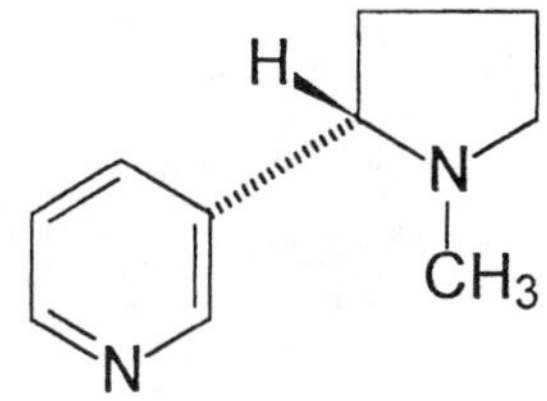

8.1

1. Can you offer any suggestions to LF as to how he might improve the "efficiency" of his nicotine chewing gum?

2. **What pertinent information should LF know about the effect of cessation of smoking on the action of other drugs?**

Case 9

Submitted by Eric J. Lien

A telephone call was transferred through our Clinical Pharmacy Department concerning the following potential chemical interaction that has not been reported in the literature: In a clinical study both ibuprofen (9.1) and penicillin VK (9.2) were used to treat endodontic pain. When either one of the two drugs was given, statistically significant relief of pain in patients was observed; when the two drugs were given together in capsules, no statistically different results from placebo were observed. The clinician asked our biomedicinal chemistry faculty to provide possible explanations.

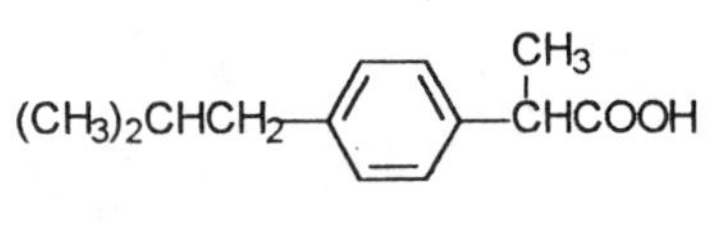

9.1

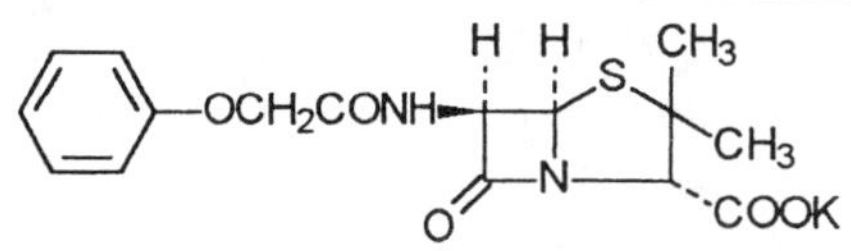

9.2

1. **Provide explanations why neither drug is effective when used in combination.**

2. **What recommendations would you make with regard to dosing schedule for these two drugs if used together in the same patient?**

Case **10**

Submitted by S. William Zito

Your company has developed a site-mutated analogue of erythropoietin, which has proved to be very effective in preliminary trials for the treatment of anemia in chronic renal failure patients, as well as in HIV-infected patients being treated with AZT. Before the analogue can be brought to market, it is discovered to quickly precipitate from solution upon standing.

1. **As the pharmaceutical chemist on the development team, you are asked to solve this problem and to make recommendations that would increase the stability of the product.**

Case 11

Submitted by Marc W. Harrold

It is 11 PM, and John has just begun a night shift at a local hospital. Before leaving his shift, Jeff tells John that patient RL in the ICU has been placed on a nitroprusside drip to control his erratic hypertension. Jeff further explains that since it had been a slow night, he prepared the nitroprusside solution that will be needed at midnight. He tells John that he just finished preparing the nitroprusside solution and has placed it in the IV room refrigerator. One hour later, ICU calls for the solution. John goes to the IV room, where he spots his technician discarding outdated IVs from the refrigerator. As he comes closer to the open refrigerator, he spots a clear glass IV bottle containing a dark blue nitroprusside solution, which is appropriately labeled for patient RL in the ICU. Disgusted, John mutters, "Thanks a lot Jeff," and starts to prepare a fresh nitroprusside (11.1) solution.

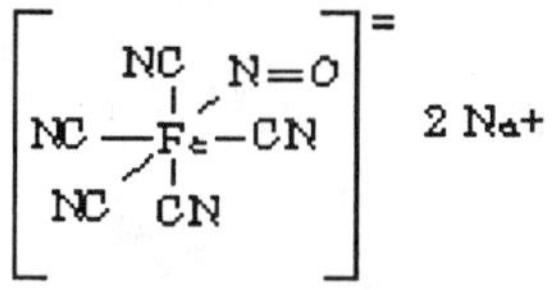

11.1

1. **Why did John prepare a fresh nitroprusside solution instead of sending the one Jeff had made?**

2. **What error(s) did Jeff make in his attempt to help John? What did Jeff do correctly?**

Case 12

Submitted by Victoria F. Roche

An oncologist calls you to consult on GS, a 69-year-old man with chronic inflammatory bowel disease who has been recently diagnosed with Stage III Hodgkin's disease. The physician wishes to initiate therapy with procarbazine hydrochloride (Matulane) (12.1), an orally active antineoplastic. The pKa of this anticancer agent is 6.8.

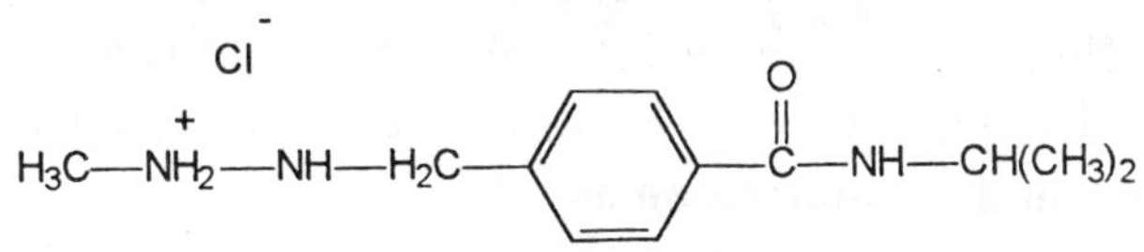

12.1

1. **Given that the pH values of gastric and intestinal fluid are 1.5 and 5.5, respectively, where would the majority of an orally administered dose of Matulane (12.1) be absorbed? Would the bowel disorder preclude po administration of this agent?**

2. The physician now plans to institute a modified MOPP multidrug regimen, with all agents to be given IV. This regimen will include mechlorethamine hydrochloride (Mustargen), vincristine sulfate (Oncovin), methylprednisolone sodium succinate (Solu-Medrol), and dacarbazine (DTIC). Could Mustargen be given in the same IV with Solu-Medrol? Why or why not?

3. UD, another Hodgkin's patient who is able to take Matulane (12.1) orally, requests that he be allowed to continue his nightly ritual of consuming a glass or two of red wine before going to bed. He often has a little fruit and/or cheese with his drink. What would be your recommendation?

Case 13

Submitted by Victoria F. Roche

Miconazole nitrate (Monistat) is an antifungal agent used topically to treat vaginal moniliasis. The pKa of this drug is 6.9. Two patients are currently in your pharmacy to pick up their prescriptions for this popular antifungal agent. One patient, SS, is 27 years old, and the other, PM, is 74 years old.

1. **What is the acid-base character of miconazole nitrate?**

2. **What is the relationship between menopausal status and vaginal pH?**

3. **What would be the ratio of ionized to unionized forms of miconazole expected in each patient?**

4. **Which patient would be more likely to experience systemic side effects after intravaginal application of Monistat?**

Case 14

Submitted by Bruce L. Currie

Bovine somatotropin or growth hormone for cows has been produced by genetic engineering (recombinant DNA technology) very successfully. When used regularly in dairy cows, this hormone can produce significant increases in milk production because of the lactotrophic (prolactin-like) action of the growth hormone. Bovine somatotropin is a single chain protein of about 20,000 MW (191 amino acids). The proposed use of this hormone to increase milk production has caused considerable controversy. One of the fears is that the growth hormone will contaminate the milk that is produced and would then be consumed by babies, children, and/or adults with potentially adverse effects.

1. **Discuss, based on the processes involved, your estimation of the relative risk to a person drinking milk produced from cows under the influence of bovine somatotropin. Discuss the scientific aspects, not the political aspects, of this case.**

Case **15**

Submitted by S. William Zito

You work for a small, entrepreneurial, biotechnology company that has been able to synthesize tissue plasminogen activator (tPA). The 527 amino acid glycoprotein was synthesized using the cDNA for natural human tPA obtained from a human melanoma cell line. The product is produced in *Escherichia coli* and as such is non-glycosylated. To increase activity, it is necessary to glycosylate the protein.

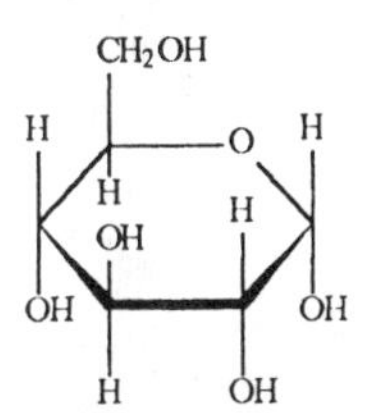

15.1

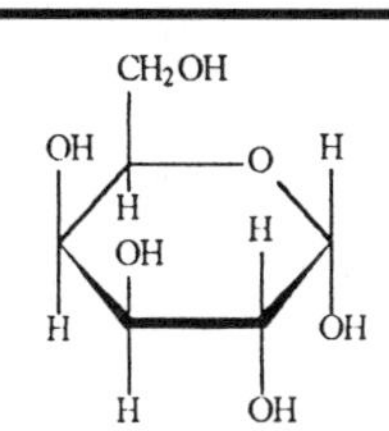

15.2

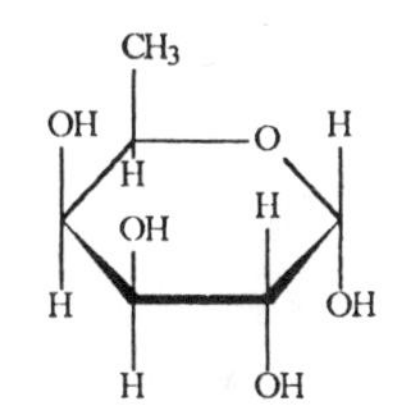

15.3

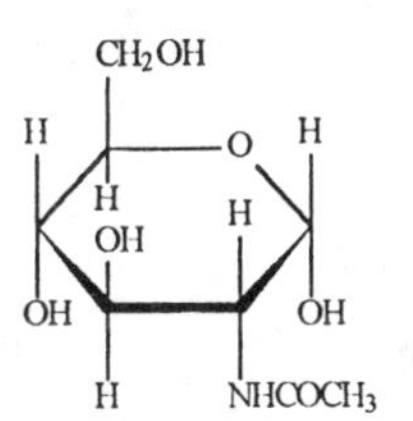

15.4

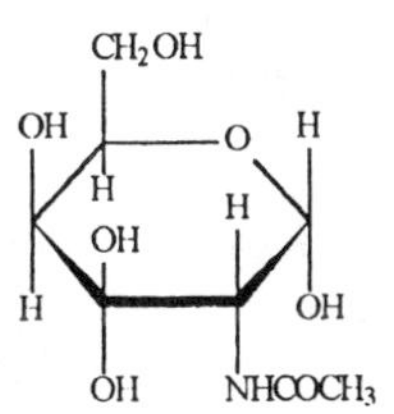

15.5

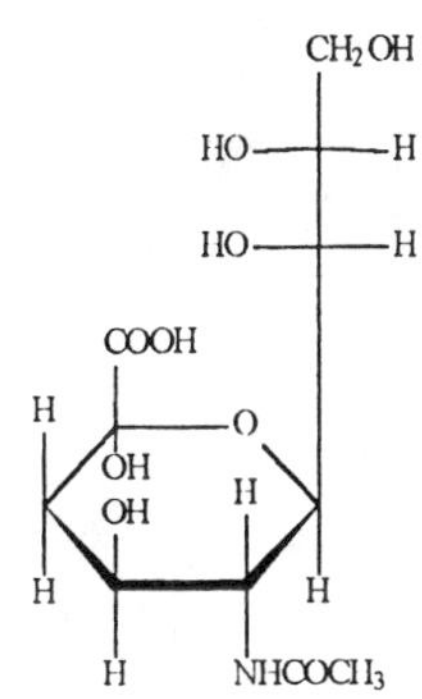

15.6

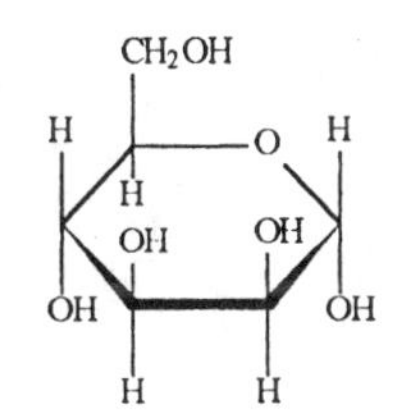

15.7

1. Discuss the importance of glycosylation.

2. **Make suggestions as to the most appropriate saccharide moieties you would use and what amino acid residues are necessary on tPA in order to carry out the chemical derivatization.**

Case 16

Submitted by Marc W. Harrold

KG was diagnosed 2 years ago with Raynaud's disease. He has no other medical problems; however, he is allergic to ampicillin (16.1), sulfisoxazole (16.2), and bee stings. Since the time of his diagnosis, he has been on several different medications, none of which have adequately decreased the number or severity of the vasospastic attacks. KG's primary physician has sent him to Dr. Brown, a specialist who has had some success in treating peripheral vascular disorders with investigational agents. Dr. Brown is currently involved in three different clinical trials for the treatment of Raynaud's disease. These trials involve pentoxifylline (16.3), an agent approved for intermittent claudication, daltroban (16.4), a thromboxane A_2 receptor antagonist, and ridogrel (16.5), a dual-acting thromboxane synthetase inhibitor/thromboxane A_2 receptor antagonist.

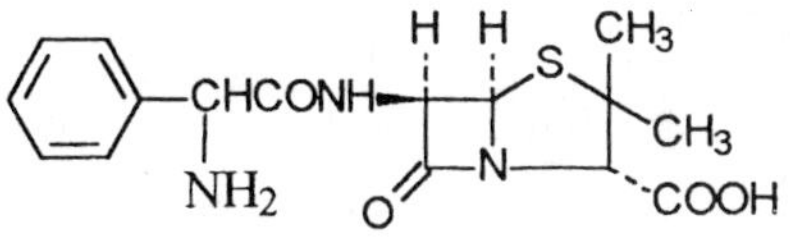

16.1

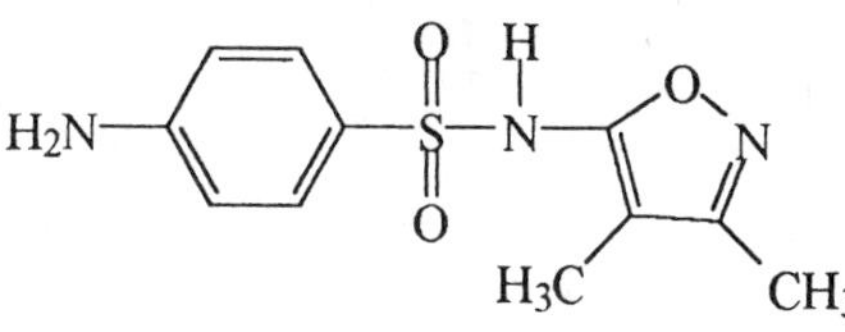

16.2

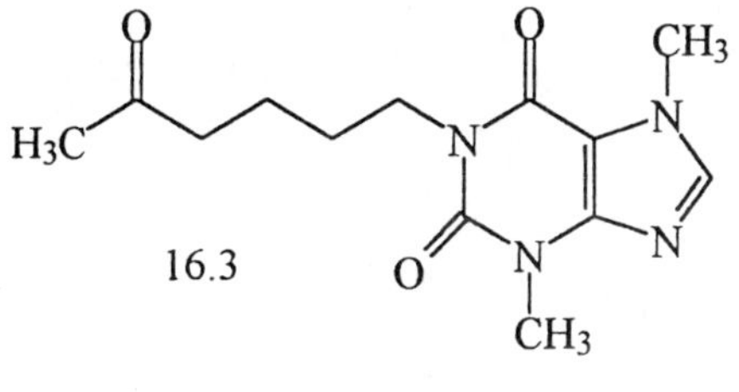

16.3

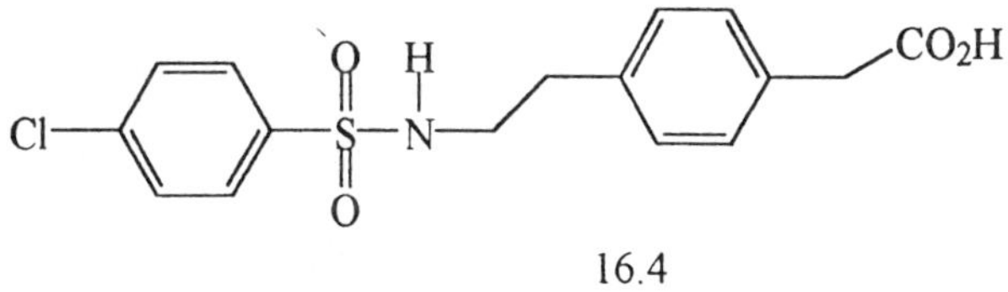

16.4

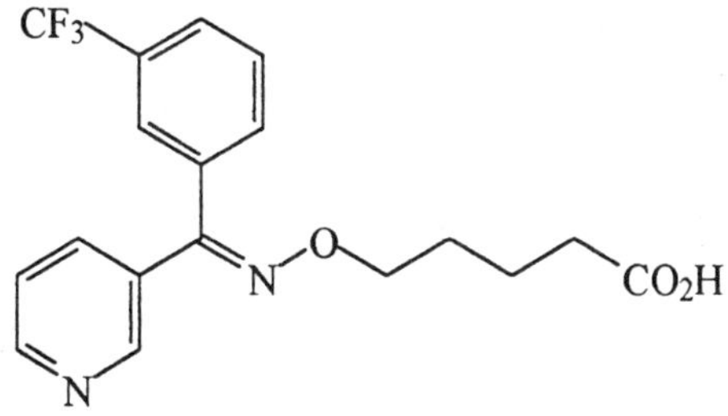

16.5

1. **Based on the information just given, which of the three clinical trials would be appropriate for KG?**

2. **To what chemical class does pentoxifylline (16.3) belong?**

Case 17

Submitted by Victoria F. Roche

You are an M.D./Ph.D. (in medicinal chemistry, of course!) practicing general medicine in the early 1930s. The medical community is all atwitter about a new anti-infective, sodium sulfanilamide, which has shown value in the treatment of urinary tract infections and infection secondary to burns. You are concerned, however, about the tendency of this agent to cause crystalluria, which can result in severe, life-threatening kidney damage. The pKa of sulfanilamide (17.1) is 10.4.

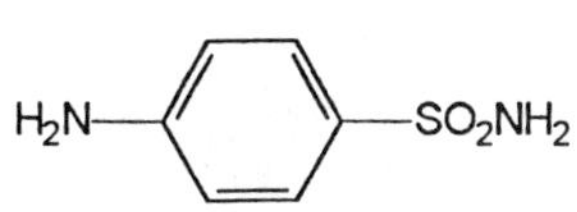

17.1

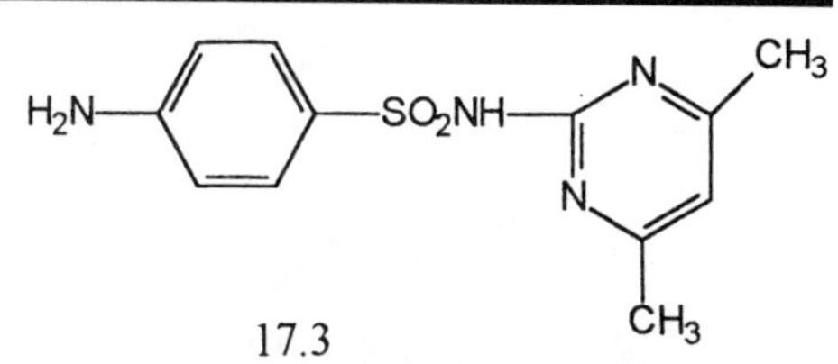

17.3

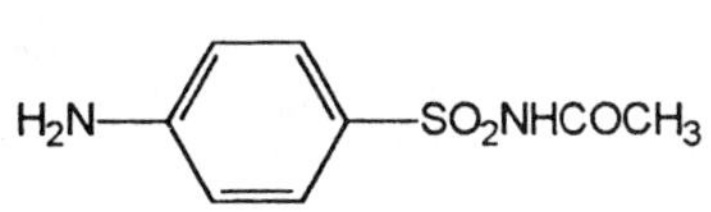

17.4

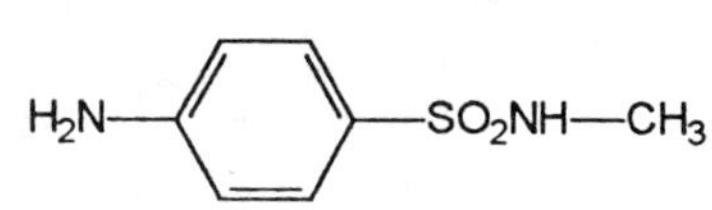

17.5

1. **Draw the structure of Sodium Sulfanilamide (17.2).**

2. **Why would a patient with a urinary pH of 5.4 be at greater risk for the development of crystalluria than one with a urinary pH of 6.4? What would be the ratio of ionized to unionized sulfanilamide drug forms at each urinary pH?**

3. **Which urinary pH modifier would you prescribe for a patient who began to show symptoms of crystalluria, NH_4Cl or $NaHCO_3$?**

4. **As a medicinal chemist, you understand the impact of structure on the therapeutic profile of drug molecules, and drug design visions are now dancing in your head. Which of the three modified sulfanilamide structures (17.3–17.5) would have a higher therapeutic index with regard to crystalluria? What is the structural basis for your answer?**

Case 18

Submitted by Victoria F. Roche

You have designed a new antinauseant for pharmacy students who must tolerate dry, boring lectures from medicinal chemistry faculty who have not yet bought into the exciting Med. Chem. case study approach to teaching. The structure of your new drug, gagamine (18.1), is provided. Note that it is an asymmetric molecule, but your synthesis is not stereospecific, so you have all possible stereoisomers in your flask.

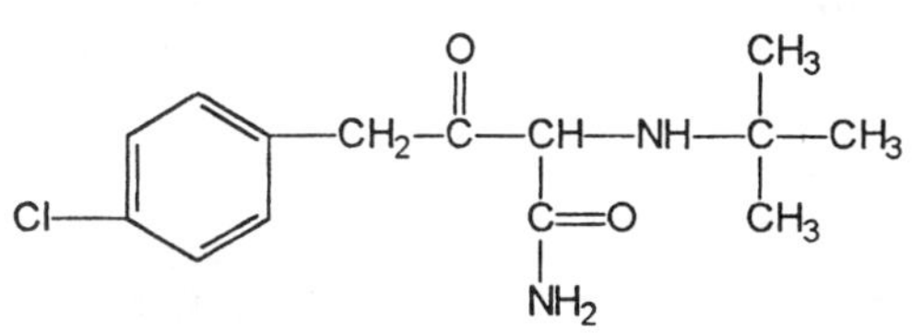

18.1

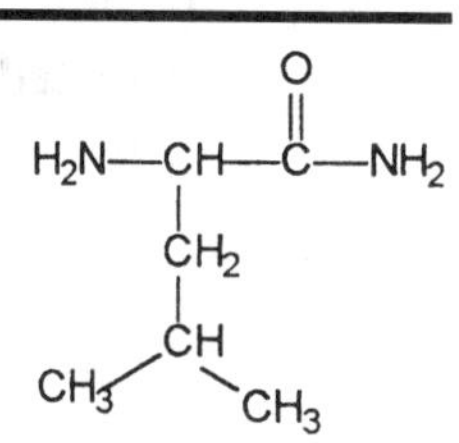

18.2

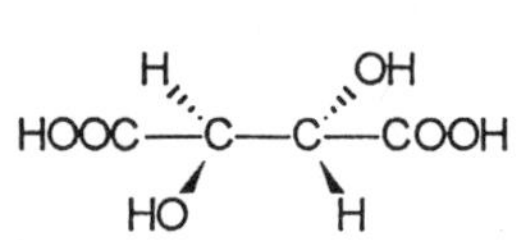

18.3

HCl

18.4

1. Circle all asymmetric centers found in gagamine (18.1). How many optical stereoisomers are possible? Could you separate these isomers directly via standard chemical methods (e.g., fractional distillation, recrystallization, or column chromatography)?

2. From pilot studies, you anticipate that the R epimer of gagamine (18.1) will be the active antinauseant isomer. Which of the compounds 18.2 to 18.4 would be of value in resolving a racemic mixture of gagamine (18.1) isomers?

3. Medicinal chemistry professors are known for their ability to drone on and on (ad nauseum...hence, the need for this new drug), so gagamine (18.1) must have a long duration of action. Discuss the ways you have creatively included structural features to retard the metabolic degradation of gagamine.

Case 19

Submitted by S. William Zito

You wish to clone the gene for human growth hormone (hGH) using the polymerase chain reaction (PCR) technique. HGH is a 191-residue polypeptide produced by the anterior pituitary. Its major biologic function is to promote overall somatic growth. Deficiency causes dwarfism, and excess causes acromegaly and giantism. The PCR technique requires that you synthesize an appropriate oligonucleotide primer. You know that oligonucleotide synthesis uses the condensation of a 3′-terminal deoxyribonucleotide with a blocked 3′-phospho deoxyribonucleotide to form a 3′-5′-phosphodiester–linked dinucleotide.

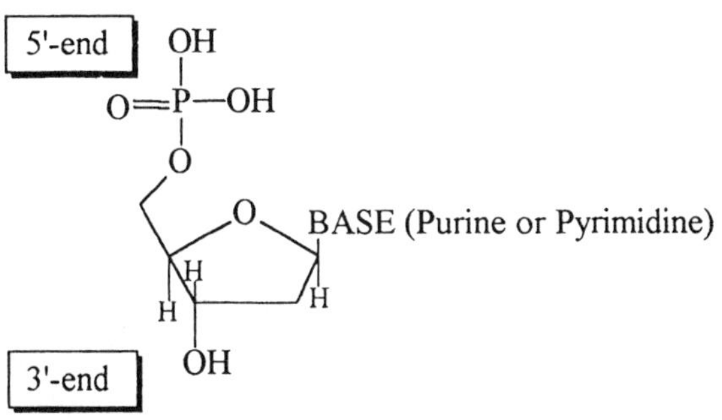

19.1

1. **Given the deoxyribonucleotide (19.1), explain how you could link it to itself so that you can add a third nucleotide? What blocking groups can be used for the 3′-terminal? What blocking groups can be used for the 5′-terminal?**

2. **What are the usual condensing reagents? How can you deblock the 5′-terminal?**

Case 20

Submitted by Denise L. Howrie and Paul L. Schiff, Jr.

Your hospital pharmacy is reviewing the use of three skeletal muscle blocking agents in order to determine the need for hospital formulary inclusion: atracurium besylate (20.1), pancuronium bromide (20.2), and pipecuronium bromide (20.3). As the hospital deals extensively with elderly patients, the anesthesiologists and also intensive care physicians have specifically requested atracurium for use in these and other patients with moderate-to-severe renal failure, particularly when multiple doses may be required, as in mechanical ventilation.

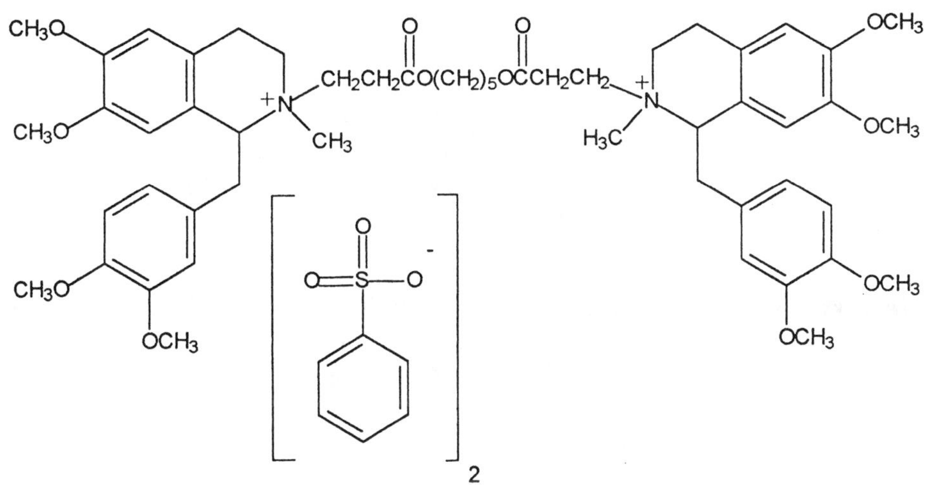

20.1

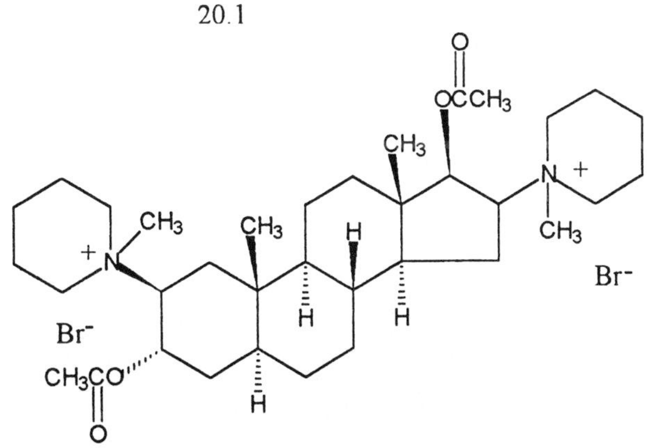

20.2

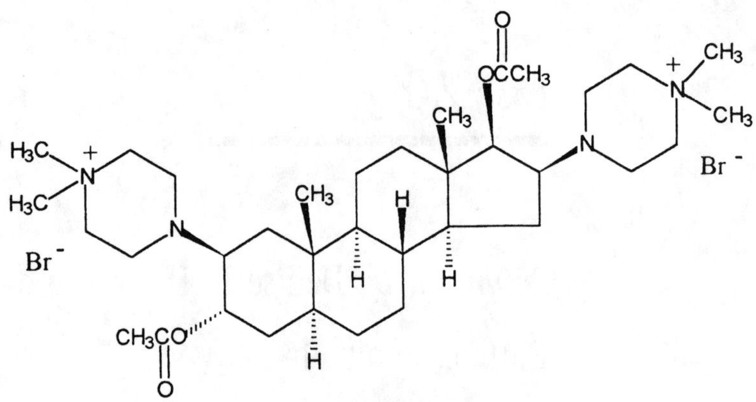

20.3

1. **Which one of the neuromuscular blocking agents (20.1–20.3) that are being considered for inclusion in your formulary would you recommend?**

2. **Discuss your reasons for this recommendation.**

Case 21

Submitted by Victoria F. Roche

AK, a 45-year-old woman, has been receiving neostigmine bromide (21.1) for treatment of myasthenia gravis for the past 5 years. She is now being admitted to the hospital for a respiratory ailment that requires intubation. The physician asks your advice on a suitable skeletal muscle relaxant to use to facilitate the intubation procedure.

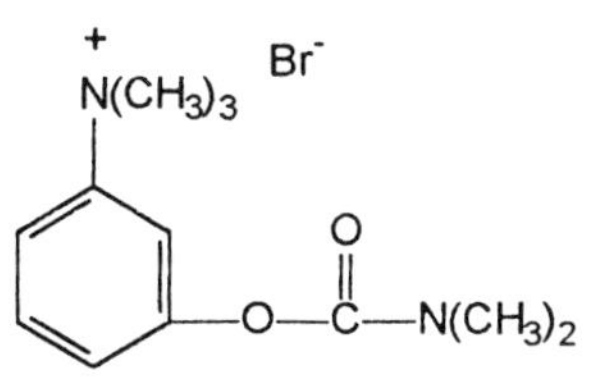

21.1

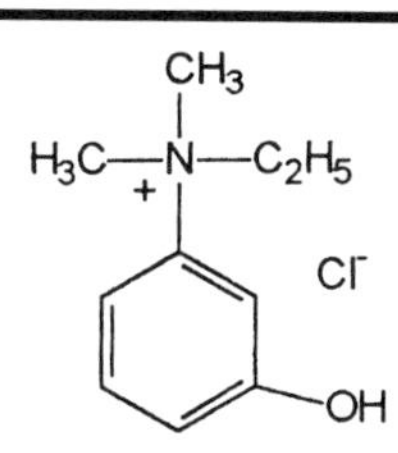

21.2

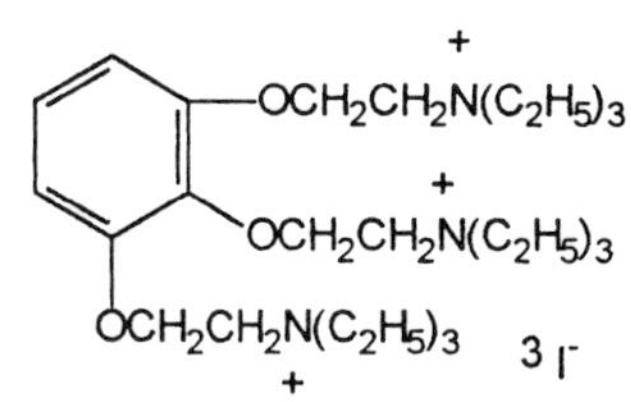

21.3

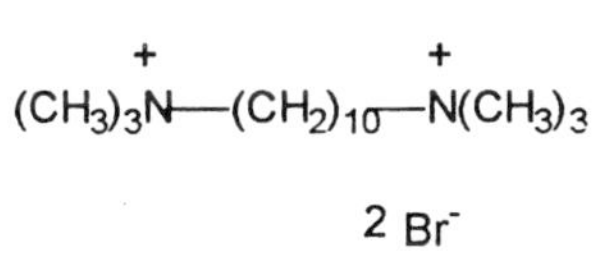

21.4

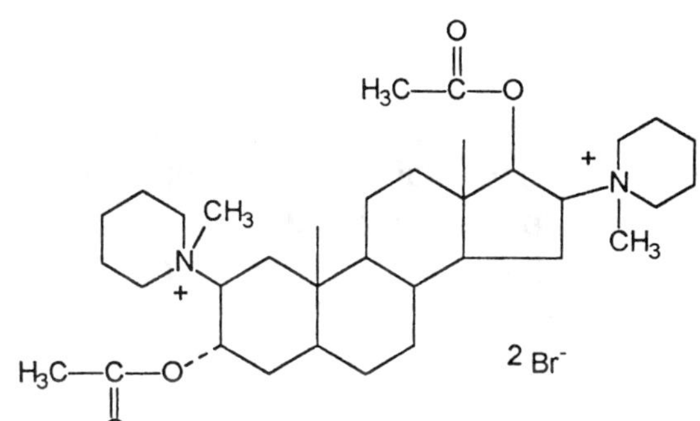

21.5

1. **Of what importance is the history of neostigmine bromide (21.1) use to the current therapeutic situation?**

2. **What is the mechanism of action of each of the neuromuscular blocking agents (21.3–21.5) shown?**

3. **Which of the three neuromuscular blocking agents (21.3–21.5) would you recommend?**

4. **If too much of your agent of choice was administered and intercostal muscles became paralyzed, could you reverse the neuromuscular blocking action by administering Edrophonium chloride (21.2)? Why or why not?**

Case 22

Submitted by Robert L. Chapman and Mary L. Chavez

The time is 1:20 AM on Christmas morning. PV presents to the emergency room with excruciating dental pain. Slowly, he explains that his temporary crown broke while he was eating a piece of fruitcake. The oral surgeon is not on the premises and is not scheduled to return to the hospital until 7:00 AM. You are called in to consult and asked to suggest a long-acting local anesthetic that will keep PV comfortable at home until he can be seen by the oral surgeon. PV is a healthy 28-year-old man who takes no medications and has no history of drug allergies.

The structure of the local anesthetic, procaine hydrochloride (22.1), is shown:

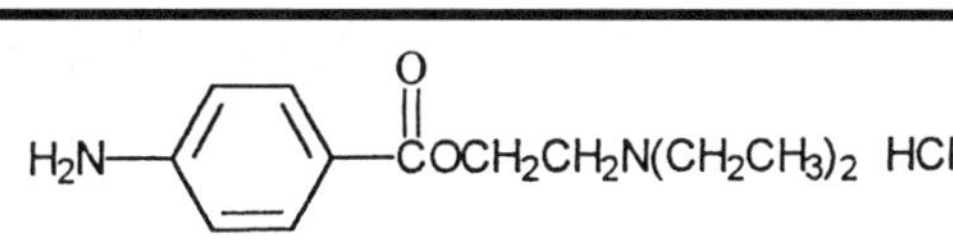

22.1

1. **Identify the pharmacophore of procaine hydrochloride (22.1). See the figure.**

2. **Based on its metabolic profile, explain why procaine hydrochloride (22.1) would or would not be an appropriate choice for the patient.**

3. **What structural modifications could be made that would increase the duration of procaine hydrochloride (22.1) as a local anesthetic?**

4. **Draw the structure of an actual long-acting local anesthetic that would be appropriate for the patient.**

Case 23

Submitted by Marc W. Harrold

RC is a 43-year-old man. He has been taking naproxen (23.1), 250 mg bid, for the past 3 years for arthritis in his left knee, which he attributes to an old football injury. He has not complained of any side effects due to the naproxen (23.1) therapy. Six days ago, he began taking co-trimoxazole (sulfamethoxazole [23.2]/ trimethoprim) tablets for treatment of an uncomplicated urinary tract infection. He has had no problems tolerating the Co-trimoxazole. Today, he is scheduled for outpatient surgery on his left knee. The physician who is to perform the surgery needs to use a local anesthetic and has the following compounds from which to choose. He has asked for your opinion.

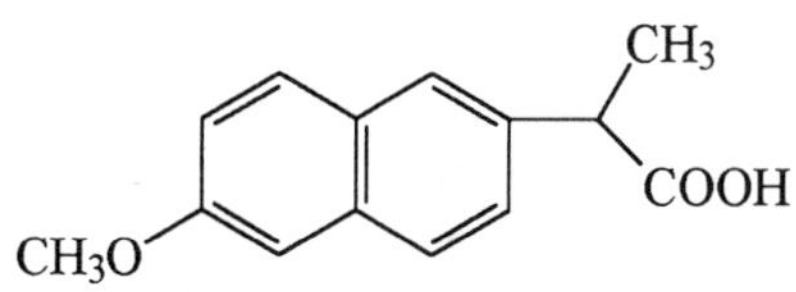

23.1

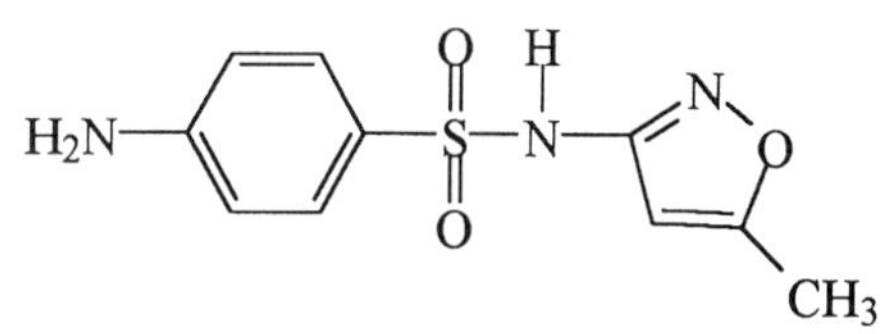

23.2

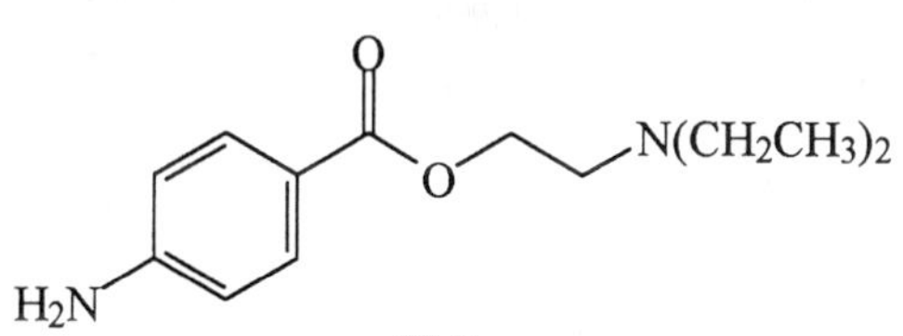

23.3

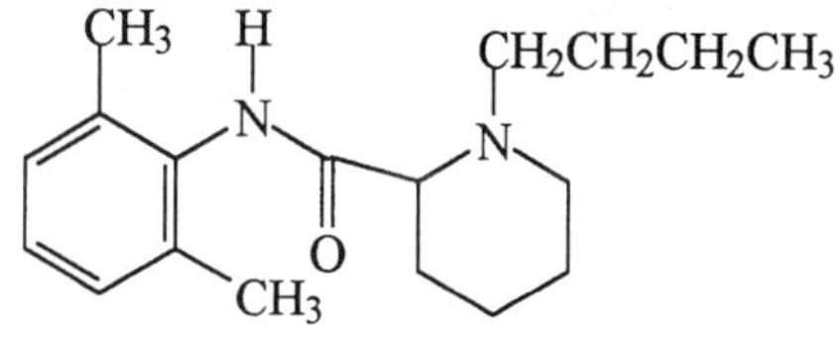

23.4

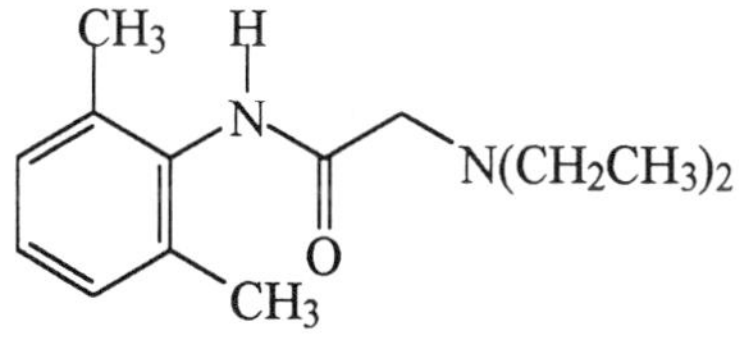

23.5

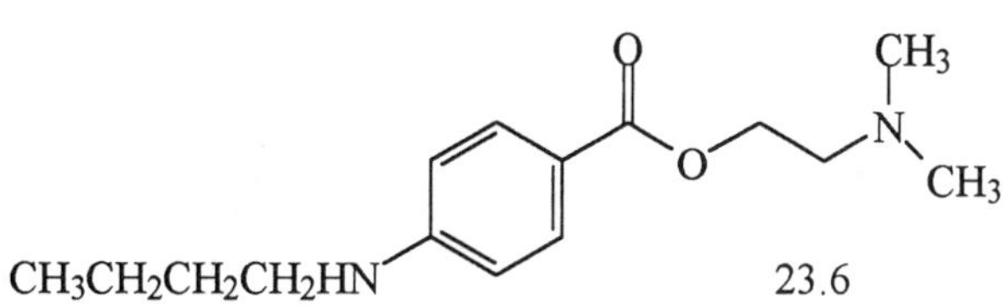

23.6

1. **Which of these local anesthetics (23.3–23.6) would be appropriate for this patient? Explain your answer.**

2. **Assume that the patient was suffering from chronic active hepatitis instead of an uncomplicated UTI. Would your recommendation change? Why, or why not?**

3. **What would your recommendation be if the patient was suffering from chronic active hepatitis in addition to an uncomplicated UTI?**

Case 24

Submitted by Matthias C. Lu and George Udeani

KC is a 42-year-old man with a history of peptic ulcers that have been previously controlled with cimetidine (24.1) 400 mg bid for the past 3 months. He is now being treated for job-related stress and insomnia. His clinic physician realizes the potential for a drug interaction between cimetidine (24.1) and diazepam (24.4).

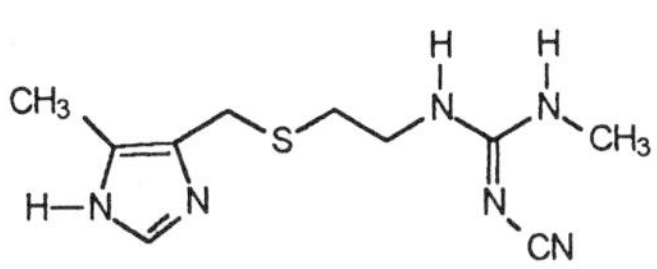

24.1

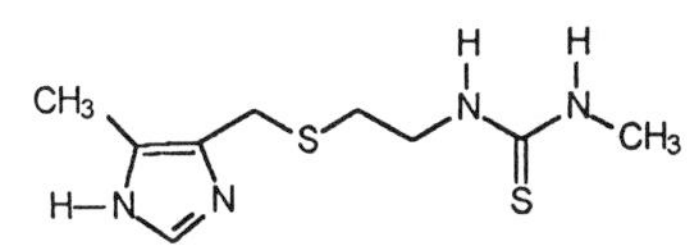

24.2

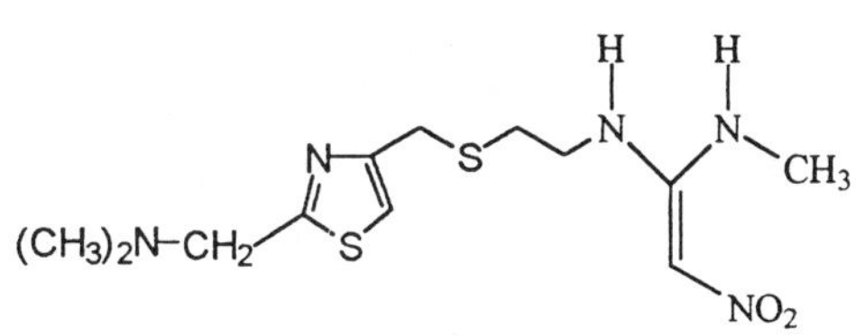

24.3

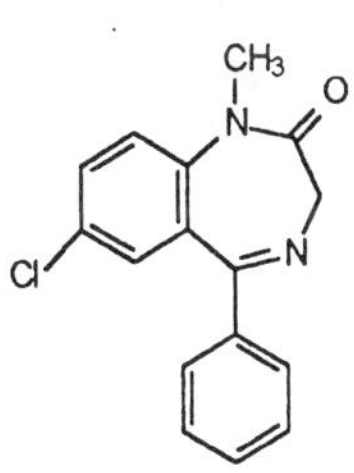

24.4

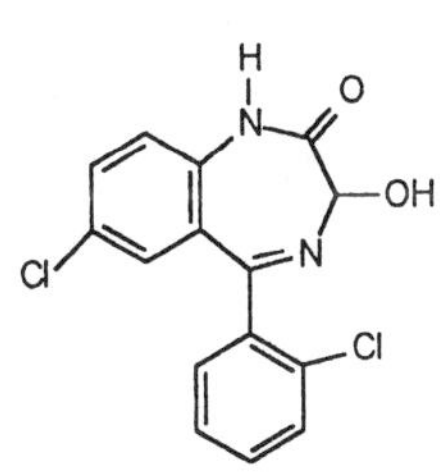

24.5

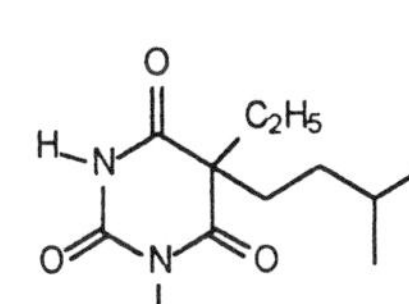

24.6

1. **Which of the following H_2 receptor antagonists (24.2–24.3) should be used as an alternative for cimetidine. Explain.**

2. **The physician also would like to know whether it is safer to use lorazepam (24.4) or amobarbital (24.6) instead of diazepam (24.4) as a sedative drug to relieve KC's insomnia. What is the risk of addiction to either of these drugs?**

Case 25

Submitted by Arnold R. Martin and Richard N. Herrier

An elderly man is brought by ambulance into the emergency ward of a local hospital after he was anonymously reported (a call to 911) to be wandering aimlessly around the neighborhood in obvious distress. Upon admission, the patient is highly disheveled, and his clothes reek of urine, vomitus, and fecal material. His symptoms are: bradycardia; shallow pulse, 60 bpm by palpation; profuse sweating, and periodic twitching. A medic alert bracelet on his wrist indicates that he is suffering from Alzheimer's disease. When contacted, the patient's caretaker reveals that the elderly gentleman takes tetrahydroaminoacridine (Tacrine) (25.1) for his condition. For treatment of his resistant reflux esophagitis, the patient has been taking metoclopramide (25.2) 10 mg qid, Mylanta prn, plus a new medication, bethanechol (25.3) 25 mg qid, which he began 5 days previous to admission.

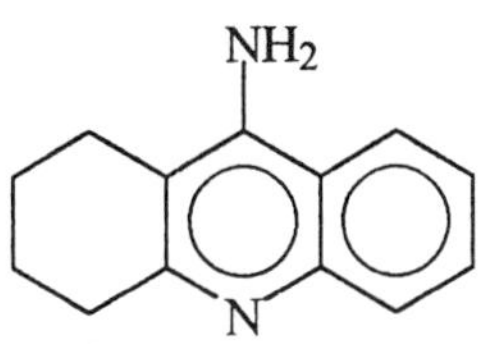

25.1

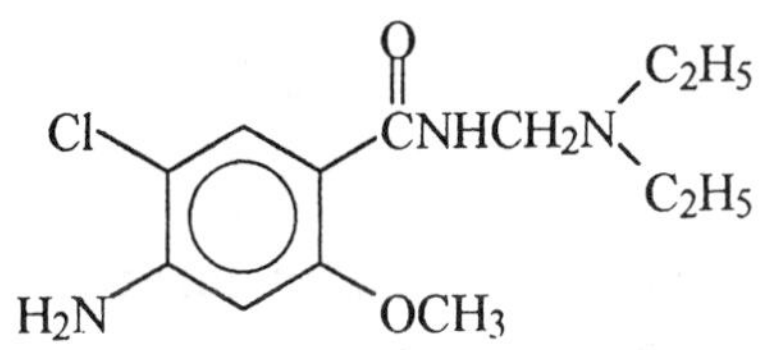

25.2

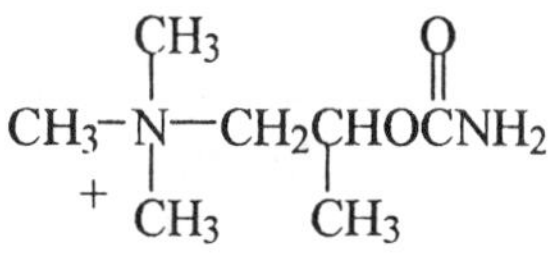

25.3

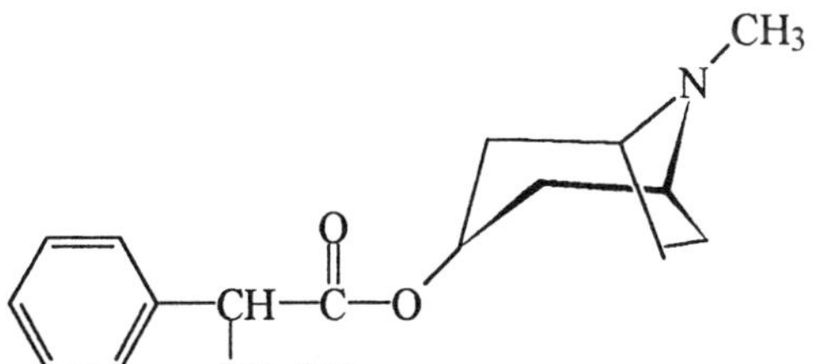

25.4

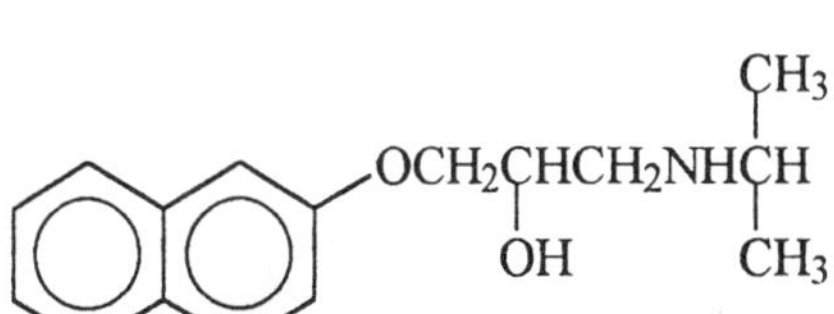

25.5

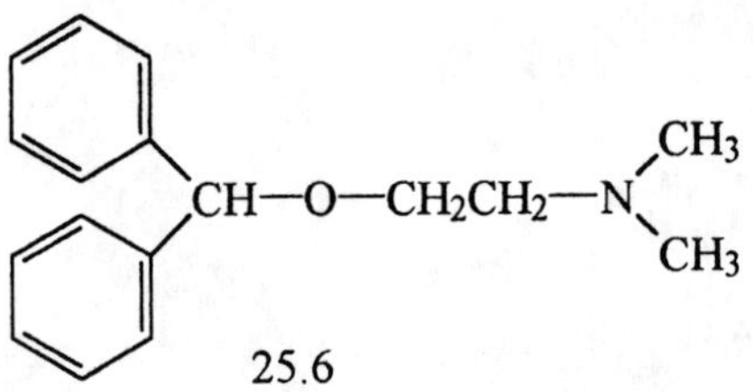

25.6

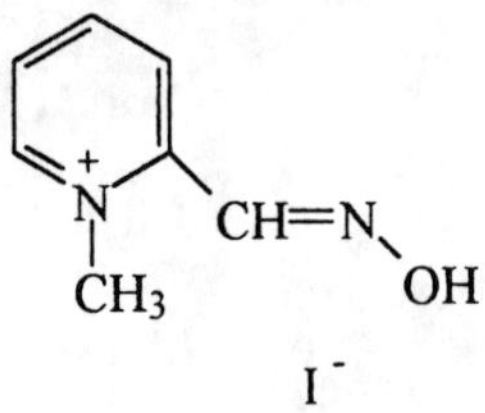

25.7

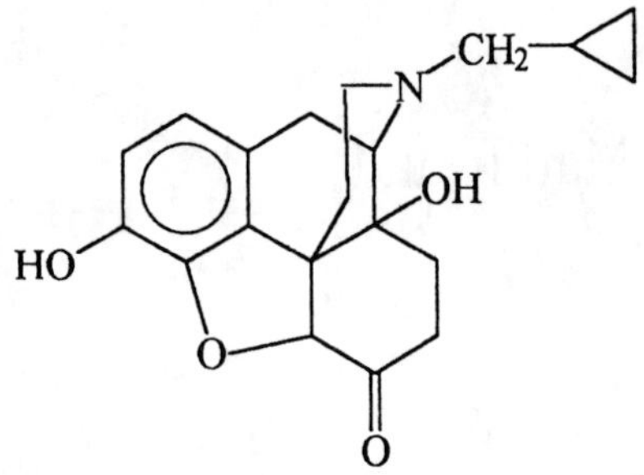

25.8

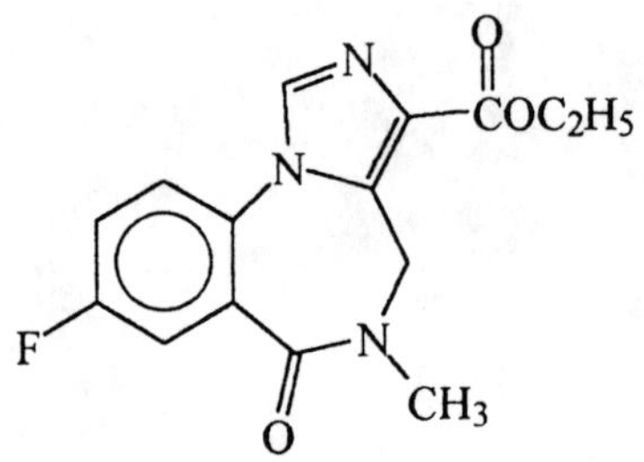

25.9

1. **Indicate the likely cause(s) of the patient's symptoms. Explain fully.**

2. **Select the appropriate drug from the structures (25.4–25.9) provided to antidote the patient's symptoms. Defend your choice.**

Case 26

Submitted by Victoria F. Roche

On the Friday before finals week, AR, a pharmacy student at the local university, develops a pruritic rash in an allergic reaction to a sulfonamide antibacterial. He comes into your pharmacy seeking an antihistamine. He has done no studying for the big Medicinal Chemistry final on Monday, and is depending on the weekend to pull it all together.

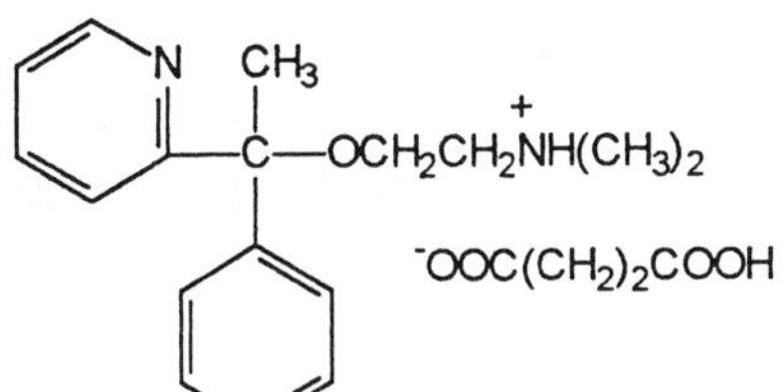

26.1

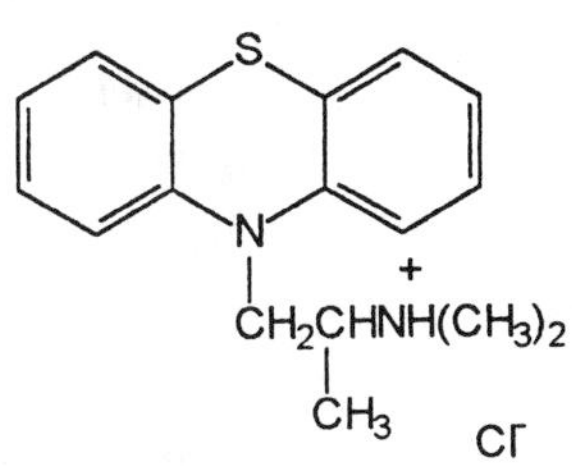

26.2

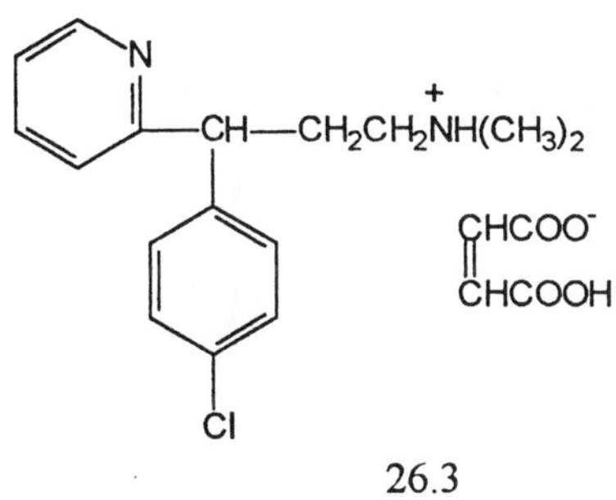

26.3

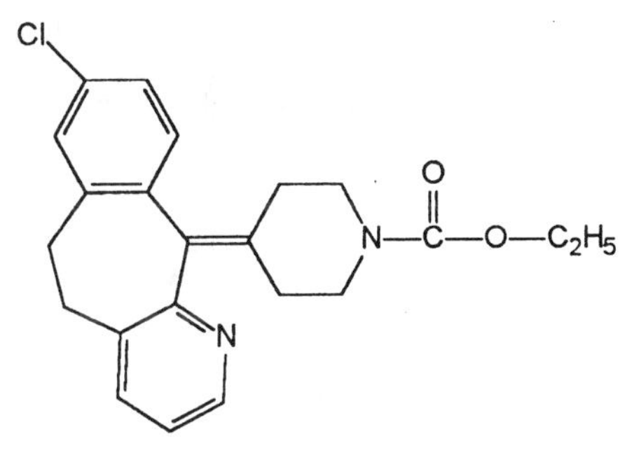

26.4

1. **Which of the antihistaminic structures (26.1–26.4) would be best suited for this patient's therapeutic and personal needs?**

Answer With all of the studying AR has before him, he cannot afford to select a highly sedative antihistamine. Molecule 26.1 can be ruled out right away. As an aminoalkylether, it is among the most sedative of all antihistamines. Structure 26.1 (doxylamine) is sedative enough to be the major ingredient in the OTC sleep aid Unisom.

Likewise, structure 26.2 would be highly sedative. Phenothiazine antihistamines are specialized ethylenediamines, and contain a hydrogen atom at position 2 of the phenothiazine ring system.

This is one structural feature that distinguishes them from antipsychotic phenothiazines. The other is the presence of a branched three or four carbon chain between the phenothiazine nitrogen and terminal amino nitrogen (antipsychotic phenothiazines have straight propyl chains). The branched chain of all antihistaminic phenothiazines provides significant sedation, so this molecule would not be appropriate for AR. In addition, it is not available OTC.

Molecules 26.3 and 26.4 are less sedative agents. Structure 26.3 (chlorpheniramine maleate) is an sp^3 propylamine antihistamine. Propylamines are among the most potent and least sedative of the traditional antihistamines, although sensitive individuals can still experience drowsiness. This agent is relatively inexpensive (good news for a college student) and available OTC. Although the p-Cl group inhibits p-hydroxylation/conjugation and prolongs the duration of action, this antihistamine is still dosed tid or qid. It would be a good first choice for AR.

Structure 26.4 (loratidine) is marketed as a non-sedative antihistamine. It can be classified as an sp^2 alkylamine (similar to an sp^2 propylamine). Many of the non-sedative antihistamines have their ionizable nitrogen (needed to anchor the drug to the anionic site of the H_1 receptor) in a piperidine or piperazine ring. This is the case with loratidine. The piperidine nitrogen atom of the intact drug is not ionizable, however, and the drug must be metabolized to provide the active descarboethoxyloratidine metabolite. The carboethoxylated amine is not basic and would not be expected to bind to histaminic receptors in either the CNS or the periphery. This, along with the aromatic chlorine atom that serves the same purpose as in chlorpheniramine (26.3), gives loratidine (26.4) a long duration of action. Once daily dosing is possible. If AR is sensitive to the sedative effects of chlorpheniramine (26.3), he might wish to obtain a prescription for loratidine (26.4), as it is not available OTC.

2. For each agent, indicate the structural features that make it either suitable or unsuitable for AR?

Answer Refer to answer for Question 1.

Case 27

Submitted by Arnold R. Martin and Richard N. Herrier

LH, a 41-year-old male independent truck driver, enters your family-owned pharmacy one evening asking you to suggest some medication for his allergy symptoms. You learn from LH that he owns his own rig, which he uses to transport merchandise on contract for clients throughout the United States. He is not certain what allergies are responsible for his symptoms (runny nose, congestion, sneezing, and severe sinus headaches), but his symptoms are usually worse in the spring. LH has tried several over-the-counter allergy remedies and has found that, although they are usually effective in curbing his symptoms, every one that he has used thus far has made him drowsy. In fact, he has twice "dozed off" and "nearly wrecked his rig" before he "came to and horsed it back on the highway." LH states that coffee does not seem to help prevent the drug-induced drowsiness.

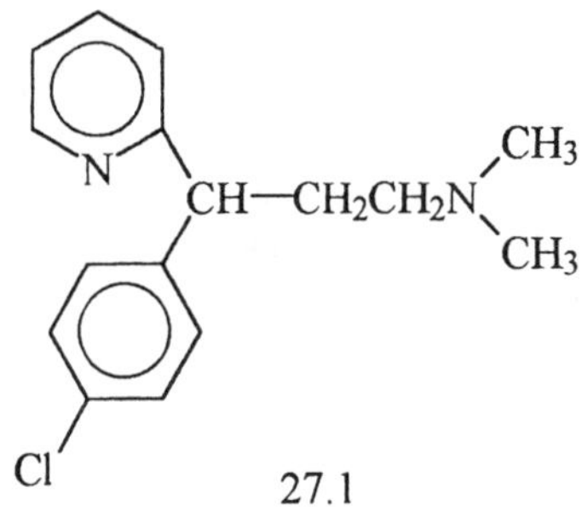

27.1

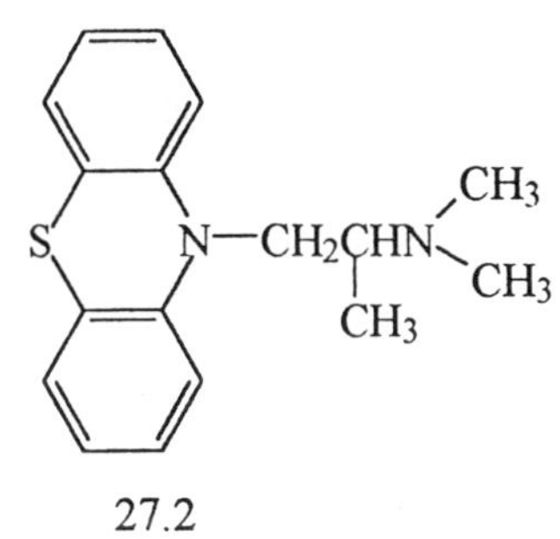

27.2

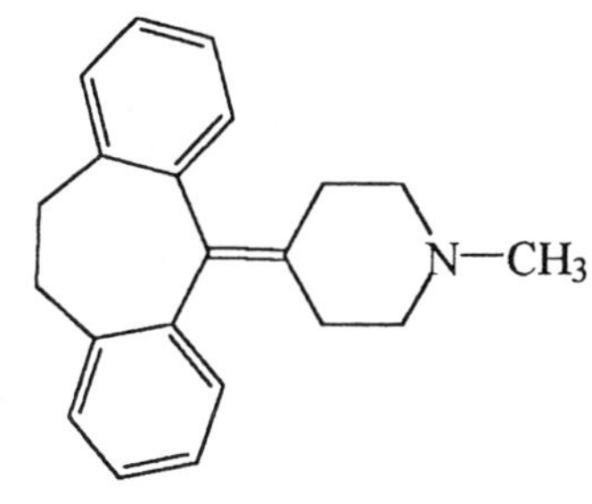

27.3

1. **From the antihistaminic structures (27.1–27.5) provided, select a suitable agent to control/prevent LH's allergic symptoms. Justify your choice on a structure-activity basis.**

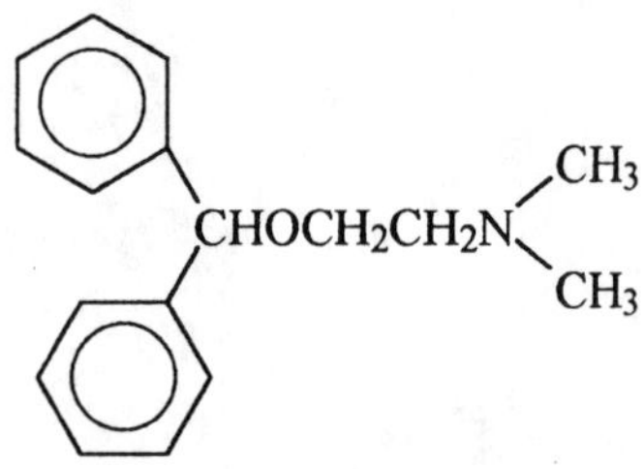

27.4

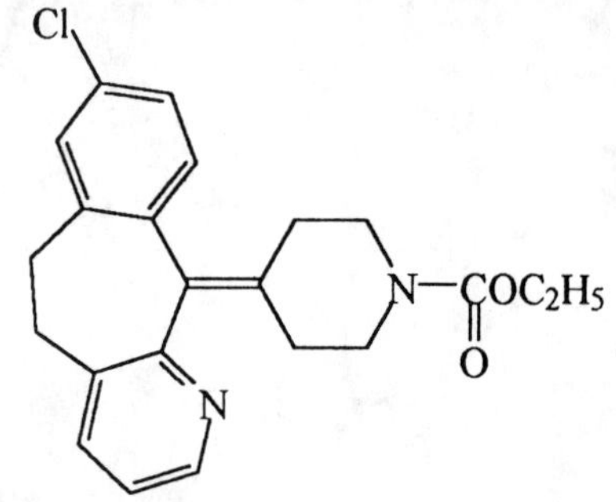

27.5

2. **Is the agent that you have selected available over-the-counter? If not, how would you instruct LH on how it should be obtained?**

Case 28

Submitted by Bruce L. Currie

LP, a prepharmacy student attending a local junior college, has been taking organic chemistry (her favorite class) and has been starting to really understand organic structures. She has been seeing her physician regularly for allergy problems that are particularly troublesome in the fall (ragweed season). She has been taking diphenhydramine (28.1) for more than a year now, but is really bothered by the dry mouth and sleepiness that she has come to associate with taking this drug. Lately, she has been drinking lots of colas to stay awake in organic chemistry and combat the dry mouth. She now thinks that she is dependent on caffeine. You are the pharmacist on duty at the drug store when LP's physician calls to ask for advice regarding an alternative medication.

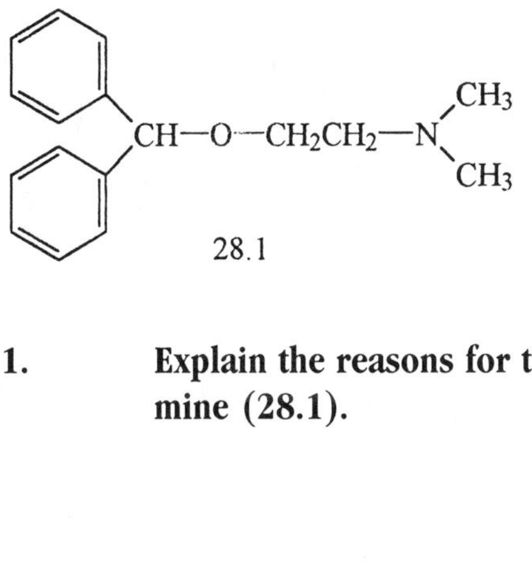

28.1

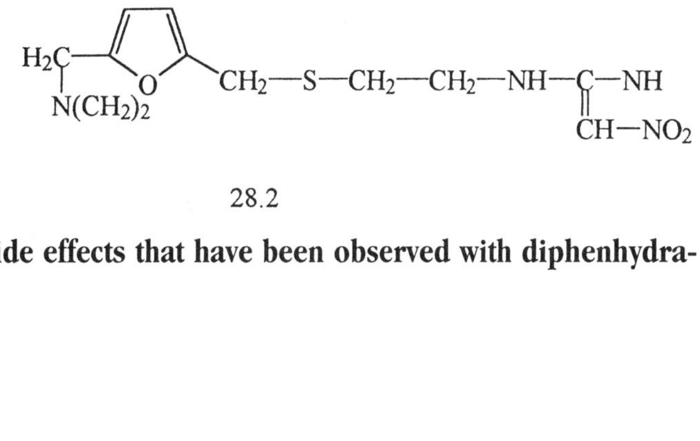

28.2

1. Explain the reasons for the side effects that have been observed with diphenhydramine (28.1).

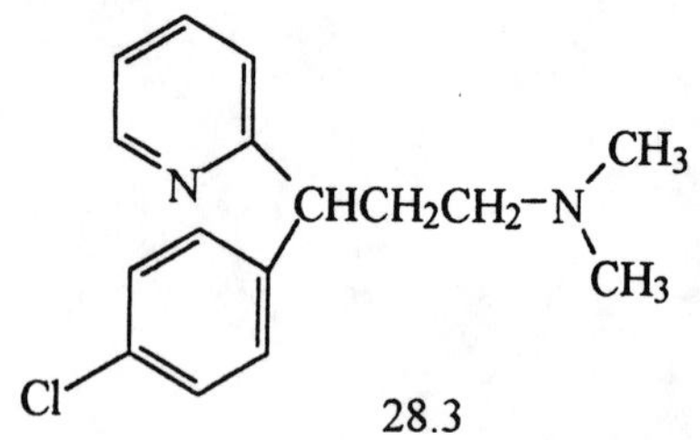

28.3

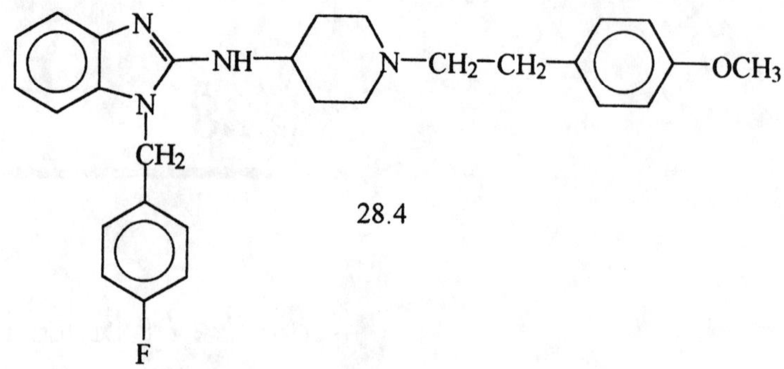

28.4

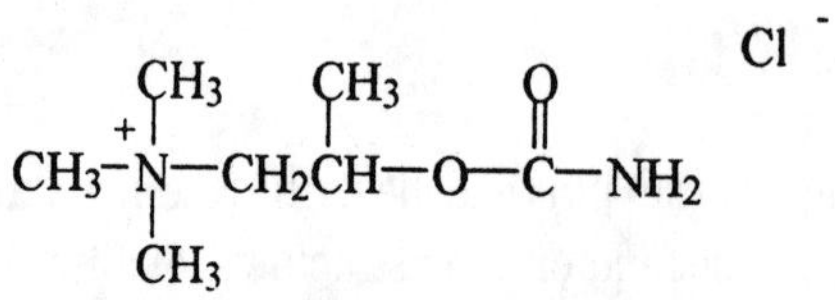

28.5

2. **Recommend one of the following drugs as an appropriate choice for this patient. Justify your choice and discuss briefly why the other choices are inappropriate.**

28.2 Ranitidine
28.3 Chlorpheniramine
28.4 Astemizole
28.5 Bethanechol

Case 29

Submitted by Matthias C. Lu and George Udeani

After school, Billy invited Johnny and his little brother Bruce over to his house to see Vendela, Billy's pet rosehair tarantula. After watching Vendela crawl up and down Billy's arm, Johnny asks if he can try too. Billy tells Johnny to hold out his arm, and Billy places Vendela on Johnny's forearm. Suddenly Bruce says, "I wanna do it too!", and grabs for Vendela. Bruce just misses grabbing Vendela, but spooks her, and she bites Johnny just above his elbow. Billy's dad gives Johnny two chlorpheniramine maleate (29.1)/phenylpropanolamine HCL (Allerest) tablets, which seem to reduce the swelling on Johnny's arm; after a short time, Johnny feels better, and Johnny and Bruce go home. After dinner, Dr. Wayne, Johnny, and Bruce's mom, notice that Johnny is experiencing a delayed allergic response to the spider bite.

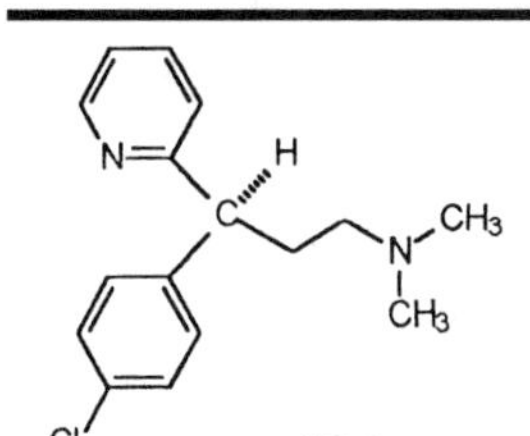

29.1

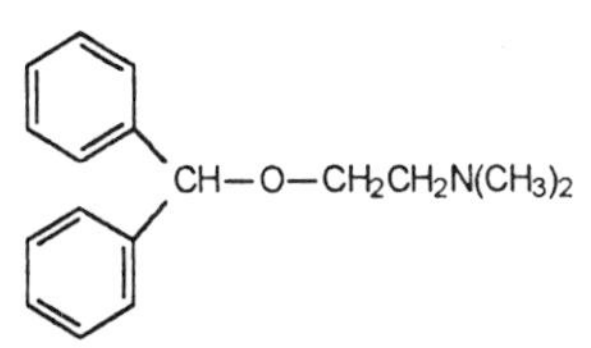

29.2

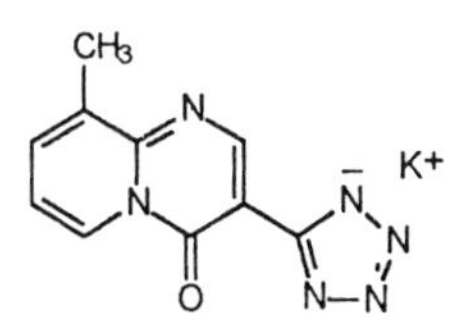

29.3

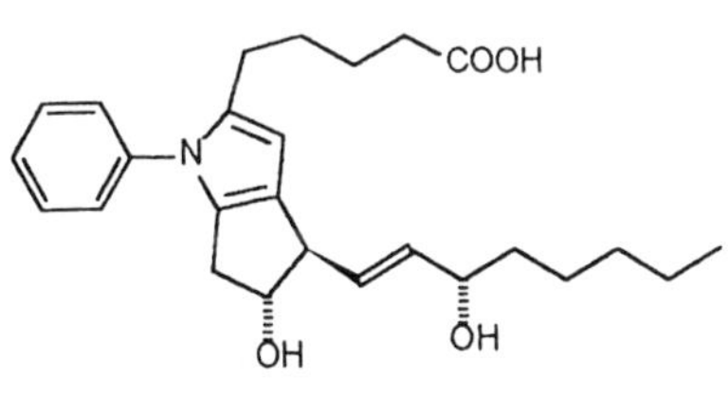

29.4

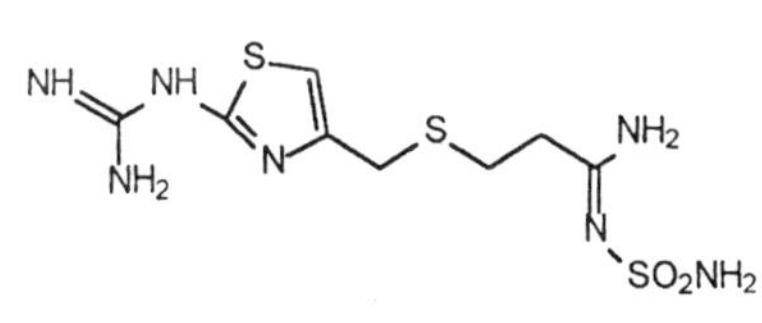

29.5

1. Which of the following antiallergic drugs, 29.2 to 29.5, could she give Johnny to relieve this symptom. Justify your rationale for selecting the appropriate drug(s) and provide reasons for eliminating the alternatives choices.

Case 30

Submitted by Victoria F. Roche

BR, a 20-year-old nature enthusiast, is halfway through a camping vacation in your southern Colorado community when she suffers multiple bee stings in a hiking mishap. Highly allergic to bee venom, she is rushed unconscious to the emergency room by her companion. Now stabilized and resting rather uncomfortably, she is ready to be transported by her friend back to her home in New York, and she is highly anxious about the long car trip. The attending physician wishes to discharge her with an appropriate antihistamine, and seeks your advice. Upon consulting her profile, you find she is currently completing a course of erythromycin therapy for mycoplasmal pneumonia, and has used scopolomine patches in the past when traveling to control motion sickness.

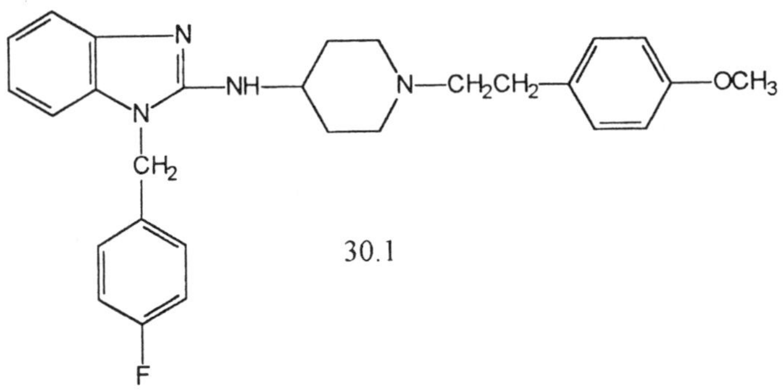

30.1

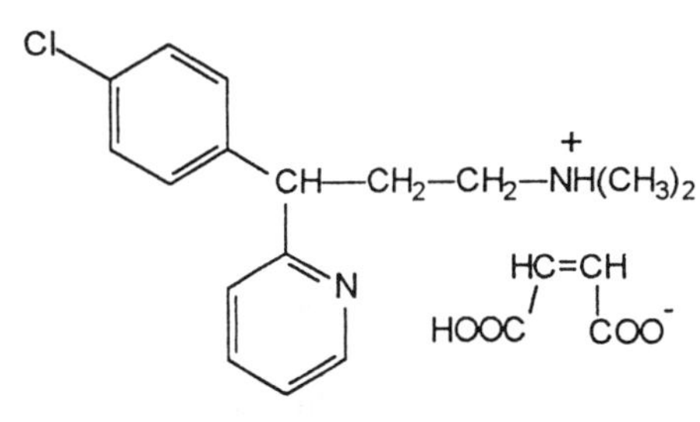

30.2

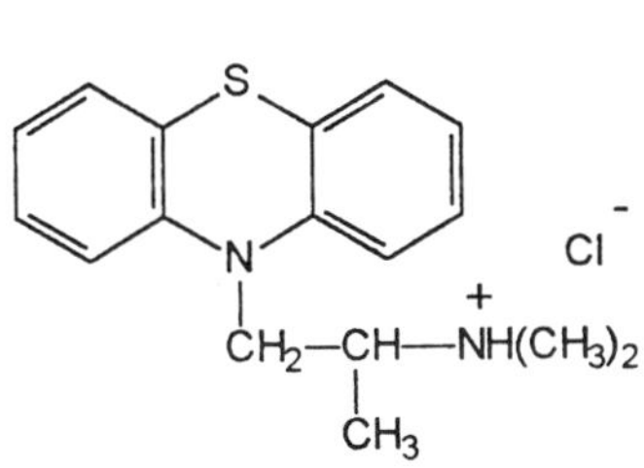

30.3

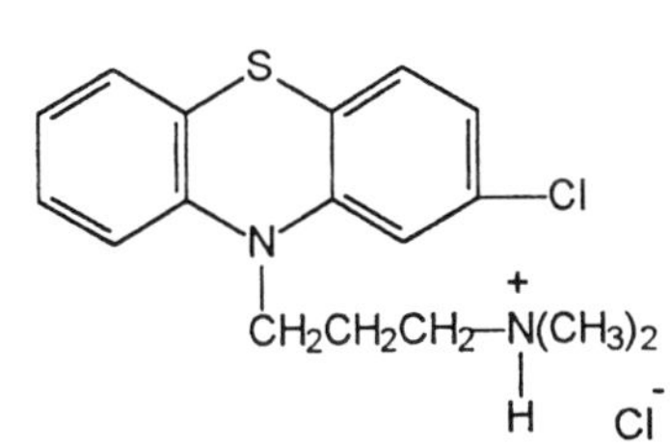

30.4

1. **What are the pros and cons of each structure (30.1–30.4) in this therapeutic situation? Are any of the agents totally unsuitable for BR's clinical condition?**

2. **What is the significance of the current erythromycin therapy? Does this concomitant therapy rule out any of your potential antihistaminic choices?**

3. **What is the significance of the past use of scopolamine patches? Is BR's need to control motion sickness important to this case?**

4. **Which of the antihistaminic structures (30.1–30.4) would you recommend?**

Case 31

Submitted by Eric J. Lien

A new (aerosol dosage form) drug of the following structure (31.1) was recently approved by the FDA as Class 1-B.

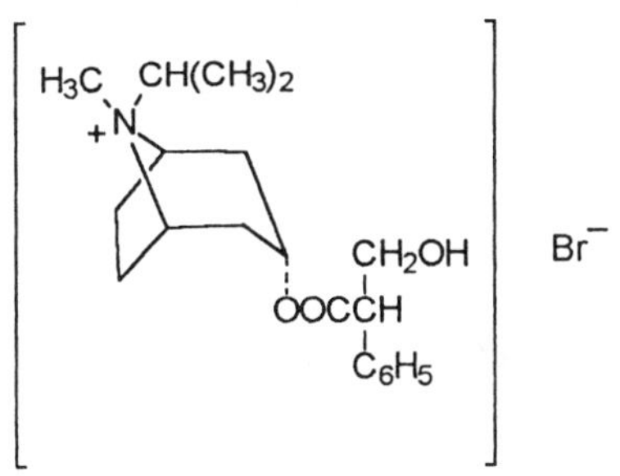

31.1

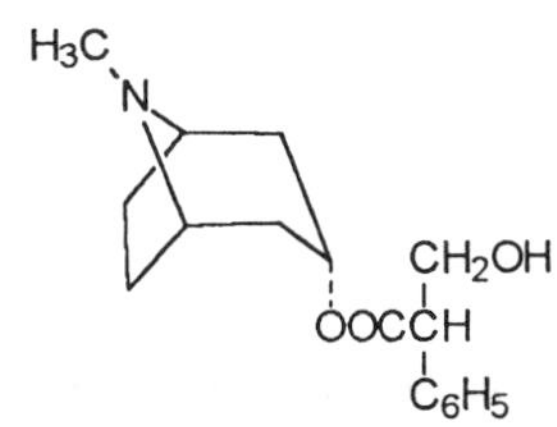

31.2

A new medical doctor asked our clinical pharmacist to answer the following questions:

1. What category does this drug belong to?

2. What is its recommended therapeutic use?

3. How does this drug differ from atropine structurally? Will it have any CNS activity?

4. Is this drug a natural product or a synthetic compound?

Case 32

Submitted by Bruce L. Currie

Late one Saturday afternoon, BC, a 26-year-old man, was playing softball with his church team. While running to second base, BC was hit on the right side of the face with a thrown ball, which resulted in a broken jaw in three places. He was taken to the emergency room of a local hospital in considerable pain. An oral surgeon was finally located at a party and asked to come to assess and repair the damage. BC was administered an analgetic by injection while waiting for the oral surgeon. The oral surgeon wired BC's jaw together, with restriction to a liquid diet for 6 weeks; BC was then sent home.

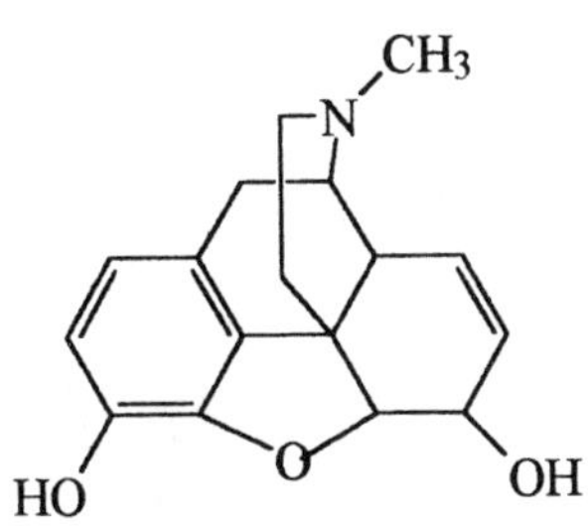

32.1

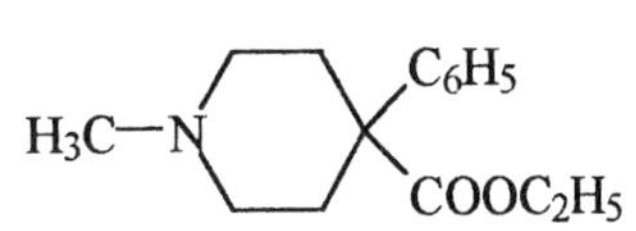

32.2

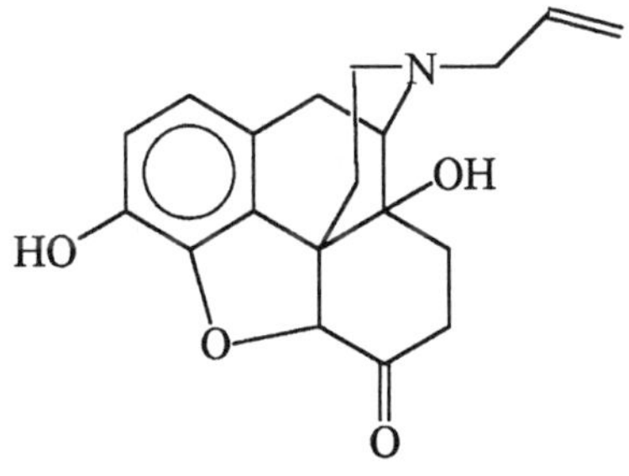

32.3

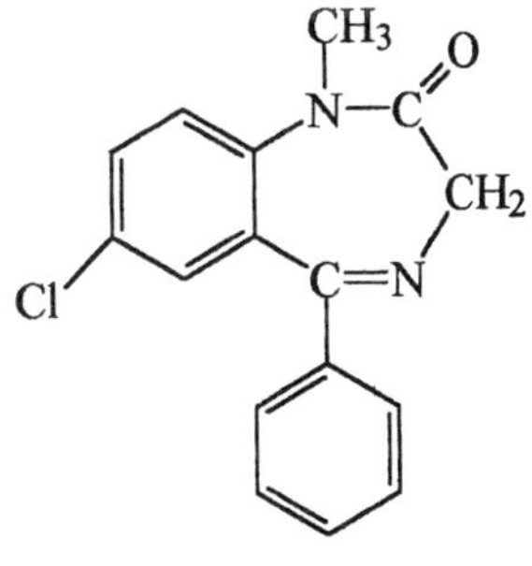

32.4

1. **Examine the four drugs shown (32.1–32.4) with regard to their suitability for analgesia in this case. Rationalize your best choice and discuss why the others were excluded.**
32.1 Morphine sulfate
32.2 Meperidine hydrochloride (Demerol)
32.3 Naloxone hydrochloride (Narcan)
32.4 Diazepam (Valium)

Answer The best choice is meperidine hydrochloride. It is a narcotic analgetic of relatively low potency, with relatively little respiratory depression effect, which would be a good choice for a patient who is being treated and sent home. Morphine is also a narcotic analgetic, but is usually used during hospitalization because of the risks associated with respiratory depression. Naloxone hydrochloride is a narcotic antagonist and would have no use in this case. Diazepam is an anxiolytic and might make the patient feel somewhat better, but does not have direct analgetic effects.

Case 33

Submitted by Victoria F. Roche

As a pharmacist specializing in home-health care, you are working with a 25-year-old terminal cancer patient, CB, who has recently moved to your city. For pain management, she had previously been stabilized on structure 33.1 (4 mg q 4-6 h), but her new physician has written for another agent, 33.2.

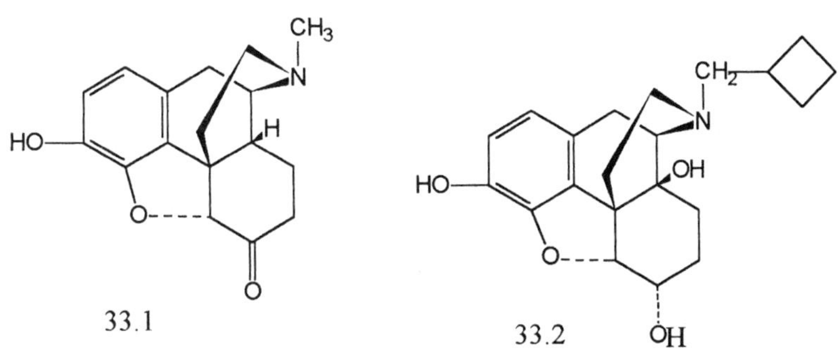

33.1 33.2

1. What is the opioid receptor activity profile of each analgetic drawn?

Answer Structure 33.1 (hydromorphone) is readily identified as a μ receptor agonist by the N-CH_3 substituent. This lipophilic CH_3 group is believed to bind hydrophobically to a site on the μ receptor that permits receptor stimulation. The N-cyclobutylmethyl group of structure 33.2 (nalbuphine), on the other hand, makes this agent a potent κ agonist and effective μ antagonist. The ring strain inherent in the cyclobutyl group permits binding to a receptor antagonist site. The 14β-OH group of molecule 33.2 enhances affinity for the μ receptor and promotes the antagonist activity. The 7,8-dihydro C ring structure of both molecules enhances molecular flexibility and allows the C_6 substituent to reach its binding site (presumably a hydrogen acceptor site) on the μ receptor.

2. What therapeutic advantages and disadvantages would be realized with a switch to agent 33.2?

Answer When contemplating a switch from a selective μ agonist analgetic to another agent, one must always evaluate the risk of precipitating withdrawal symptoms. As a rule, commercially available κ analgetics are not cross tolerant with μ analgetics because the nitrogen substituents that promote κ agonism also promote binding of the antagonist conformer to the μ receptor. In this case, CB has been stabilized on μ agonist 33.1, so tolerance and dependence on μ receptor occupancy is assumed. Abrupt discontinuation of μ receptor stimulation by switching to compound 33.2, which would antagonize the μ receptor even though it stimulates the κ receptor, would precipitate a withdrawal response in CB. The switch should not be made.

The use of κ agonist analgetics in patients not dependent upon μ agonists has a number of clinical advantages. Stimulation of the κ receptor produces a profound analgetic response, often providing pain relief equal to or greater than that obtained from a similar dose of morphine. The euphoria often associated with opioid analgetic use is derived primarily from μ receptor stimulation and, although a patient can become psychologically dependent upon any analgetic, addiction is more common with μ agonists than with κ agonists. Likewise, respiratory depression is associated more with μ than with κ receptor stimulation. One drawback with the use of κ analgetics is that some patients will experience dysphoria.

3. **Would you fill this new prescription? Why, or why not?**

Answer Refer to solution to question 2.

4. **Should you be worried about the possibility of opioid addiction in CB?**

Answer Although the subject is certainly open to debate, opioid addiction in a terminally ill patient is not viewed as a problem by many health-care professionals. Rather, allowing a dying individual as much freedom from pain as possible so that quality of life at its end can be maximized is the primary therapeutic objective.

Case 34

Submitted by Victoria F. Roche

JC, a 59-year-old postal employee, is in your community pharmacy to pick up some Maalox tablets for an upset stomach initiated by job stress. A talkative guy, he tells you that he had learned in "Reader's Digest" about the benefits of aspirin in preventing "stroke and heart attacks" in men. He thinks his boss is going to cause him to have one or the other, but he has trouble taking aspirin (34.1) because of GI distress. Therefore, he has been taking Tylenol (acetaminophen)(34.2) qd because "all those pain killers are alike."

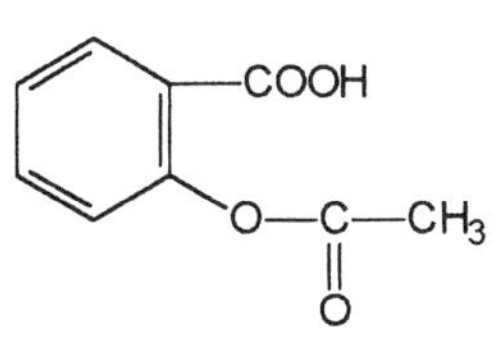

34.1

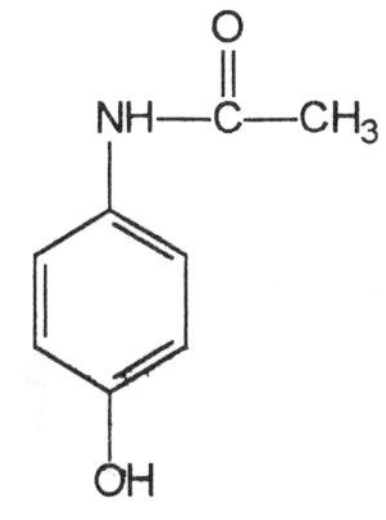

34.2

1. **What is the mechanism by which aspirin (34.1) decreases risk of cardiovascular disease?**

2. **What functional group(s) on aspirin (34.1) are crucial to this mechanism?**

3. **How would you counsel this patient?**

Case 35

Submitted by Victoria F. Roche

SO, an athletic faculty member who teaches in the business administration college of your local university, has developed a nagging case of tennis elbow. After self-medicating unsuccessfully for a couple of weeks with Tylenol (acetaminophen) (35.1) he bought at the grocery store, he made an appointment with his physician, who is now seeking your advice on appropriate nonsteroidal anti-inflammatory drug (NSAID) therapy. SO complains that his busy schedule makes taking medication problematic. He is currently on Athrombin K (warfarin potassium) (35.2) to control a thromboembolic disorder and, although he claims to take this Rx medication faithfully as prescribed, he admits to being significantly less diligent with OTC drugs. The physician is toying with the NSAID structures indicated.

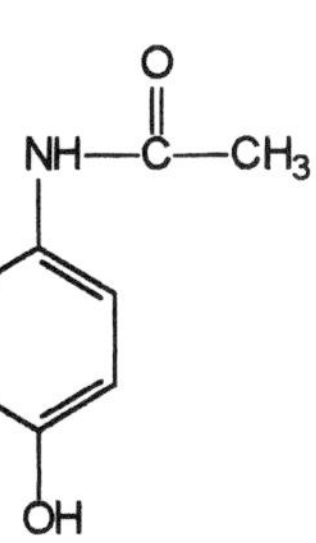

35.1

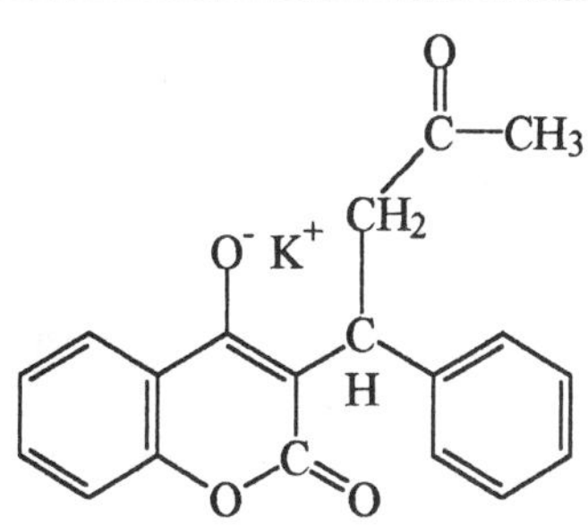

35.2

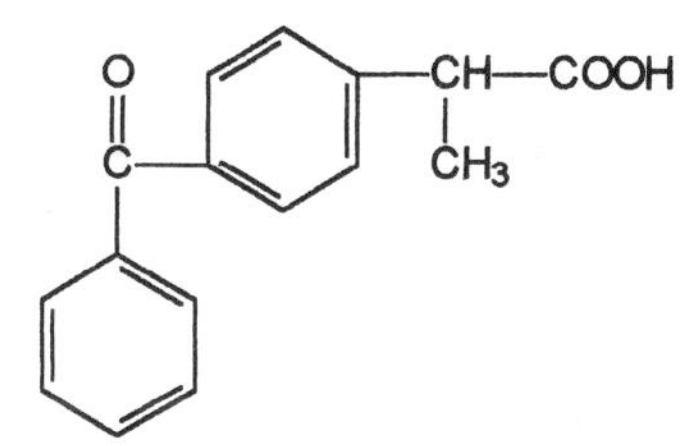

35.3

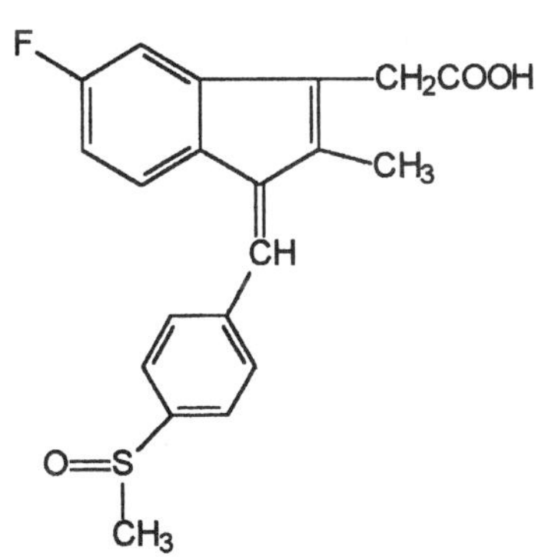

35.4

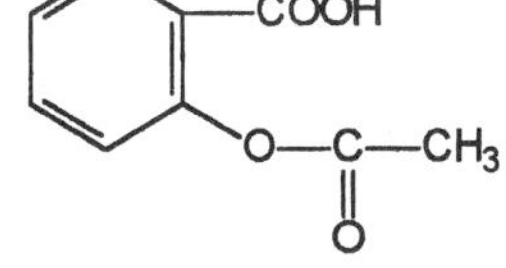

35.5

1. **Which of the following would be a sound therapeutic recommendation in this clinical situation?**

   a. **Explain that Tylenol (35.1) purchased from the grocery store is not bioequivalent to that purchased from a pharmacy. Recommend that the patient purchase good Tylenol from *your* pharmacy, kick up the dose, and wait it out for a couple of additional weeks.**

   b. **Recommend that the expensive Rx NSAIDs under consideration (structures 35.3 and 35.4) be avoided in favor of aspirin (35.5), which will have the same effect at a lower cost.**

   c. **Recommend Rx NSAID 35.3.**

   d. **Recommend Rx NSAID 35.4.**

2. **What structural considerations support your therapeutic recommendation?**

3. **Would you recommend that NSAID therapy be rejected in favor of short-term treatment with a corticosteroid?**

Case 36

Submitted by Marc W. Harrold

DJ suffers from periodic migraines. The migraines usually occur without aura and have been therapeutically managed using subcutaneous sumatriptan (36.1). Six months ago, DJ began playing racquetball during his lunch hour with one of his colleagues. Today, he visited his physician complaining of a sore elbow. His physician has diagnosed the problem as bursitis, and wants to prescribe treatment for the problem.

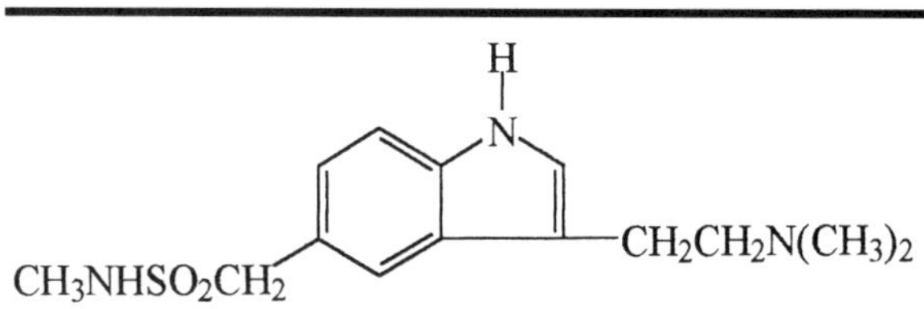

36.1

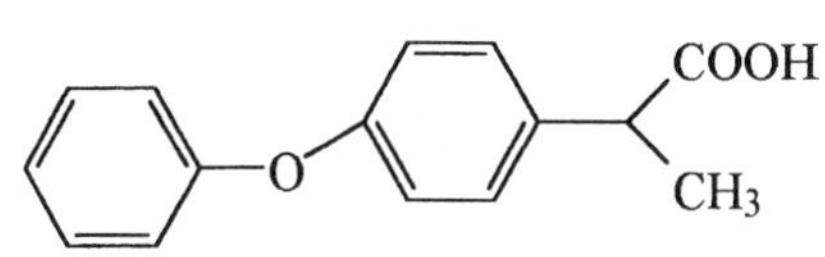

36.2

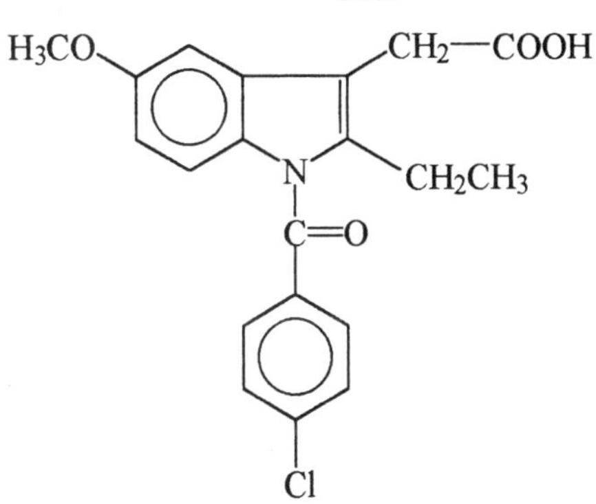

36.3

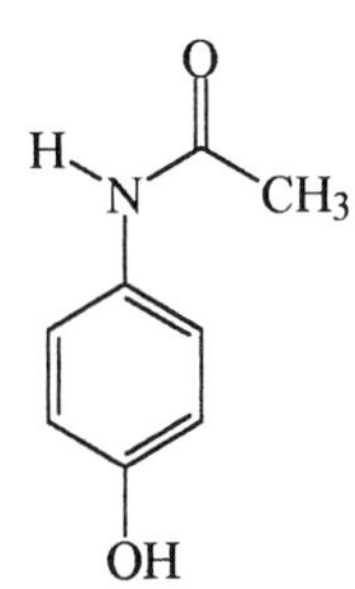

36.4

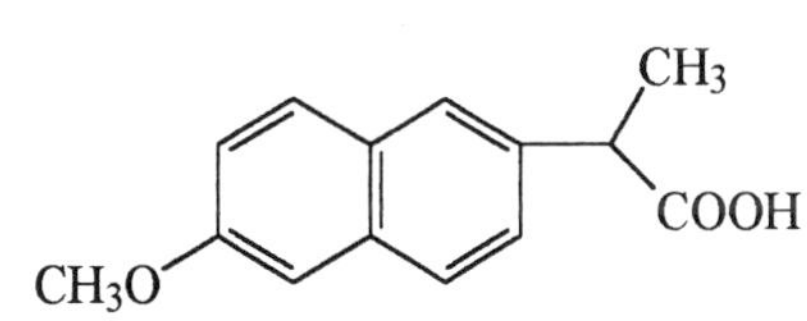

36.5

1. **Which of the agents (36.2-36.5) should not be used in this patient? Explain your answer using the chemistry of the compounds.**

Case 37

Submitted by Victoria F. Roche

A two-year-old child is rushed to the emergency room of your hospital by his distraught mother. She tells the E.R. staff that her boy ate approximately half of a full bottle of Tylenol (acetaminophen) (37.1) tablets. Along with gastric lavage, you recommend the po administration of a 5% solution of structure 37.2.

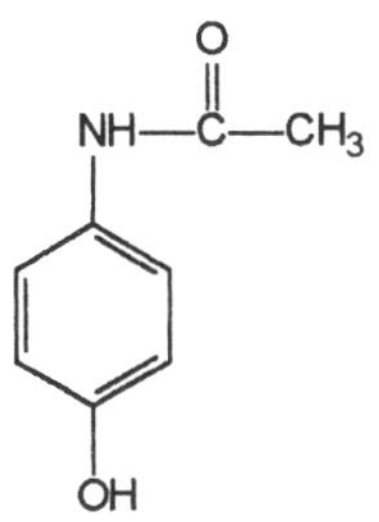

37.1

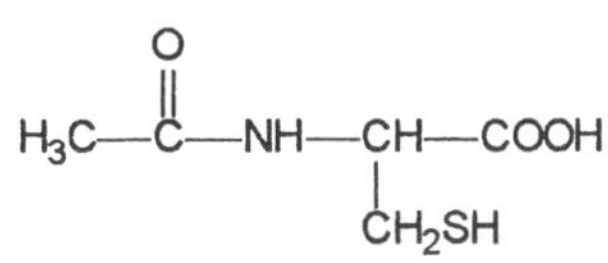

37.2

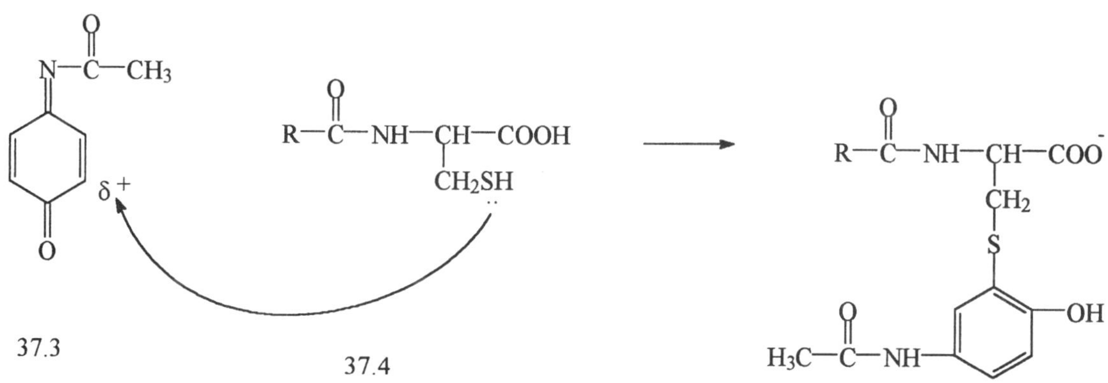

37.3 37.4 37.5

1. **How is acetaminophen (37.1) normally metabolized? Based on this metabolic route, what medical emergency is facing this child?**

2. **What is the chemical rationale for the administration of structure 37.2?**

Case 38

Submitted by Victoria F. Roche

You are the medicinal chemist in charge of a research group employed by a British pharmaceutical company. Your company markets prodine (structure 38.1, an analogue of the opioid analgetic meperidine), and you are discussing the chemistry of this drug with a newly hired technician. Specifically, you tell the technician that the diasteriomeric form that places the C_3 CH_3 group and the C_4 phenyl group cis to one another provides the greatest antinociceptive potency. This form is known as *beta* prodine, and it has an analgetic action approximately nine times that of morphine. You ask him to build a molecular model of *beta* prodine, and he provides structure 38.2.

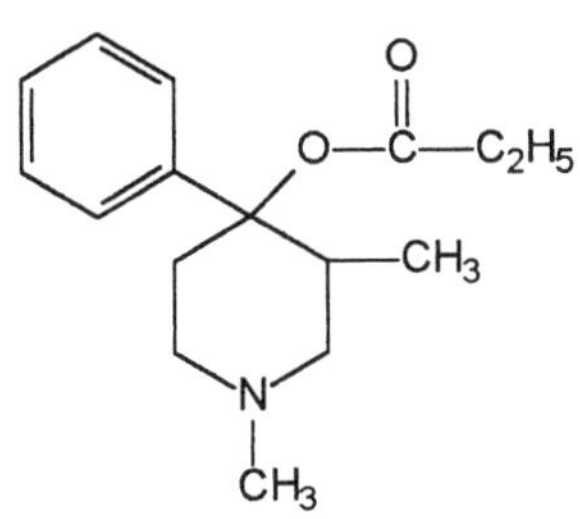

38.1

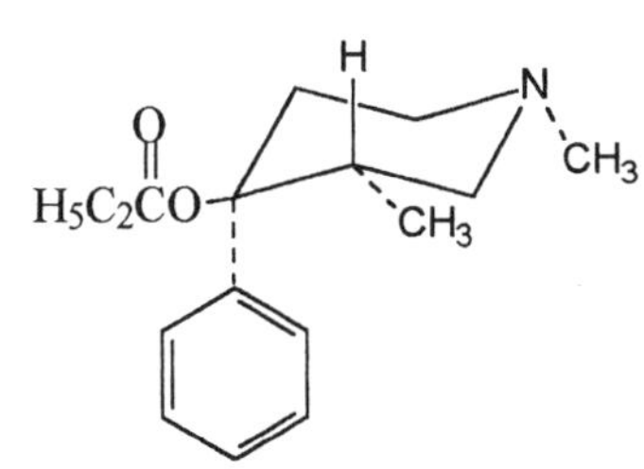

38.2

1. **Do you congratulate this technician on his sound understanding of stereochemistry, or send him to the local university for a refresher organic chemistry course?**

2. How could you convert the isomer built by the technician to a model of ***alpha*** prodine (which has the C_3 CH_3 and the C_4 phenyl groups trans to one another, and is only twice as active as morphine)? Is the conversion of ***beta*** prodine to ***alpha*** prodine a configurational or a conformational change?

3. The literature suggests that flexible 4-phenylpiperidine analgetics like prodine may interact at the opioid receptor with the phenyl group in the equatorial position (rather than in the axial orientation demanded by the more rigid fused-ring analgetics like morphine). Does the model built by the technician have the phenyl ring oriented axially or equatorially to the piperidine ring?

4. What would you need to do to the technician's model to get the desired equatorial phenyl ring? What would happen to the C_3-CH_3 when you did this? Have you made a configurational or a conformational change in your model? Would you classify this new structure as ***beta*** prodine or ***alpha*** prodine?

Case 39

Submitted by S. William Zito

Your company has developed a new enkephalin derivative (39.1) that shows excellent parenteral antinociceptive activity. They are ready to begin clinical trials when it is discovered that the peptide very quickly and significantly hydrolyzes in aqueous buffered media. As the formulation expert, you are called in to help solve this problem.

---

1. **What is the chemical cause of this instability?**

2. **Suggest synthetic solutions to this problem.**

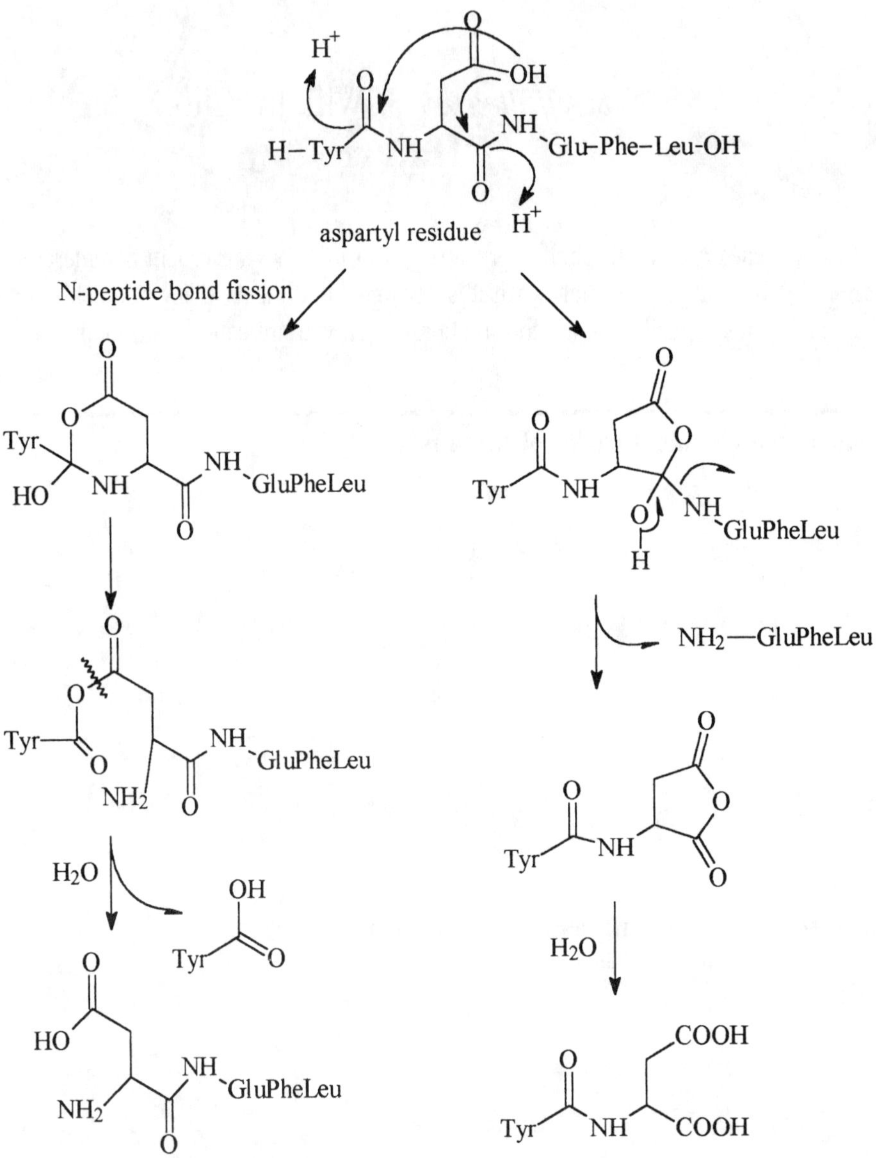
H⁺
OH
NH
H–Tyr
NH
Glu–Phe–Leu-OH
H⁺
aspartyl residue
N-peptide bond fission
Tyr
HO
NH
NH
GluPheLeu
Tyr
NH
NH
GluPheLeu
H
NH₂—GluPheLeu
Tyr
NH
GluPheLeu
NH₂
Tyr
NH
H₂O
OH
Tyr
H₂O
HO
NH
GluPheLeu
NH₂
COOH
Tyr
NH
COOH

39.1

Case 40

Submitted by Victoria F. Roche

As a service oriented pharmacist, you are making a presentation on designer drugs to a group of physicians in your clinic. One physician asks specifically about how to treat an individual highly addicted to the currently popular MPPP (4-hydroxy-1-**m**ethyl-4-**p**henyl**p**iperidine**p**ropionate) (40.1). You ascertain that his patient has never acutely overdosed on MPPP and contemplate each of the following possible responses.

1. Immediately administer structure 40.2.
2. Abruptly switch patient to drug 40.3 and gradually taper dose until the patient is drug free.
3. Abruptly switch patient to drug 40.4 and gradually taper dose until the patient is drug free.

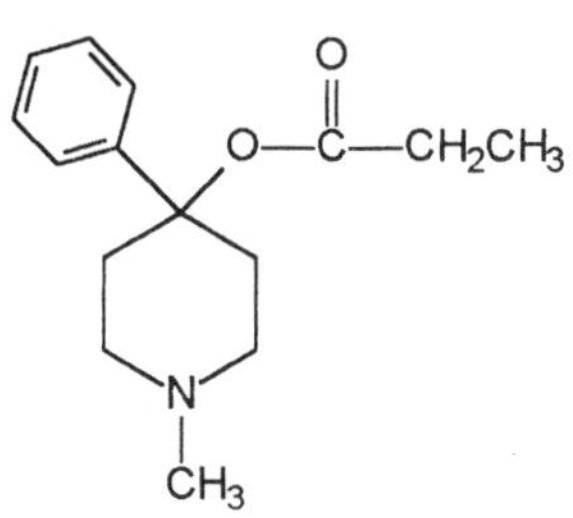

40.1

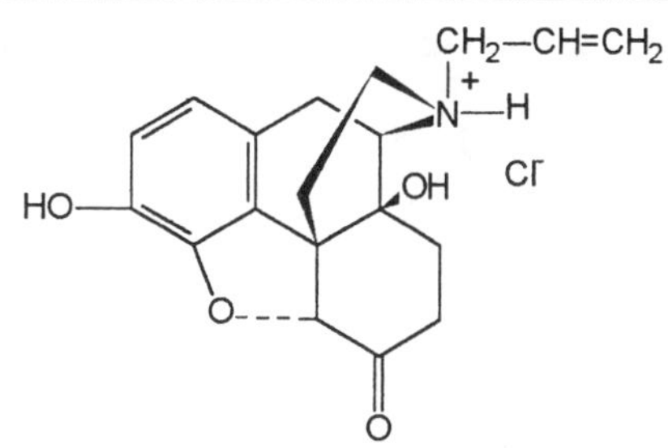

40.2

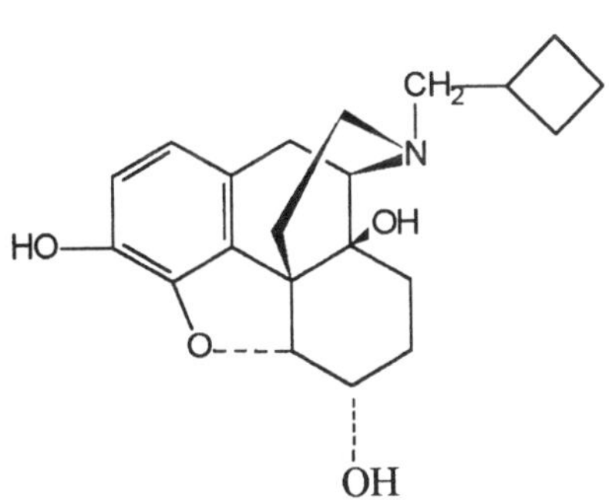

40.3

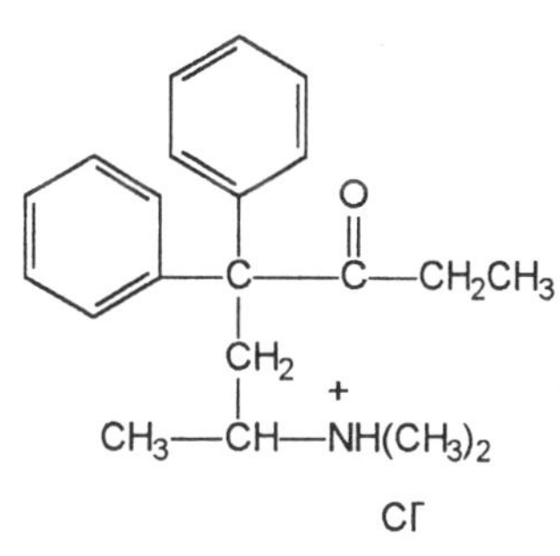

40.4

1. What is the opioid receptor activity profile of each of the drug molecules provided in this case?

2. **What is your therapeutic goal?**

3. **How would you answer this physician?**

Case 41

Submitted by Victoria F. Roche

A semiconscious individual is brought into the emergency room of your hospital by some good samaritans who found him on a downtown sidewalk. He has severely depressed respiration and pinpoint pupils.

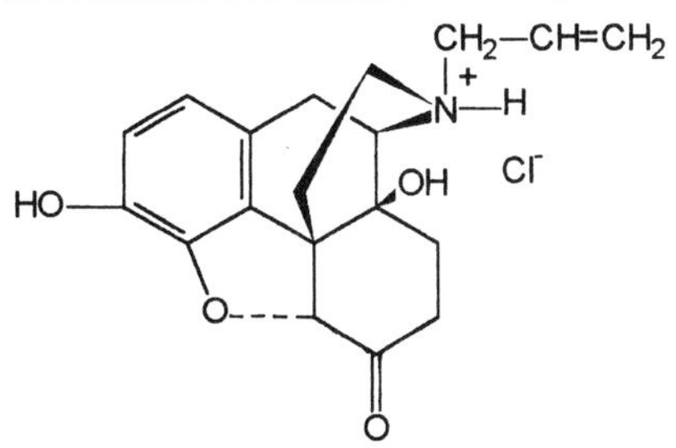

41.1

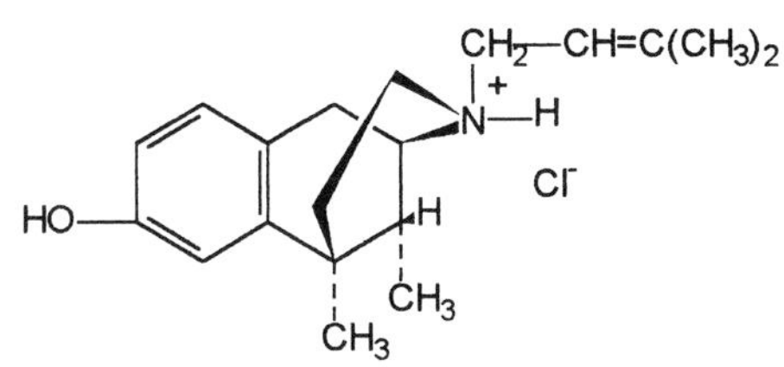

41.2

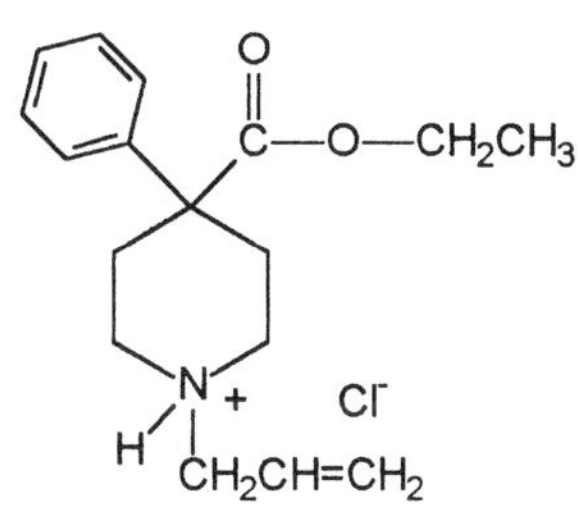

41.3

1. What might be this individual's problem?

2. **Identify the opioid receptor activity profile of each of the structures drawn (41.1–41.3). Which drug structure would be appropriate to administer to this unidentified man?**

Case 42

Submitted by Victoria F. Roche

A physician consults you on a newly transferred hospital patient, JL, who was the victim of a serious motorcycle accident. At the hospital from which he was transferred, JL had been given Dilaudid (hydromorphone)(42.1) regularly for 4 weeks. This physician is concerned about drug-induced dependence, and has been reading about the newer analgetics whose manufacturers claim provide greater clinical safety. The physician wishes to change therapy, and seeks your opinion.

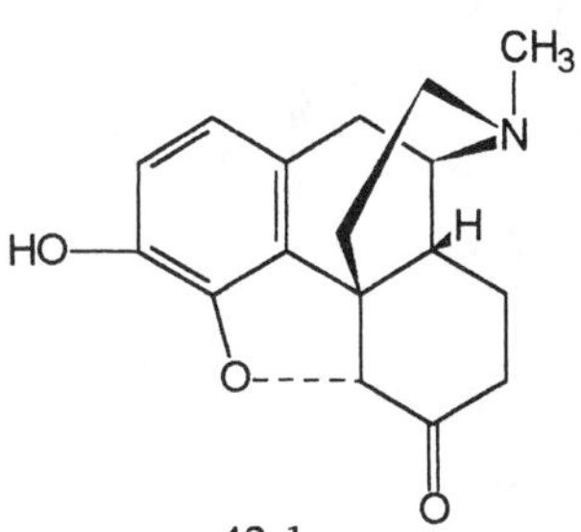

42.1

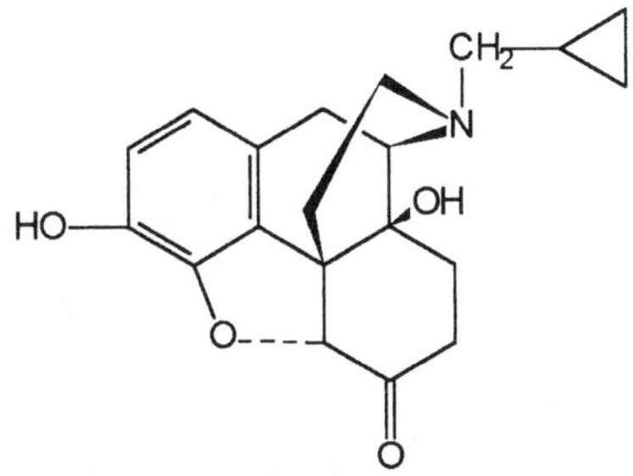

42.2

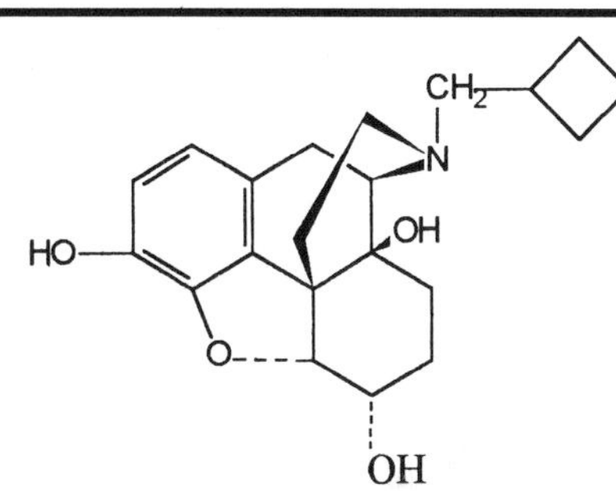

42.3

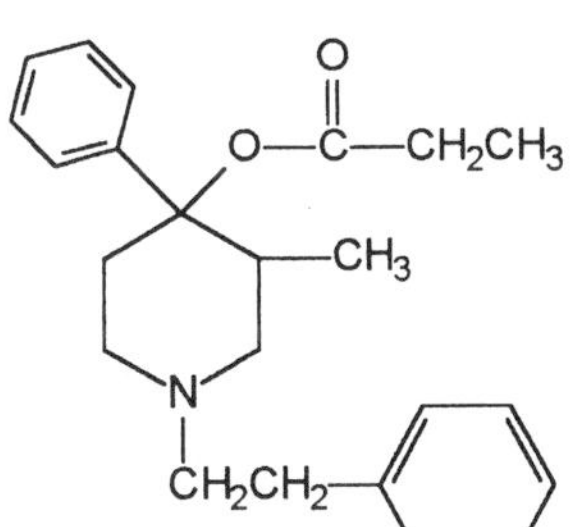

42.4

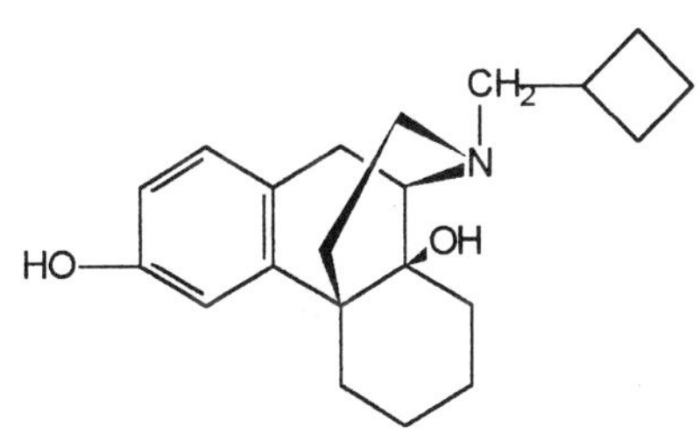

42.5

1. What is the opioid receptor activity profile of each agent drawn (42.1–42.5) in this case?

2. What is your therapeutic goal?

3. Which, if any, of the alternative agents would you recommend?

Case 43

Submitted by Victoria F. Roche

Your community has started a halfway house where heroin (43.1) addicts can continue rehabilitation after being released from health-care institutions. This facility used po naltrexone (43.2) in drug-free addicts as a deterrent to heroin abuse. When asked, an addict who didn't show up for his last naltrexone (43.2) dose says he has been heroin free since last receiving naltrexone (43.2) 2 weeks ago. He is given his 150-mg dose of naltrexone (43.2), and in about 30 minutes begins to experience severe tremor, painful muscle spasms, and abdominal cramps. He becomes hypertensive and tachycardic.

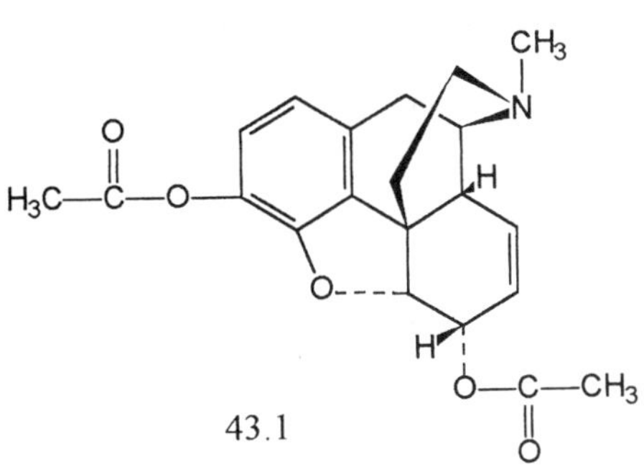

43.1

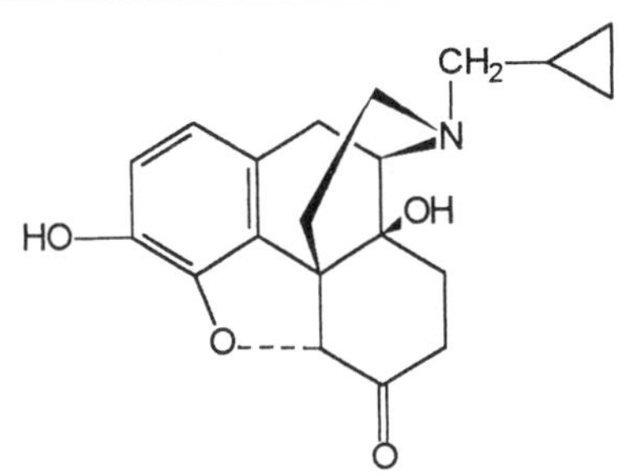

43.2

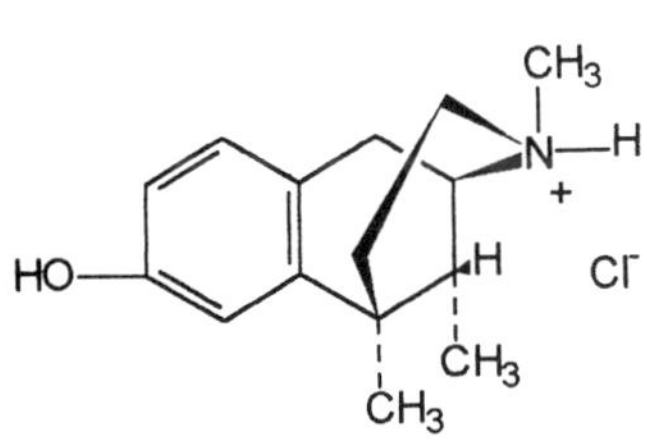

43.3

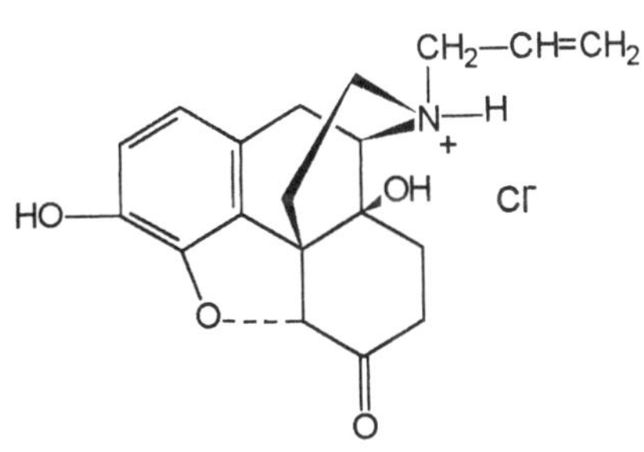

43.4

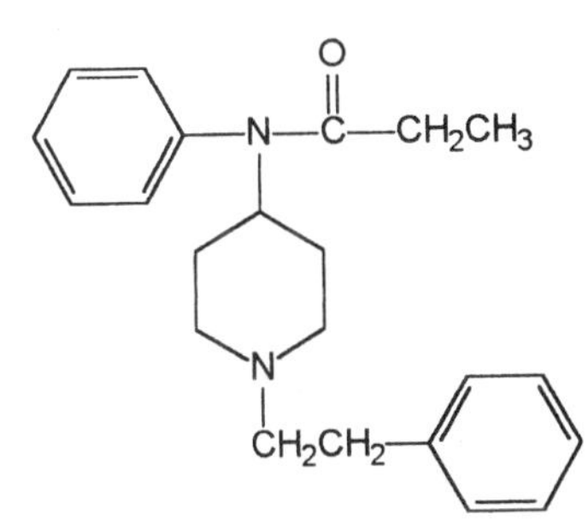

43.5

1. What is the matter with this patient? What are his current therapeutic needs?

2. **What is the opioid receptor activity profile of each of the structures drawn (43.2–43.5) in this case?**

3. **Which of the three agents (43.3–43.5), if any, would you recommend be administered to this patient?**

Case 44

Submitted by Arnold R. Martin and Richard N. Herrier

The chief of surgery calls the pharmacy service with a difficult problem. He has a 32-year-old male surgery patient with an open fracture of the left humerus. The problem is that the patient is 5'7" and weighs 475 lb. He has no other medical problems other than the severe obesity. The patient needs pain relief with an opiate, but the surgeon is not sure which drug would be safest and most effective and whether he should use lean body mass (ideal body weight) or actual weight as the basis for dosing the patient. You perform a literature search and fail to find a study that addresses the use of narcotics in obese patients. The narcotics in your formulary are: morphine (44.1), codeine (44.2), meperidine (44.3), oxycodone (44.4), and methadone (44.5). Based on your knowledge of drug structures and medicinal chemistry, answer the following:

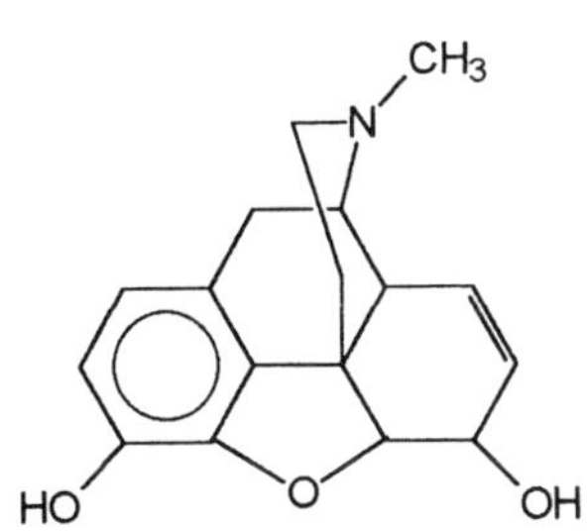

44.1

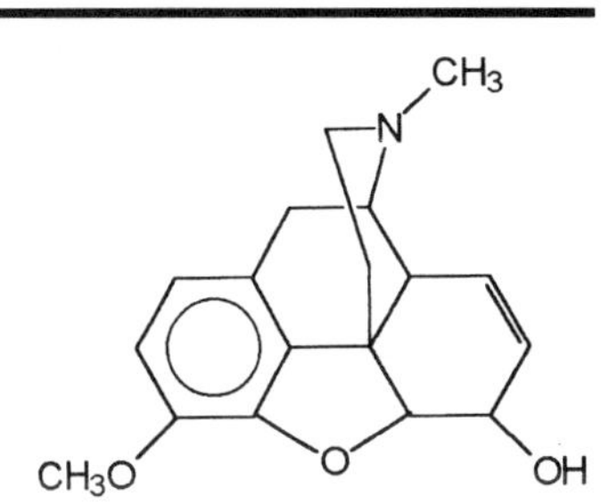

44.2

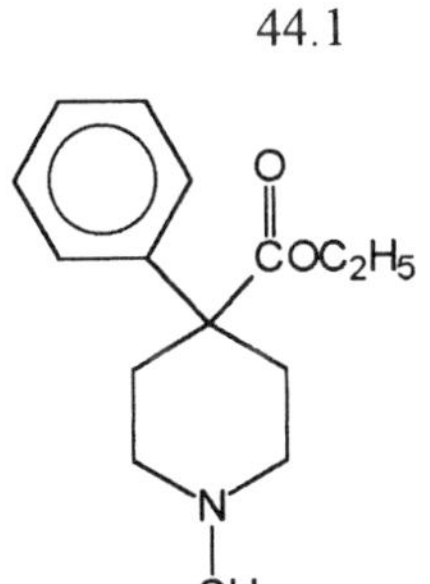

44.3

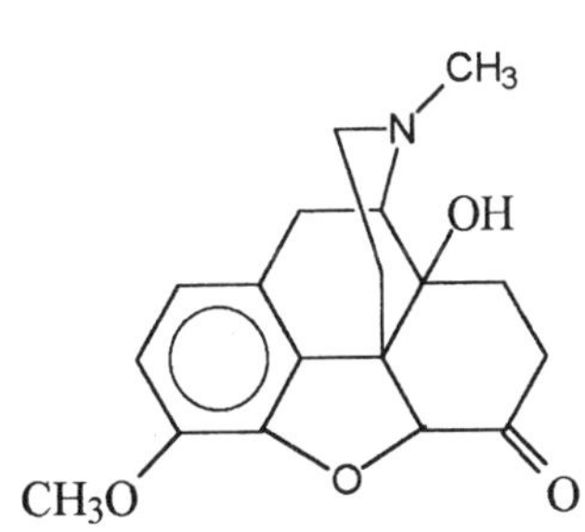

44.4

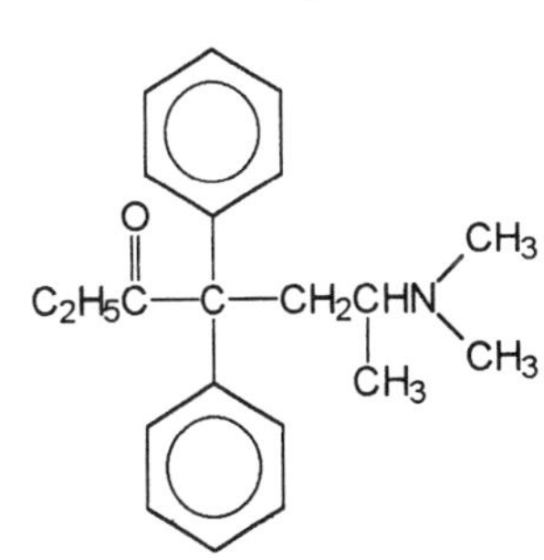

44.5

1. **Which available opiate (44.1–44.5) would be the best to use in this patient? Explain your rationale.**

2. **Should ideal or actual body weight be used as the basis for dosing the patient?**

Case 45

Submitted by Victoria F. Roche

NF, a 57-year-old woman with Alzheimer's disease and myasthenia gravis, is being cared for at home. Her husband is still employed full time, and volunteers from their church help with her care during the day. The woman has become increasingly combative, and will no longer accept medication from non-family members. You are asked to comment on the merits of three cholinesterase inhibitors that could be used to treat NF's myasthenia gravis. NF is not currently being medicated for the Alzheimer's disease.

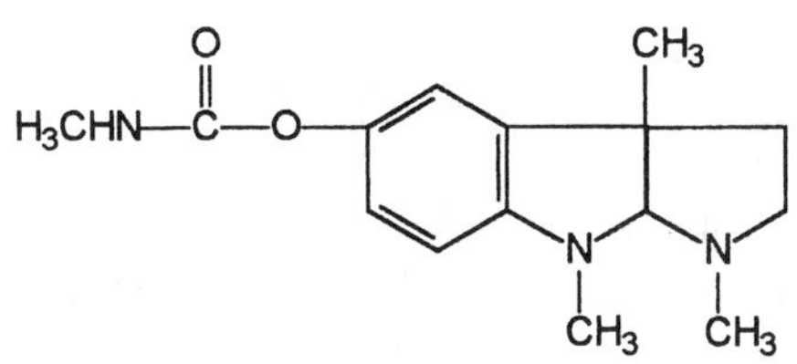

45.1

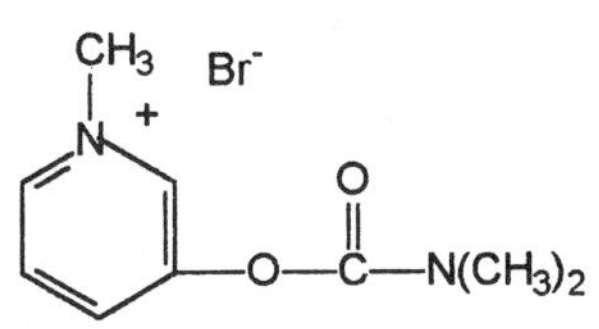

45.2

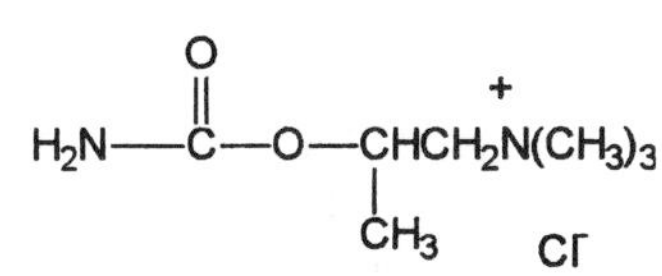

45.3

1. What are the therapeutic and personal needs of NF and her family?

Answer NF requires a drug that provides nicotinic cholinergic action of a long duration. Cholinesterase inhibitors provide both muscarinic and nicotinic activity because they reversibly inhibit cholinesterase enzymes found in both synapses. The endogenous transmitter acetylcholine, since it is not being destroyed as rapidly by acetylcholinesterase (AChE), can act longer at receptors.

2. **How well do the structural features of the drug candidates provided (45.1–45.3) meet these needs?**

Answer The requirement for nicotinic action of long duration would rule out drug 45.3 which, although an effective cholinesterase inhibitor, has a short duration of indirect cholinergic action due to the primary carbamate group. The lack of CH_3 groups on the carbamate nitrogen will minimize steric hindrance to attack of the carbamylated AChE serine residue by activated water, and the enzyme will be regenerated faster than would be seen if the carbamate structure was secondary or tertiary. Molecule 45.3 is a simple carbamate derivative of the neurotransmitter acetylcholine and has direct as well as indirect cholinergic action. The carbamate group of 45.3 is not conjugated with an aromatic ring, so the limited resonance delocalization of carbamate electron density is centered over the carbonyl carbon. Along with the steric and inductive hindrance provided by the β methyl group, this dramatically decreases its affinity for the AChE serine. It is very slow to be destroyed by AChE and would have a prolonged duration of direct cholinergic action. But the direct cholinergic activity produced by 45.3 is selectively muscarinic, because the β methyl group essentially destroys affinity for nicotinic sites. This drug (bethanechol chloride) (45.3) would not be of tremendous value to NF.

3. **Which of the three cholinesterase inhibitors (45.1–45.3) would you recommend?**

Answer A cholinesterase inhibitor with a long duration is needed so that traumatic confrontations with care providers can be avoided. Both remaining structures 45.1 and 45.2 are aminophenolcarbamates and act only by inhibiting the cholinesterase enzyme. Their affinity for postsynaptic muscarinic and nicotinic receptors has been destroyed by the presence of aromatic ring systems, which are too bulky to permit access to receptor surfaces. The extensive resonance delocalization of the carbamate's electrons with the aromatic ring increases the electrophilic character of the carbonyl carbon, and it is readily attacked by the AChE serine. When it hydrolyzes, it leaves behind an AChE serine carbamylated with either a secondary (45.1) or a tertiary (45.2) carbamate group. The methyl groups on the carbamate nitrogen will provide significant electronic and steric hindrance to the attack by activated water (the enzyme's mechanism of regeneration). Until the water can free the AChE serine, the enzyme is inhibited and will not hydrolyze endogenous acetylcholine. The more methyl groups attached to the carbamate nitrogen, the greater the hindrance to activated water and the longer the duration of therapeutic action. As a tertiary carbamate, structure 45.2 (neostigmine bromide) would have a longer duration of action and could be dosed less frequently than secondary carbamate 45.1 (physostigmine). It would be the best one to try in NF.

It should be noted that the secondary carbamate (45.1) is the only tertiary amine of the compound choices provided. It alone could penetrate the blood-brain barrier and inhibit acetylcholinesterase in the CNS, which is proposed to be beneficial in Alzheimer's disease. Physostigmine (45.1) is currently being used in some Alzheimer's patients and is dosed 0.5 mg q 2 h.

Case 46

Submitted by Jeffrey J. Christoff

Imagine yourself as a drug information specialist at a poison control center. A technician from the coroner's office is investigating a case and requires assistance in identifying the possible sources of benzodiazepines (BZDs) in the toxicology profile of a particular corpse. The technician has identified four distinct BZDs in this blood sample. She believes that the major component is diazepam (46.1) (72% of the identified BZDs) and that the remaining three compounds are metabolites (**NOTE:** the assay identifies only active compounds).

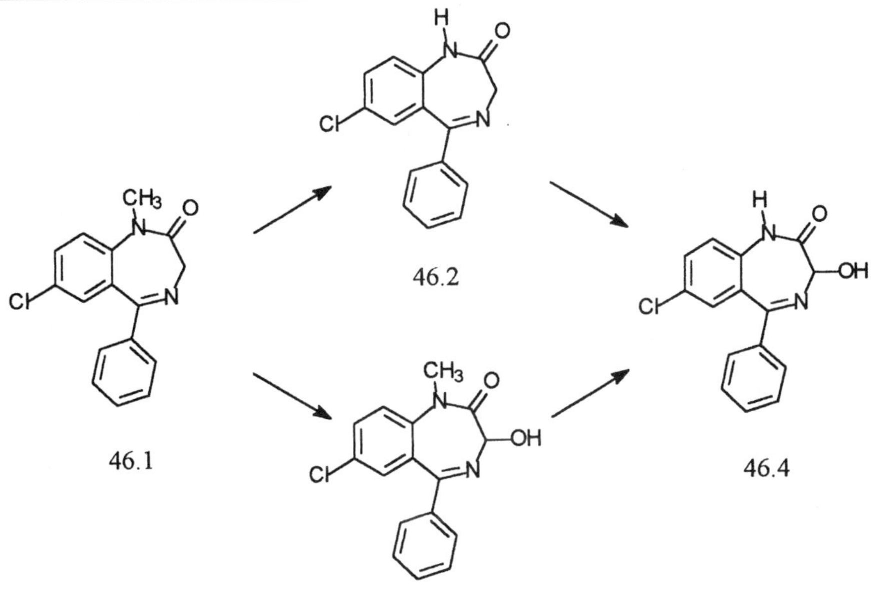

46.1 46.2 46.3 46.4

1. **What are the three structures of potential ACTIVE metabolites for diazepam (46.1).**

Answer Diazepam (46.1) can undergo N-dealkylation to the active metabolite desmethlydiazepam (46.2), or diazepam (46.1) can undergo oxidation at the 3 position to form temazepam (46.3). Both desmethyldiazepam and temazepam (46.3) can be metabolized further by oxidation or N-dealkylation respectively to produce oxazepam (46.4). Both temazepam (46.3) and oxazepam (46.4) are marketed along with diazepam (46.1) as CNS depressants.

Case 47

Submitted by Arnold R. Martin and Richard N. Herrier

MG is a 72-year-old widowed grandmother of ten living in Phoenix, Arizona in a planned retirement community. She likes to visit her two sons, their wives, and her 10 grandchildren in Seattle, Washington and Portland, Oregon whenever she can afford it. She doesn't like to drive because the trip is so long and exhausting, and she is becoming increasingly less confident of her driving skills. Her only choice is to fly, but she had a very bad experience on her one and only airplane trip, which has made her exceedingly anxious. She does not want to be "zonked out" during the flight, because she wants to see the snow-capped mountains en route and arrive at her destination reasonably alert.

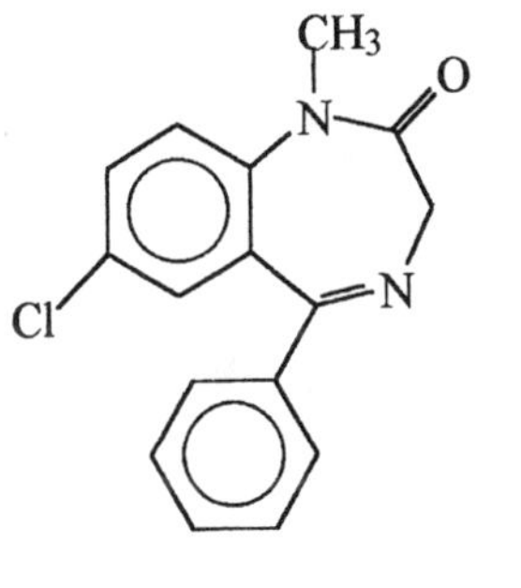

47.1

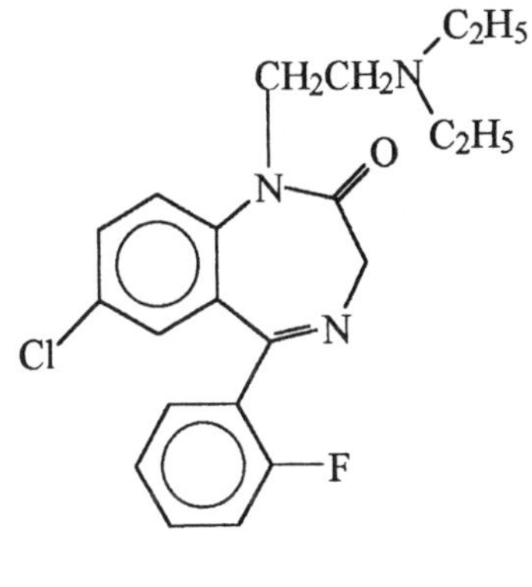

47.2

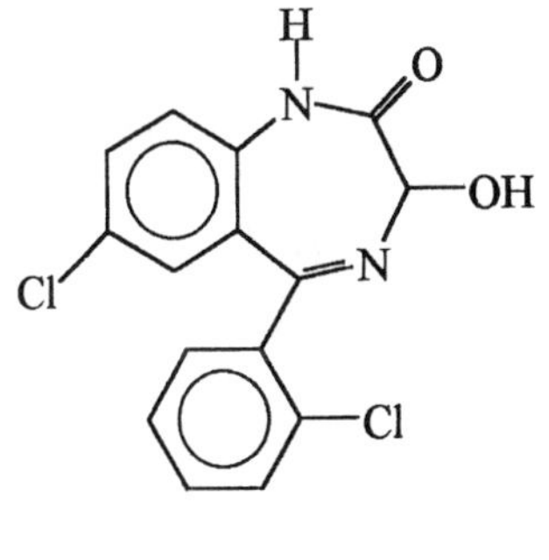

47.3

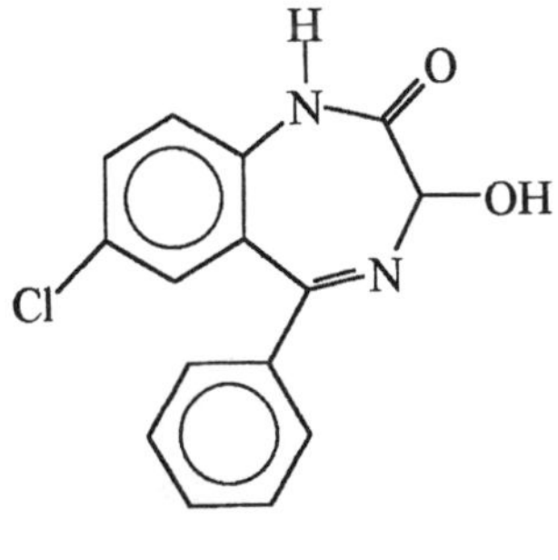

47.4

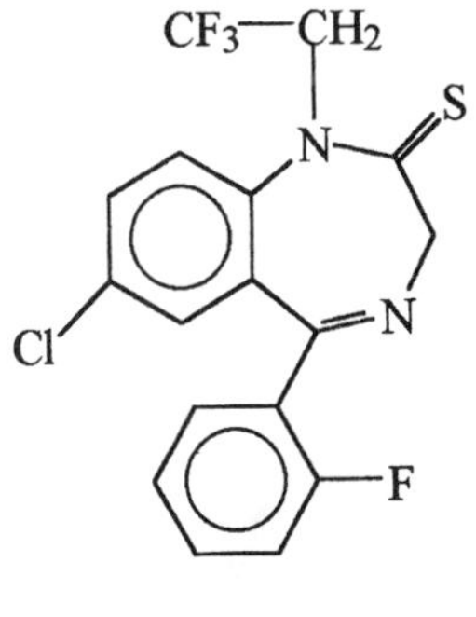

47.5

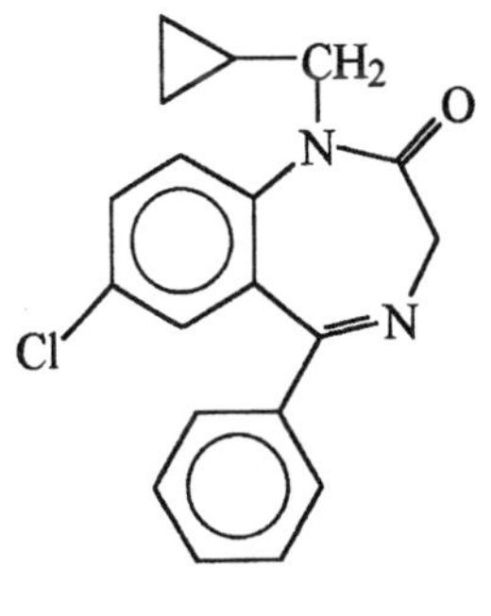

47.6

1. **Select a benzodiazepine from the structures provided (47.1–47.6) that will best fit the needs of MG. Justify your choice on the basis of comparative potencies and pharmacokinetics of the structures.**

Answer The anxiolytic of choice is oxazepam (47.4). It is the least potent and shortest acting of the structures listed.

2. **Rank the benzodiazepines given in the order of comparative potencies (most to least potent) and durations of action (shortest to longest).**

Answer Potencies of benzodiazepines correlate with their lipid solubilities, primarily. The order is flurazepam (47.2) > quazepam (47.5) > lorazepam (47.3) ≥ prazepam (47.6) ≥ diazepam (47.1) >> oxazepam (47.4). Durations of action correlate with metabolic inactivation (O-glucuronidation) *versus* bioactivation (N-dealkylation). The order: oxazepam ≤ lorazepam << diazepam ≤ prazepam ≤ quazepam << flurazepam.

Case 48

Submitted by Victoria F. Roche

JB, a 34-year-old woman, was the recent victim of a car accident in which she sustained a severe head injury. She has come to the emergency room after suffering a severe, generalized tonic-clonic convulsive episode. She is hospitalized and, within the next few days, two more generalized seizures are experienced. It is decided to initiate chronic anticonvulsant therapy, and your advice is solicited.

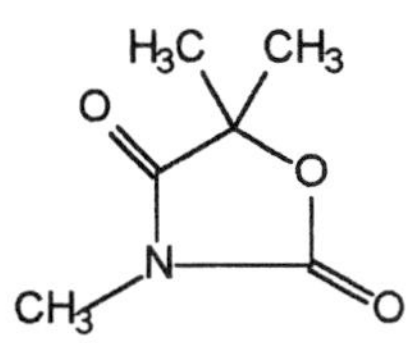

48.1

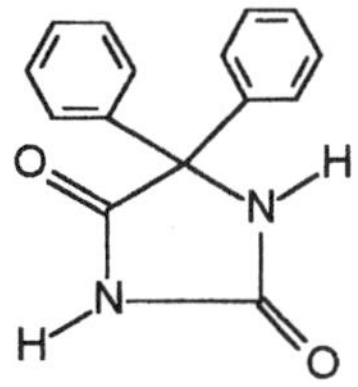

48.2

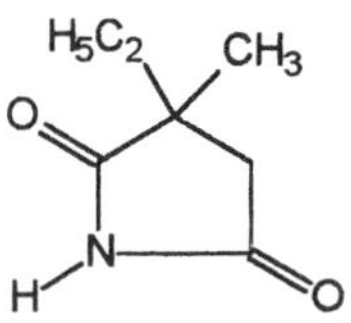

48.3

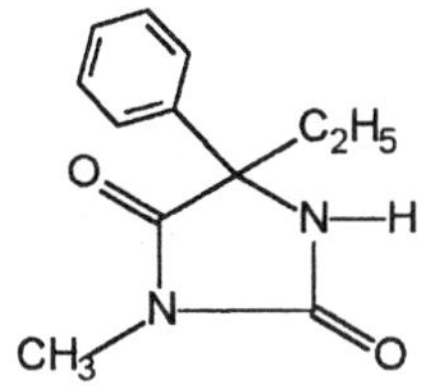

48.4

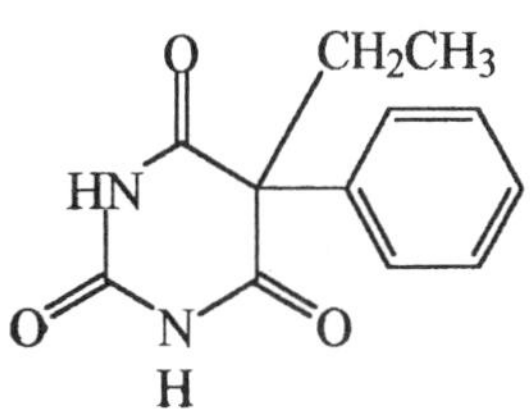

48.5

1. Which of the anticonvulsant structures (48.1–48.5) should be administered to JB?

Answer Compound 48.2 (Phenytoin) would be the drug of choice in this case. Phenobarbital (48.5), a ring expanded analogue of phenytoin, has been used for prophylactic anticonvulsant therapy following head injury. The drug can be given qd hs and has been shown to have no effect on intracranial pressure.

2. **What structural features of your drug of choice are responsible for your answer?**

Answer The substitution pattern at C_5 of the hydantoins (48.2, 48.4) and oxazolidinediones (48.1) (or C_2 of the succinimides [48.3]) is an important determinant of the type of antiseizure activity exhibited by the compound. Hydantoins with at least one C_5 phenyl moiety are the drugs of choice in generalized tonic-clonic (grand mal) seizures. Oxazolidinediones, such as compound 48.1, are generally substituted at C_5 with small alkyl chains (e.g., methyl or ethyl), which make them less effective against grand mal seizures and more effective against absence (petit mal) seizures. The oxazolidinediones are toxic structures and succinimides, like structure 48.3, were developed as safer alternatives for absence therapy. The most potent anti–petit mal succinimides have small alkyl groups at C_2. Regardless of ring type, disubstitution at C_5 (or C_2) provides more potent anticonvulsant agents than monosubstituted derivatives.

Given the preceding, one can rule out structures 48.1 and 48.3. These structures are an oxazolidinedione and a succinimide, respectively, with small alkyl groups at C_5 or C_2. They would be appropriate for treating absence seizures, but would not be effective in the grand mal seizures being experienced by JB.

The hydantions 48.2 and 48.4 differ in substitution pattern at C_5. The diphenyl substitution pattern of molecule 48.2 not only provides more potent anti–grand mal activity than the single phenyl substituent of molecule 48.4, it is also associated with a decrease in toxicity.

Case 49

Submitted by Bruce L. Currie

GM, a 58-year-old woman, has been taking diazepam (49.1) to help her sleep for the past 2 years since her husband died. Her physician has recently prescribed cimetidine (49.2) for her because her ulcer has been acting up. She has come to you for a refill on her Valium (49.1) prescription and complains to you that she has a difficult time waking up in the morning.

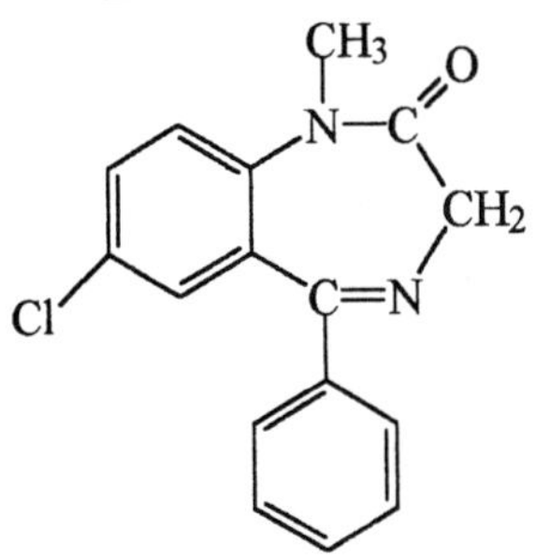

49.1

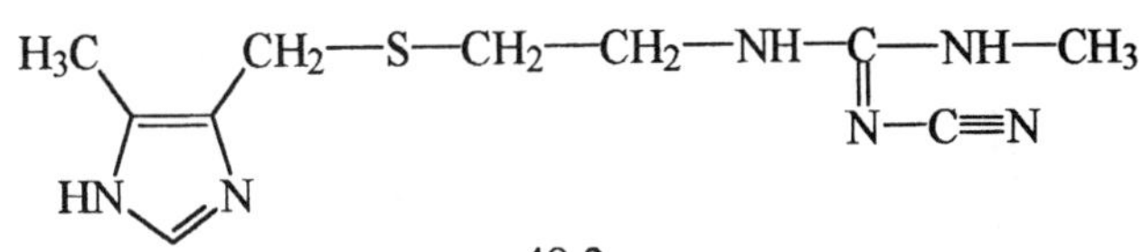

49.2

1. **Identify any problems with the medications.**

2. **What should you, the pharmacist, recommend to the patient's physician before filling the diazepam (49.1) prescription? Make a recommendation and explain your rationale.**

Case 50

Submitted by Arnold R. Martin and Richard N. Herrier

SA is a 20-year-old university coed who presents to the student health service complaining of extreme nervousness, irritability, and difficulty concentrating. For the past year, SA has constantly worried about achieving sufficiently high grades in her prepharmacy courses to qualify for entry into the Doctor of Pharmacy program. She is also concerned about her skin condition, which has worsened since she entered the University 2 years ago. SA has suffered from atopic dermatitis since high school, a condition which until recently had been kept under control using topical hydrocortisone (0.5% cream). As the time for her interview with the College of Pharmacy admissions committee approaches, SA's worry increases, and she has become increasingly agitated, is unable to concentrate on her studies, and has shortness of breath and palpitations. To make matters worse, the itching rash on her face and arms has worsened.

The student health physician, in consultation with SA's family doctor, decides on a diagnosis of generalized anxiety disorder (GAD). The two physicians agree that a benzodiazepine is indicated.

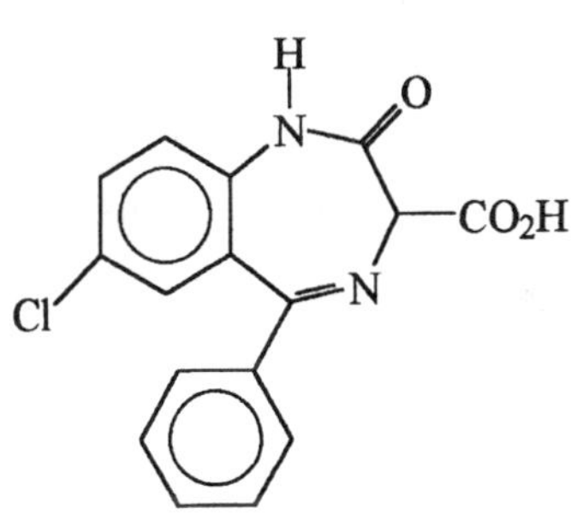

50.1

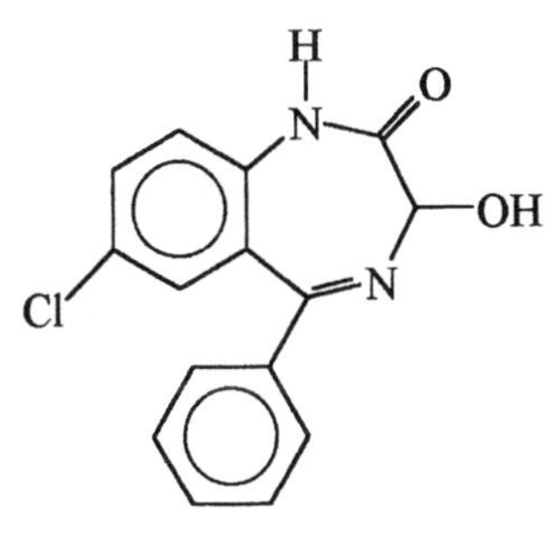

50.2

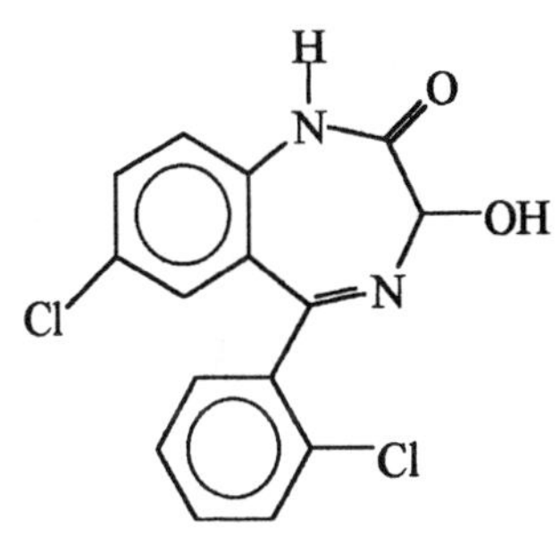

50.3

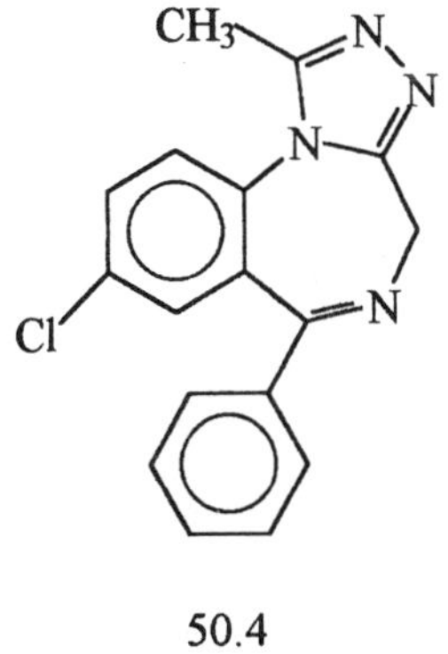

50.4

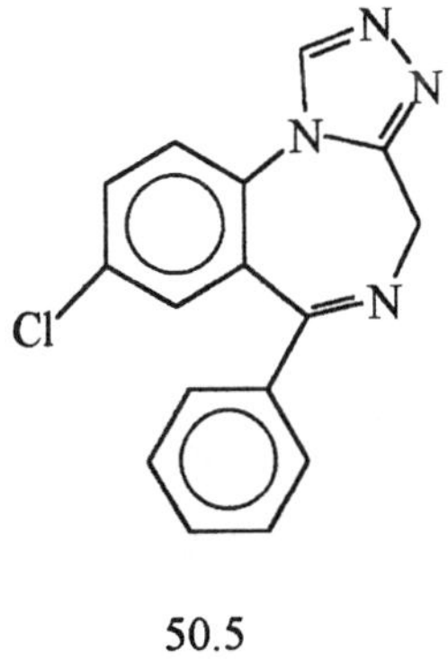

50.5

1. **Select a benzodiazepine from the structures (50.1–50.5) provided that would be the most appropriate for treating SA's anxiety. Justify your choice on the basis of structure/potency and duration of action/relationships. Eliminate inappropriate choices on the same basis.**

2. **How long should SA be maintained on anti-anxiety medication? Explain fully.**

3. **What are you going to do for the patient's atopic dermatitis?**

Case 51

Submitted by Arnold R. Martin and Richard N. Herrier

A psychiatrist at the health maintenance organization where you are employed as a staff pharmacist calls you for advice concerning the choice of medication for one of her patients. The patient, YO, is a 27-year-old male schizophrenic who has been treated for the past 6 years using chlorpromazine (51.1). It has been necessary for the psychiatrist to gradually increase the dosage of chlorpromazine through the years to maintain control of YO's symptoms. During the past 6 months, YO began to complain of excessive thirst, constipation, and blurred vision. At a checkup last week, the psychiatrist observed that YO was severely sunburned and had greyish brown splotches on his arms, neck, and forehead. On consulting his *Physician's Desk Reference*, the psychiatrist discovered that chlorpromazine (51.1) can cause abnormal skin pigmentation and photosensitivity. The psychiatrist thus decided to change YO's medication.

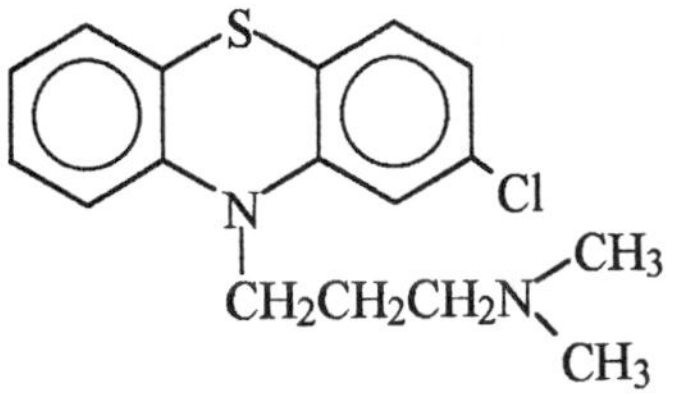

51.1

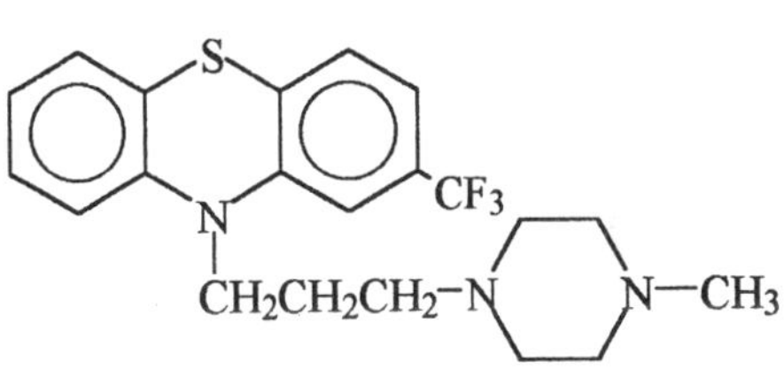

51.2

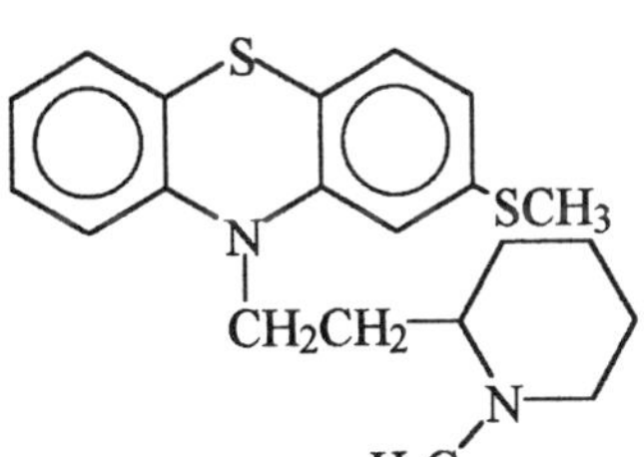

51.3

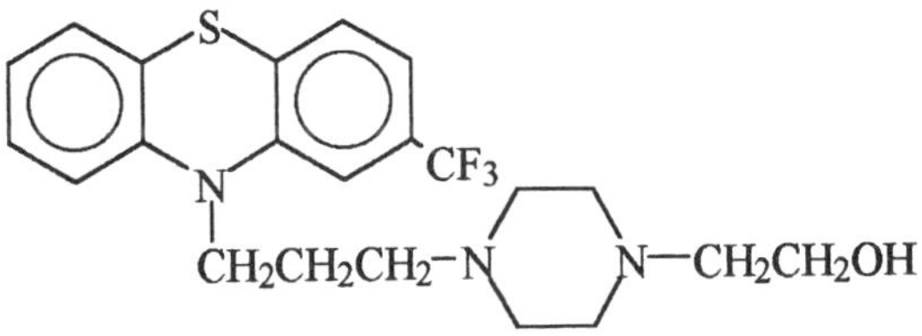

51.4

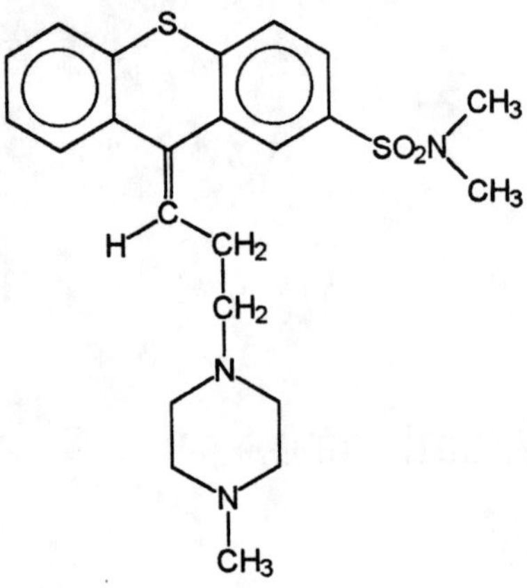

51.5

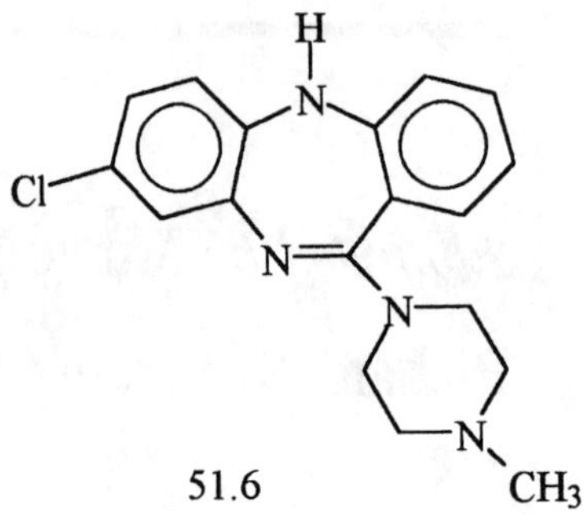

51.6

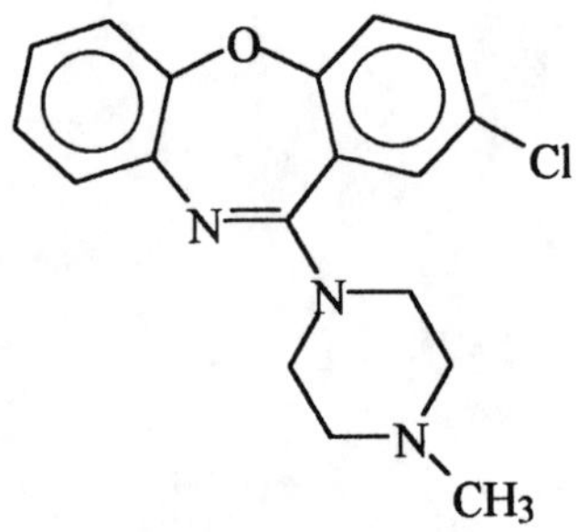

51.7

1. **Select a suitable antipsychotic agent from the structures (51.2–51.7) provided that will: (1) maintain control of YO's psychotic symptoms, (2) reduce the intensity of side effects that YO is experiencing, and (3) not exacerbate YO's skin problems. Justify your choice on the basis of the comparative structure-activity relationships among the available structures.**

Case 52

Submitted by Jeffrey J. Christoff

NP is a 6-year-old boy with attention deficit disorder. He has been treated with methylphenidate (52.1) for 6 weeks with substantial improvement in behavior. This drug has multiple effects on various neurotransmitters.

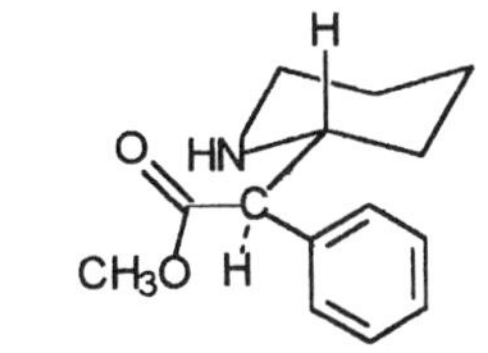

52.1

1. **Based on its structural resemblance to other phenylethylamines, describe one effect this drug (52.1) can have on a particular neurotransmitter (include a brief structure-activity relationship of methylphenidate [52.1]).**

2. **Even though 80 to 90% of the drug is absorbed following oral administration, only about 4% of methylphenidate (52.1) actually penetrates the CNS. Why would this be the case?**

Case 53

Submitted by Jeffrey J. Christoff

GC is a 56-year-old man who suffers from insomnia. He has tried sleep therapy and self-treatment with Sominex (diphenhydramine), but neither has been successful. His physician requests your assistance for drug selection because GC is a former drug abuser (heroin, cocaine, and alcohol) and has attempted suicide in the past. GC has no prior history of any chronic medical conditions and is an active member of Alcoholics Anonymous. The following drugs are currently available hypnotics.

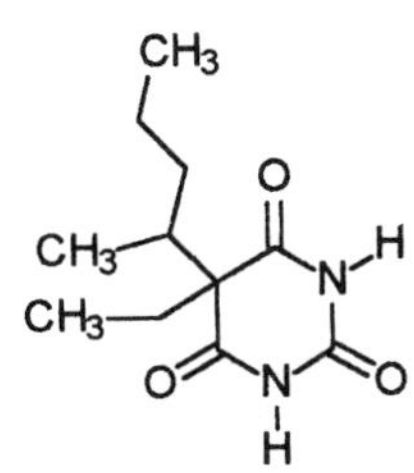

53.1

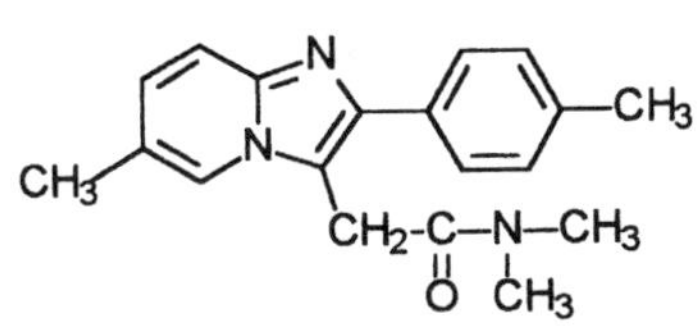

53.2

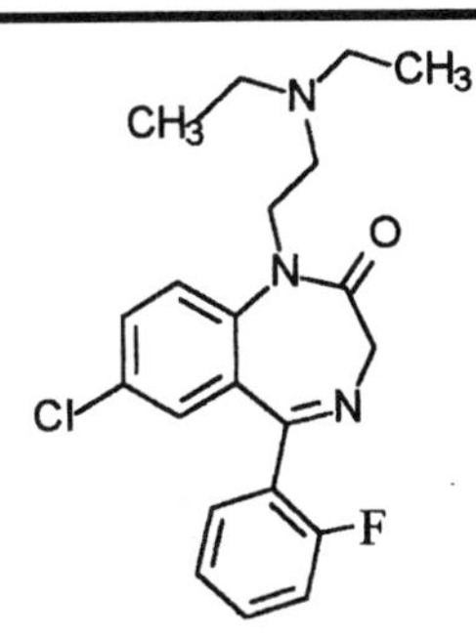

53.3

1. Rank these hypnotic agents (53.1–53.3) according to their addictive liabilities.

________________	________________	________________
least potential for addiction	**intermediate potential for addiction**	**highest potential for addiction**

2. Name the basic pharmacophore for each of the drugs shown.

Case 54

Submitted by Arnold R. Martin and Richard N. Herrier

A 68-year-old woman was admitted to the oncology unit for a colonoscopy to determine if her rectal and colonic polyps have become cancerous. Because she is highly anxious about the procedure, the attending physician has decided to perform the colonoscopy while the patient is conscious, but highly sedated. Since the procedure requires only a few minutes to perform, the doctor desires a drug with a rapid onset and a speedy recovery for his patient.

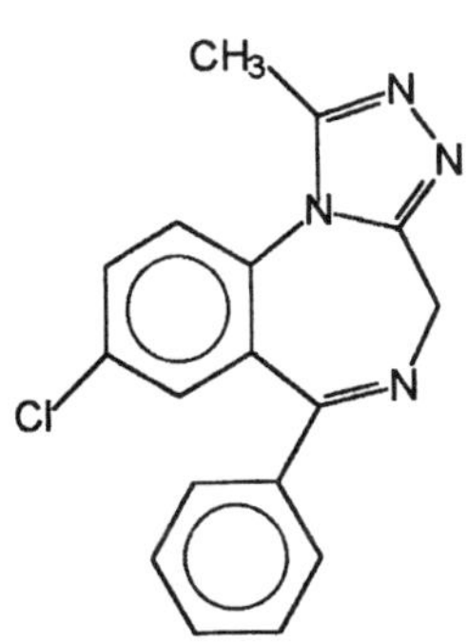

54.1

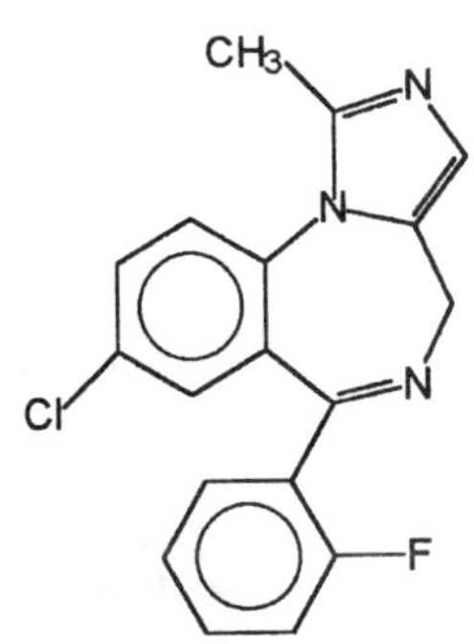

54.2

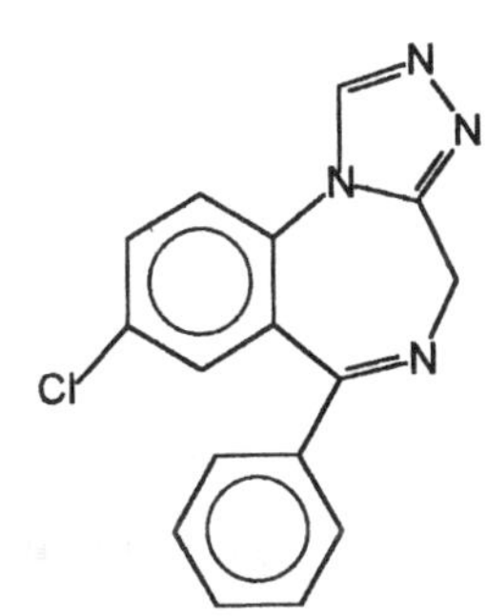

54.3

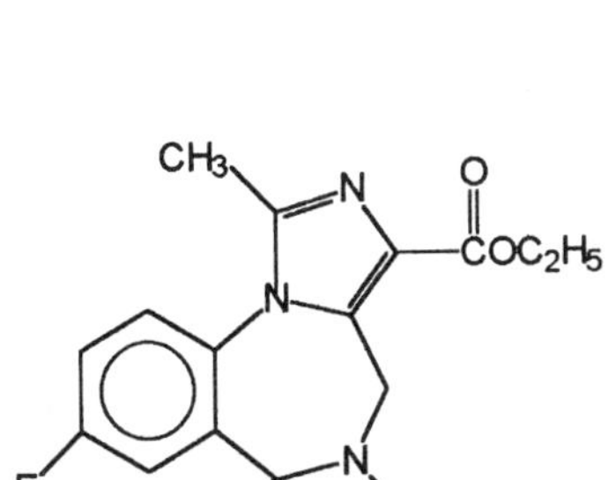

54.4

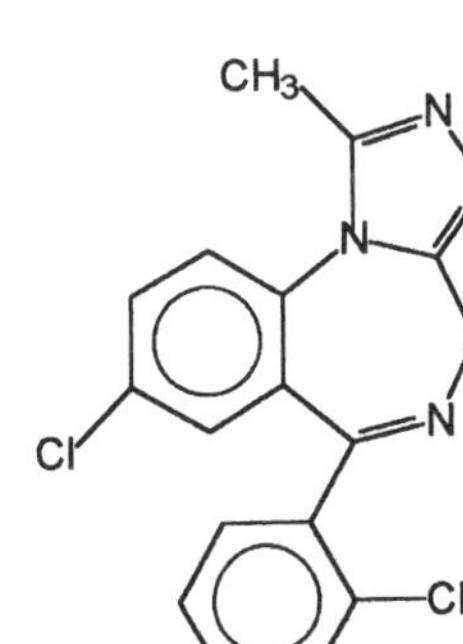

54.5

1. **Select the best agent from the benzodiazepine structures (54.1–54.5) provided to serve as an anxiolytic/sedative for the colonoscopy. Justify your choice on the basis of the chemical properties and comparative potencies and pharmacokinetics of the structures.**

2. **Outline the biotransformation of each of the benzodiazepines listed and rank them in the order of their increasing durations of action.**

3. **One of the structures is a benzodiazepine antagonist. Identify the structure and the functional group that confers antagonist properties.**

Case **55**

Submitted by Arnold R. Martin and Richard N. Herrier

JN is a 44-year-old male conservative radio/TV talk show host with a type A personality. Although he has always been a trifle overweight and moderately hypertensive, JN worried little about his health until the chest pains started to occur. The pains were particularly common just before, and even during, the stress of his radio and TV broadcasts. JN was given a prescription for nitroglycerin by his doctor with the advice to "stay on your diet and slow down, particularly now that there are republican majorities in both the house and senate." However, JN finds it impossible to curtail his activities, particularly now that his fans are touting him for the governor's chair (currently held by a liberal democrat). Convinced that JN usually doesn't follow medical advice, his physician elects to prescribe an anxiolytic benzodiazepine, hoping that it will relax JN and thereby relieve some of his symptoms (if not moderate his type A behavior). It should be pointed out that JN is somewhat careless about taking his medication, except as a part of his morning ritual.

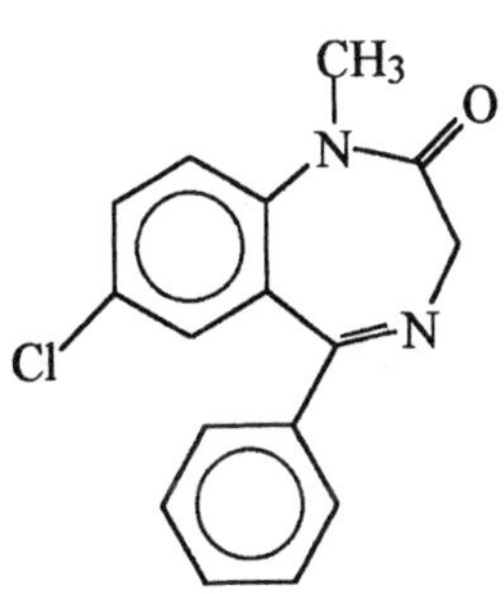

55.1

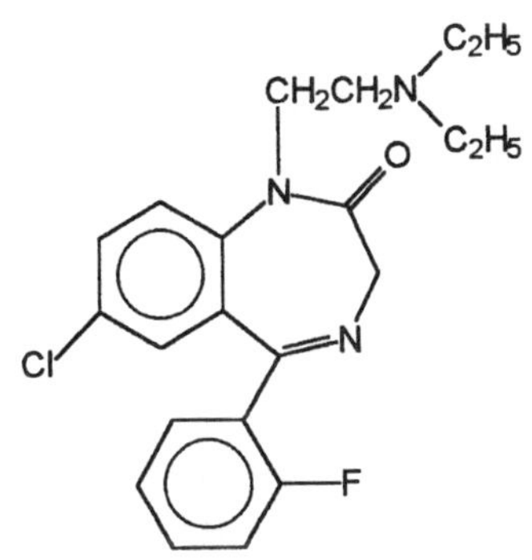

55.2

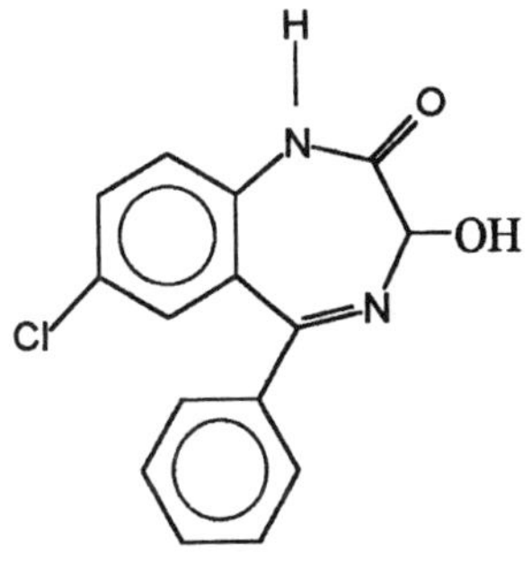

55.3

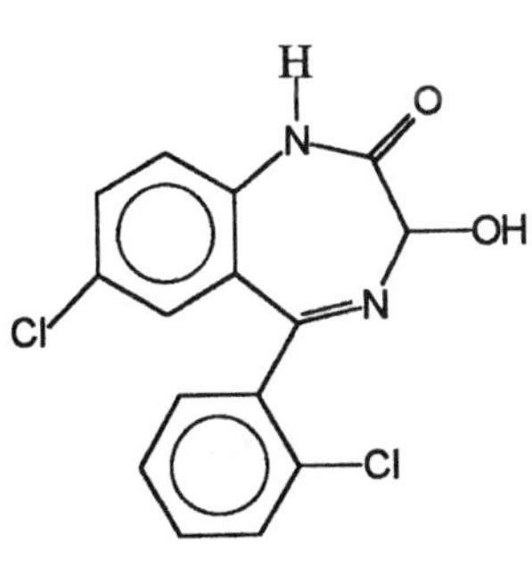

55.4

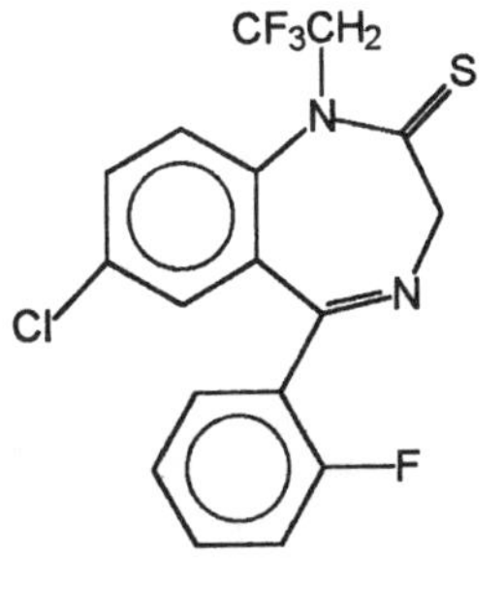

55.5

1. **Select the most appropriate anxiolytic benzodiazepine for JN from the structures given (55.1–55.5). Justify your choice on structure/potency/duration of action considerations.**

Case 56

Submitted by Victoria F. Roche

You are taking a night off from the pharmacy and have rented an old gangster movie from the video store to watch with friends. At a critical point in the show, a shady looking character slips a capsule into an innocent lady's martini. She sips on her drink unknowingly through several minutes of boring dialogue and, a short time later, slumps over on the bar, unconscious. Your friends, obviously not pharmacists, ask about the chemistry behind this dramatic scene. You know that the capsule contained chloral hydrate (56.1).

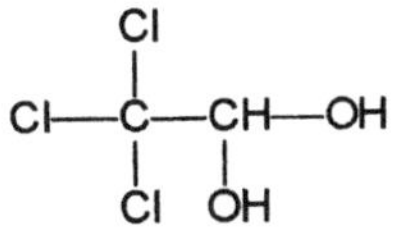

56.1

1. **What is chloral hydrate? What is its mechanism of pharmacologic action?**

2. **Why couldn't the innocent victim tell something was fishy with her drink?**

Case 57

Submitted by Victoria F. Roche

VR, a medicinal chemistry professor whose students were unable to draw the structure of acetylcholine after a 2-week lecture series on cholinergic agonists, has been brought, straight-jacketed and screaming, into the emergency room of your hospital. Years of trying to teach everything there is to know about the chemical basis for drug action in five credit hours has induced a persistent tachycardia in this professor, and medical history reveals pigmentary dispersion syndrome (a predisposition to glaucoma). You are asked to recommend initial therapy for this professor, and find the following phenothiazine (or related) structures on hand.

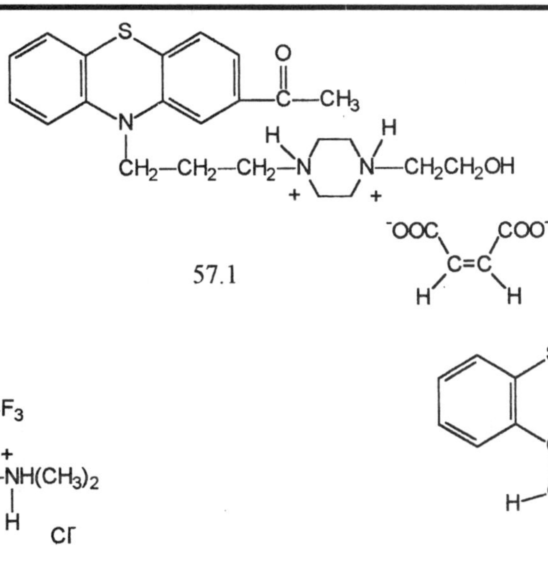

57.1

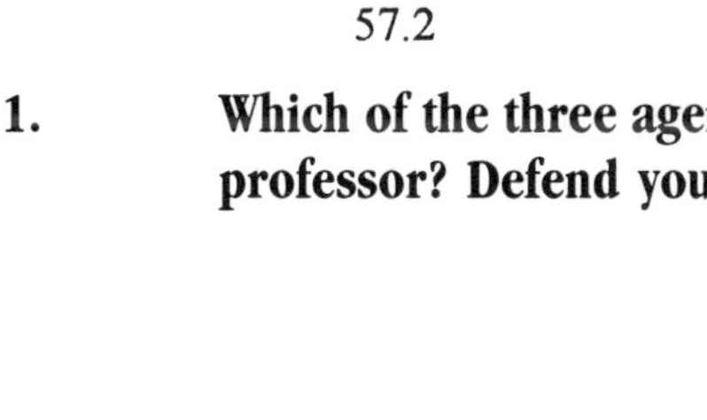

57.2

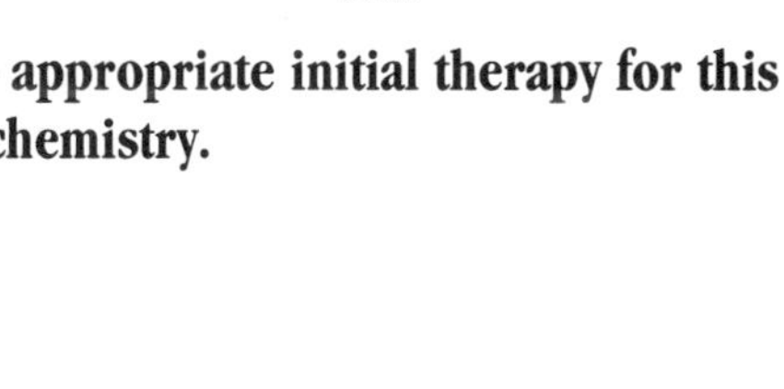

57.3

1. **Which of the three agents (57.1–57.3) would be appropriate initial therapy for this professor? Defend your choice based on drug chemistry.**

2. **This professor is argumentative and non-compliant with her medication regimen. Could you modify the structure of your antipsychotic of choice to address these therapeutic and behavior issues? If so, how?**

Case **58**

Submitted by Victoria F. Roche

A physician asks your advice on therapy for JF, a newly diagnosed schizophrenic who has just been admitted to the psychiatric hospital. In a review of the case, you find he is exhibiting mild psychotic symptoms, and has a history of intermittent atrial tachycardia, along with mild parkinsonism currently controlled with L-dopa/carbidopa (Sinemet). You consider the four phenothiazine structures (58.1–58.4) drawn as potential candidates to treat JF's schizophrenia.

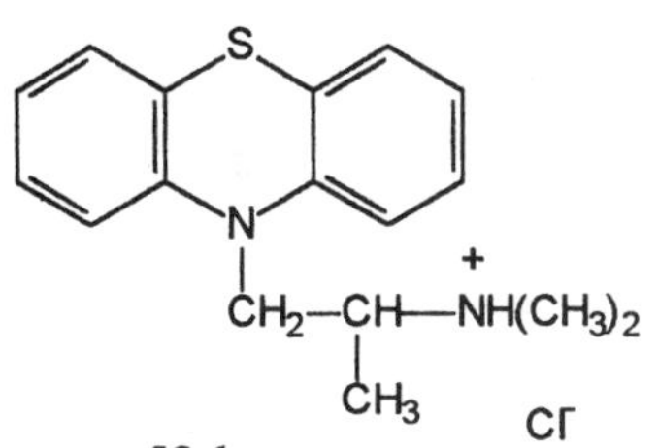

58.1

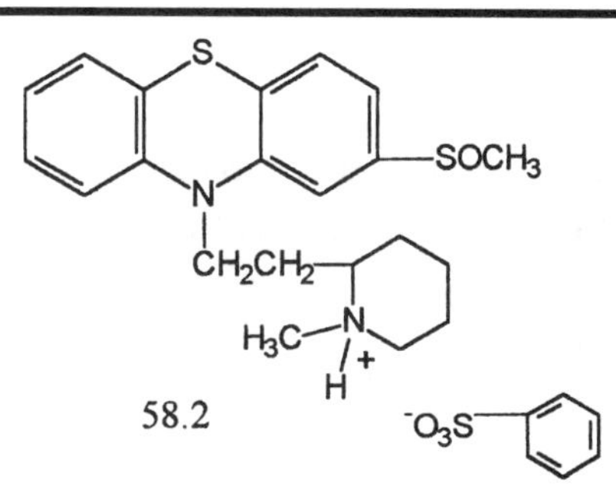

58.2

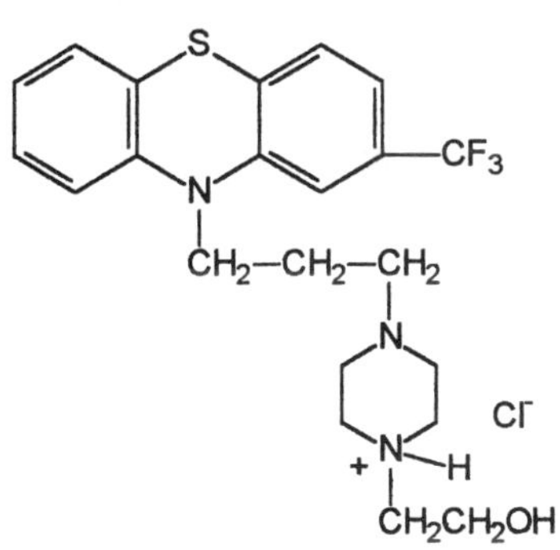

58.3

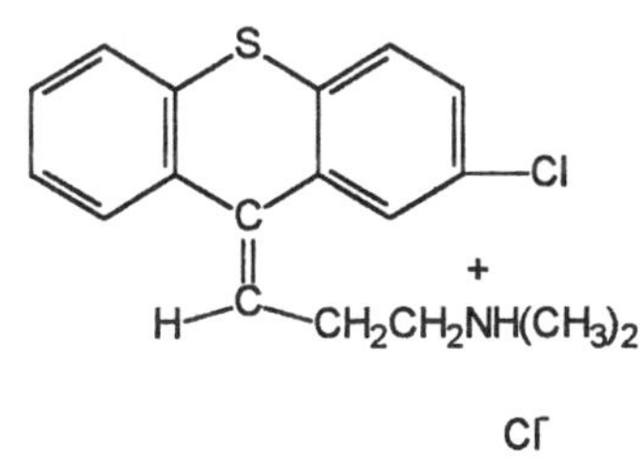

58.4

1. How do JF's other disease states complicate the treatment of his psychiatric disorder?

2. **What structural features on each of the four phenothiazine molecules (58.1–58.4) would promote antipsychotic activity? What structural features would cause problems for JF?**

3. **What would be the pros and cons of each of the four phenothiazine molecules in this therapeutic situation. Which one, if any, would you recommend?**

Case 59

Submitted by Victoria F. Roche

You are the medicinal chemist in charge of an industrial research team involved in synthesizing a new antimuscarinic agent for use in treating motion sickness. Your plan is to make an aminoalkylester using the naturally occuring 8-azabicyclo[3.2.1]octane skeleton as your receptor anchor. You want to provide a compound that is substituted for optimal potency and plan to market your new drug in patch form for transdermal drug delivery. The azabicyclooctane skeleton is drawn (structure 59.1).

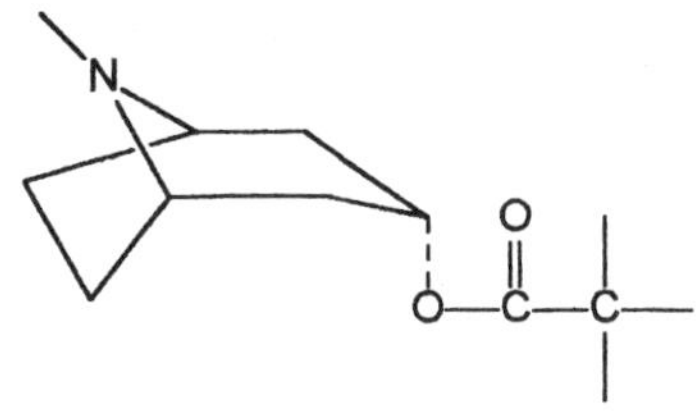

59.1

1. **How would you substitute this structure at the amino nitrogen and the alpha carbon to accomplish your therapeutic aims?**

2. **What salt form, if any, would be of value in this therapeutic agent?**

3. **Provide a discussion as to why your substitution pattern would be optimal for both effective delivery and high potency.**

Case 60

Submitted by Victoria F. Roche

You are the medicinal chemist in charge of an industrial research team involved in synthesizing a new antimuscarinic agent for use in Parkinson's disease. Your plan is to make an aminoalkylether using the naturally occuring 8-azabicyclo[3.2.1]octane skeleton as your receptor anchor. You want to provide a compound that is substituted for optimal potency. The azabicyclooctane skeleton (in salt form) is provided (structure 60.1).

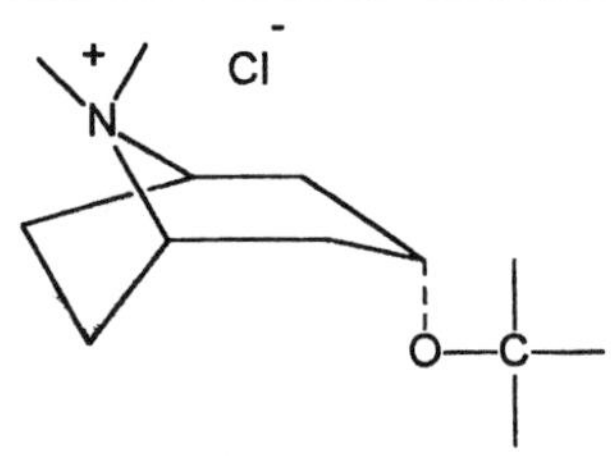

60.1

1. **How would you substitute this structure (60.1) at the amino nitrogen and the alpha carbon to accomplish your therapeutic aims?**

2. **From a receptor interaction standpoint, why is your substitution pattern optimal?**

Case 61

Submitted by Jeffrey J. Christoff

MR is a 26-year-old man who weighs 420 pounds. He has sought counseling for weight loss, but modifications of diet and exercise alone have never really been successful for him. He recently watched a "20/20" program (aired 2/3/95) that described the use of prescription drugs as part of a weight loss program. The drug therapy used a combination of fenfluramine (61.1) and phentermine (61.2).

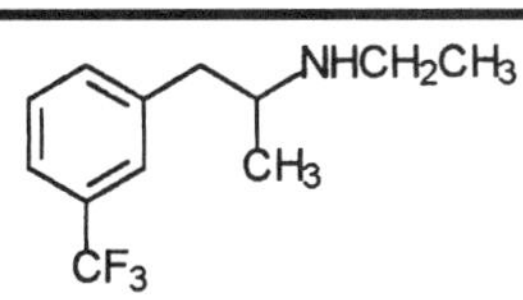

61.1

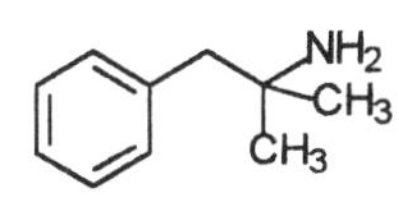

61.2

1. **Explain the mechanism of action of EACH drug (including a brief structure-activity relationship).**

2. **Would the addition of a β-hydroxyl group to either of these drugs improve CNS penetration? Why, or why not?**

Case 62

Submitted by Bruce L. Currie

FM, a 75-year-old man, has been treated for Parkinson's disease for the past 5 years. He has been taking L-dopa (62.1)/carbidopa (62.2) plus amantadine (62.3), but has been experiencing decreased function, cog-wheel rigidity, and renewed onset of hypokinesia. At the last clinic visit, FM was placed on reserpine (62.4) and a diuretic as an approach to treating his hypertension. Over the last 6 weeks, he has had decreased mobility and increased mental status changes, which can be attributed to the medication additions.

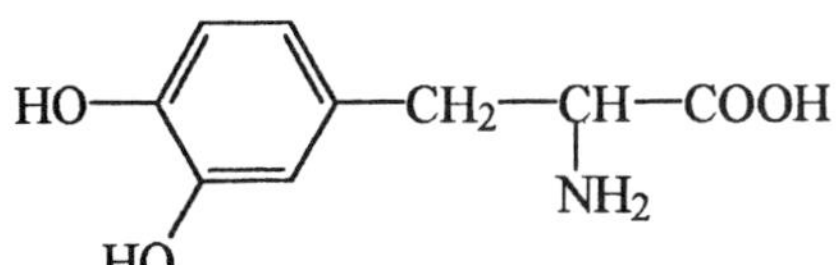

62.1

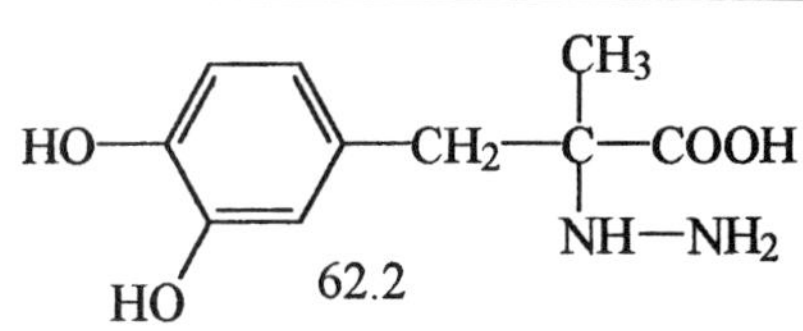

62.2

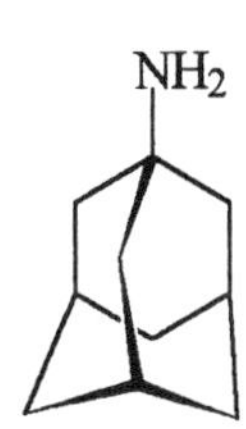

62.3

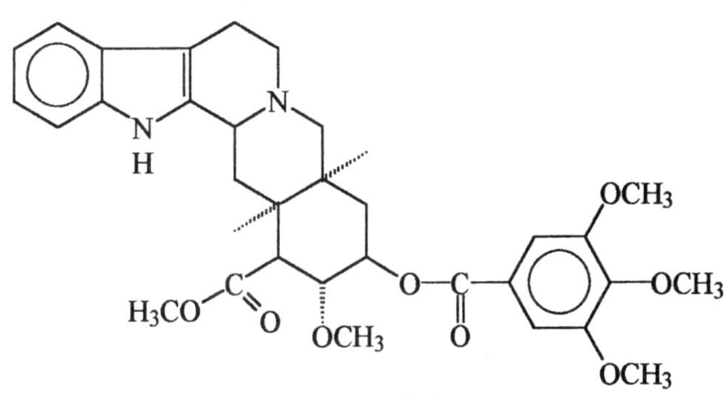

62.4

1. Outline what you think the problems are with the management of FM.

2. Propose a solution or an approach to therapy and defend your approach briefly, but clearly.

Case 63

Submitted by Victoria F. Roche

An 87-year-old woman, newly widowed, has been placed in a nursing home in your community. Her grief is currently being expressed as a rather severe depression, and the attending physician wishes to initiate antidepressant therapy. This patient has osteoporosis and, due to intermittent vertigo, uses a cane to steady her gait. She has a history of cardiac arrhythmia, and is stabilized on atenolol (63.5) (a β-adrenergic blocking agent).

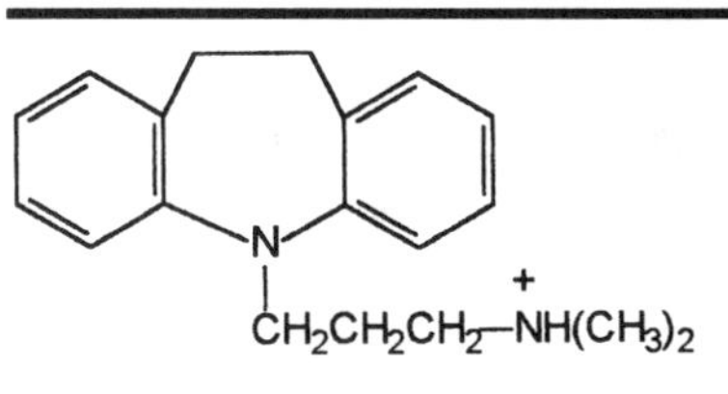

63.1

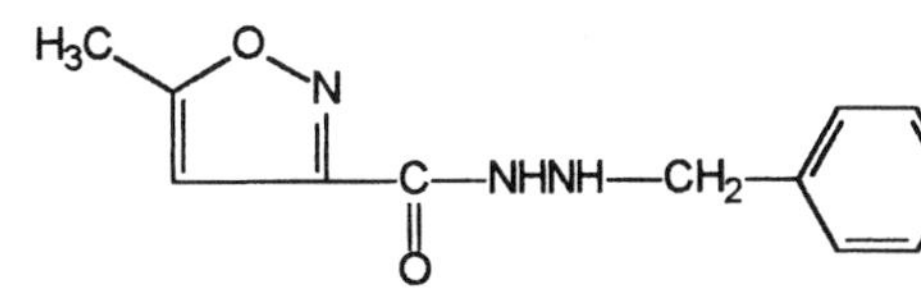

63.2

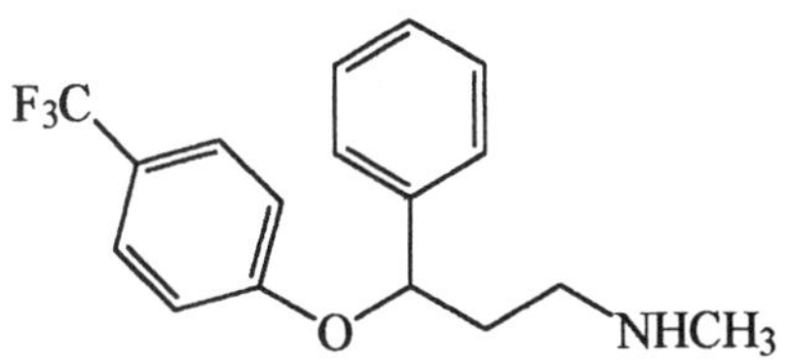

63.3

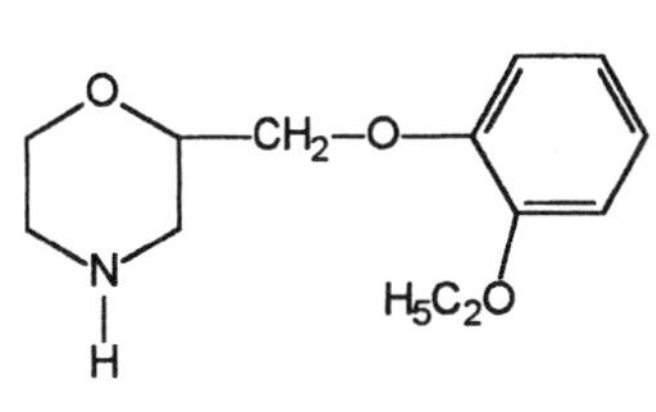

63.4

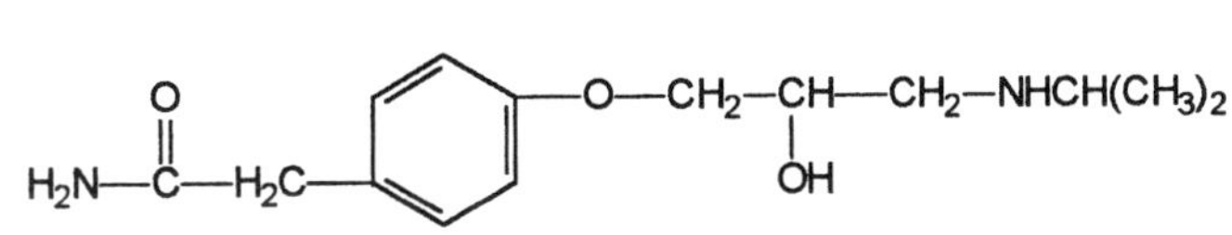

63.5

1. **What are the mechanism(s) of antidepressant action of the drug structures (63.1–63.4) provided in this case?**

2. **What would be the pros and cons of each agent in treating this patient?**

3. **Which of the four antidepressant structures (63.1–63.4) would you recommend?**

Case 64

Submitted by Victoria F. Roche

PW, a 28-year-old single mother of three, works as an elementary school bus driver during the academic year. She has been seeking counseling for depression that has been worsening over the past 6 months, and has now turned to her physician for help. Medical history indicates current use of ophthalmic epinephrine dipivaloate ester (64.5) (for glaucoma) and diphenhydramine (64.6) (for allergies). Her blood pressure is routinely slightly elevated, but she is not currently on antihypertensive medication. You have the four antidepressant molecules (64.1–64.4) available.

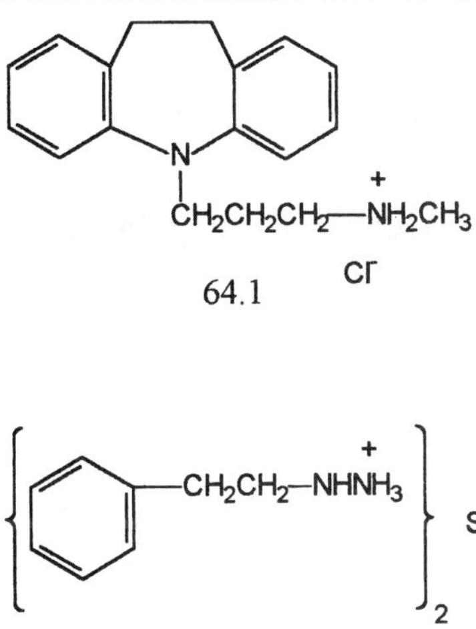

64.1

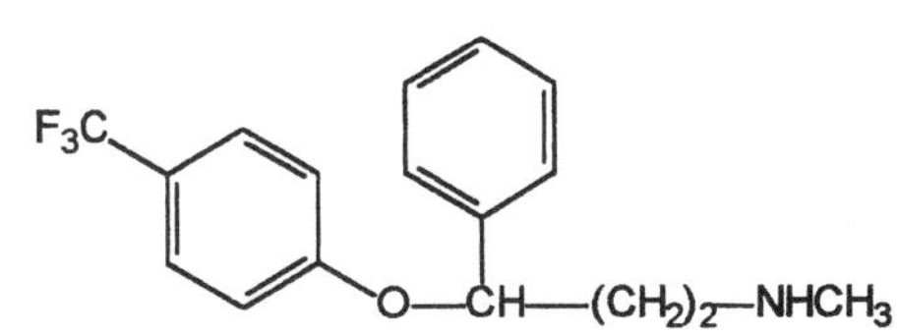

64.2

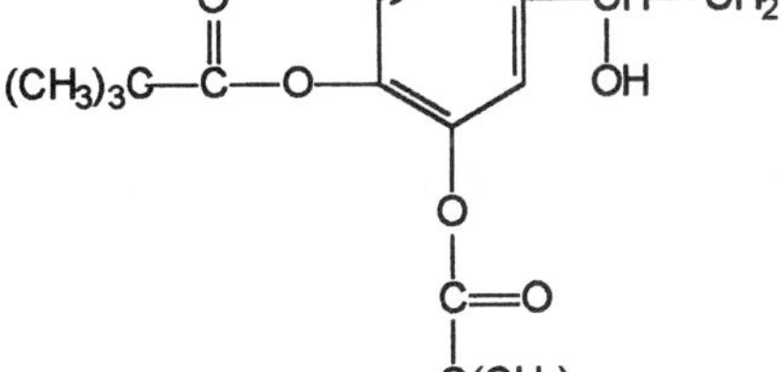

64.3

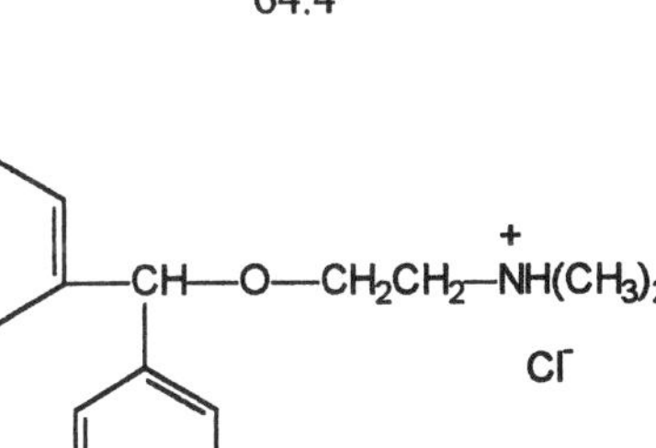

64.4

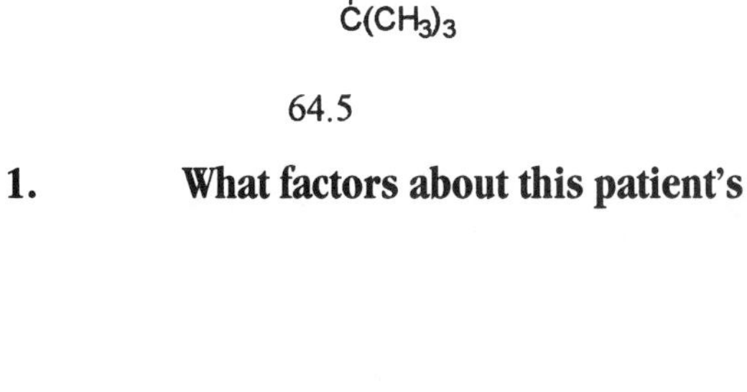

64.5

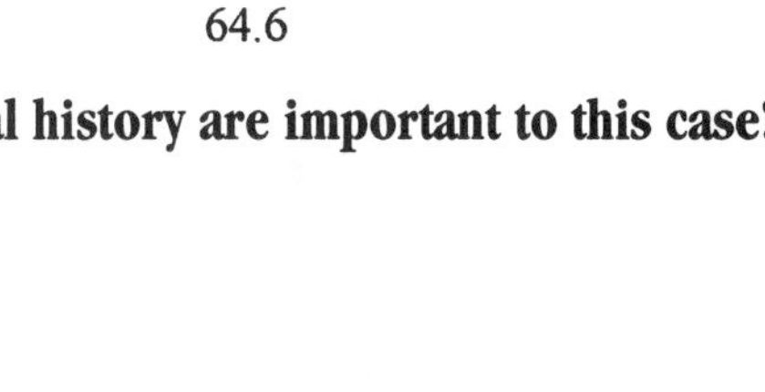

64.6

1. What factors about this patient's job and medical history are important to this case?

2. What is (are) the mechanism(s) of antidepressant action of each antidepressant agent (64.1–64.4) shown?

3. What are the pros and cons of selecting each of these agents for this patient?

4. Which of these four agents (64.1–64.4) would you recommend?

Case 65

Submitted by Bruce L. Currie and

Kimberly A. Linowiecki

RJ, a 10-year-old child, total body weight 30 kg, presents to the neurology clinic accompanied by his mother with a chief complaint of uncontrolled seizure disorder. The patient's mother states that her son has not experienced improvement in seizure frequency since starting primidone (65.1). The patient's mother adds that her child has also been "acting tired" and "sleeping more."

Past medical history: (1) Complex partial seizure disorder since childhood; (2) slight mental retardation.

Medication history: Current medication, primidone (65.1) 250 mg q 12 h (takes at 9 AM and 9 PM, last dose yesterday PM, started 6 months ago, patient states compliance with the regimen). Past medications: (1) carbamazepine; (2) valproic acid; (3) phenytoin. (The patient experienced therapeutic failure with each of the previously mentioned antiepileptic drugs [monotherapy] tried in the past.) Labs (today 8 AM): Primidone serum conc. = 15 μg/mL (5–15 μg/mL)

Physician assessment: Therapeutic failure with primidone (65.1) monotherapy despite primidone (65.1) serum concentration at the upper end of the usual therapeutic range in a patient with signs and symptoms of toxicity.

Physician plan: Begin polytherapy by adding phenobarbital (65.2) to the patient's current primidone (65.1) regimen.

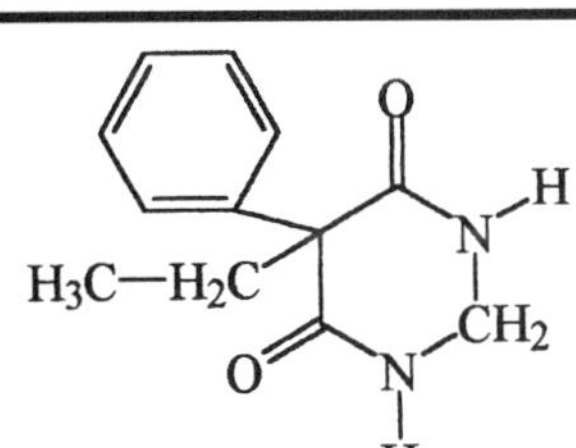

65.1

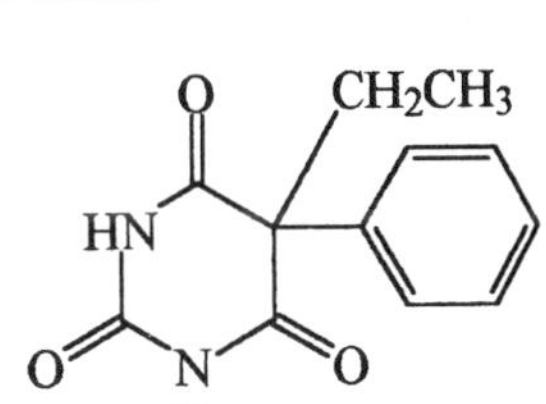

65.2

1. **Identify any problems that you see with the addition of phenobarbital (65.2) to this patient's antiepileptic drug regimen and explain briefly each problem.**

2. **What recommendation would you as the pharmacist make to the prescribing physician after receiving this prescription for RJ? Briefly explain the reasons for your recommendation.**

Case 66

Submitted by Victoria F. Roche

WD has been taking sodium phenytoin (66.1) for years to control generalized tonic-clonic (grand mal) epilepsy. He is now being admitted to the hospital with acute thrombophlebitis that developed 3 weeks after successful coronary bypass surgery. The resident physician has ordered the standard dose of sodium warfarin (66.2) to be administered intravenously to arrest further clotting.

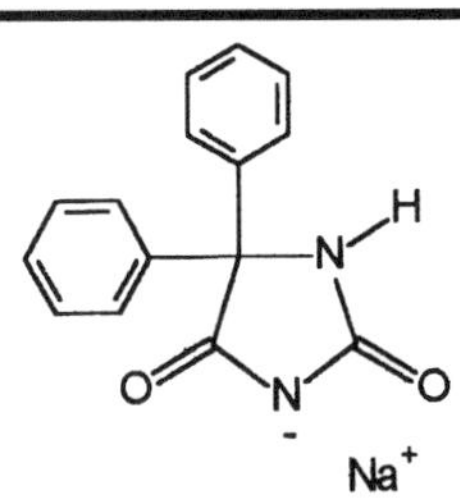

66.1

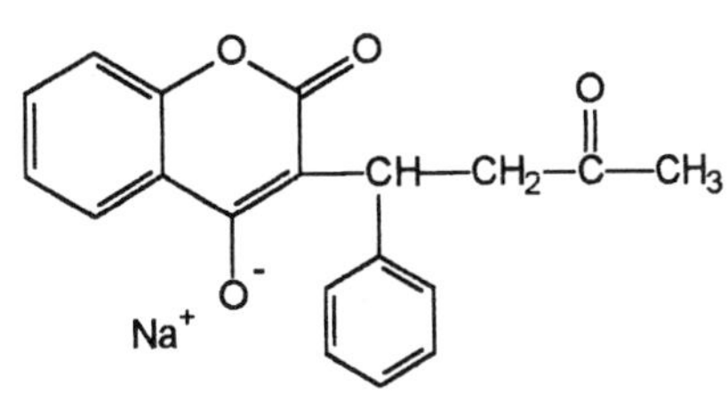

66.2

1. **Would you fill this drug order? Why, or why not?**

Case 67

Submitted by Matthias C. Lu and

George Udeani

JR is a 35-year-old woman with a history of generalized tonic-clonic seizures. Her treatment for this condition was initiated with carbamazepine (67.1) 200 mg bid; however, her dose was gradually increased over the following 8 weeks to 1200 mg daily (600 mg, bid) due to persistent seizures and low serum carbamazepine concentrations.

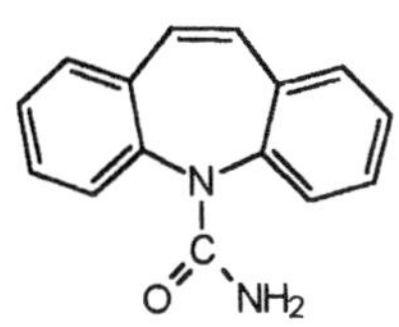

67.1

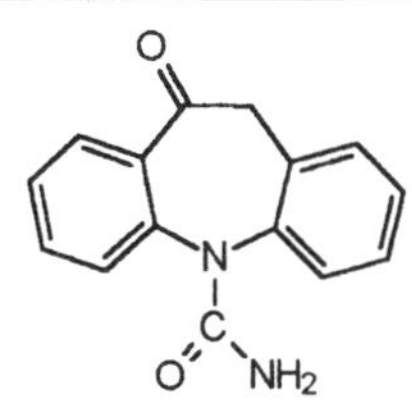

67.2

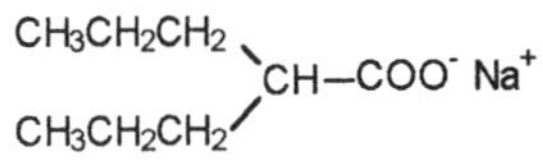

67.3

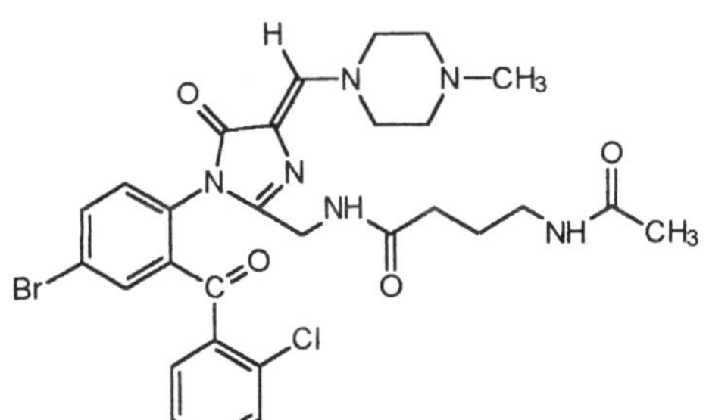

67.4

1. Should an alternative anticonvulsant therapy be considered, and why?

2. **Routine laboratory tests revealed that JR has low L-glutamic acid decarboxylase (GAD) activity in the brain, which prohibits using Sodium Valproate (67.3) for her therapy. Her physician has recently come across a report describing the use of an investigative anticonvulsant drug, Seizure-Free (67.4). Explain whether or not this new anticonvulsant drug might be used by JR to control her seizures.**

Case 68

Submitted by Victoria F. Roche

An agitated man comes into your pharmacy with a homemade capsule he confiscated from his teenage son's room. The son confessed that he bought the drugs at school, and that they were supposed to contain methamphetamine. After having the capsule analyzed by a cooperative local school of pharmacy chemistry faculty member, you find that it contains the structure shown (68.1).

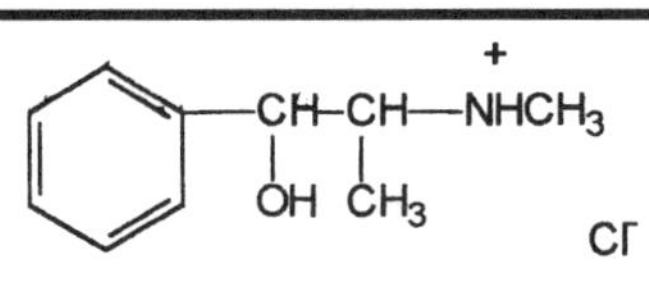

68.1

1. **Could this agent cause any CNS stimulation? What structural features of the molecule support your answer?**

2. **The boy who sold the capsules to the man's son claimed they would provide the desired CNS stimulation if swallowed, or the contents could be dissolved in water and injected. Do you agree? What chemical features of the structure support your answer?**

Case 69

Submitted by Victoria F. Roche

You are the consultant pharmacist to an area nursing home that cares for a large number of Alzheimer's patients, and you are having a discussion with the facility's chief physician about the lack of adequate drug therapy for this disease. The physician is recalling some clinical trials with reversible cholinesterase inhibitors that gave disappointing results, when a senior pharmacy student on rotation comes into the room. You decide to test his understanding of cholinesterase chemistry by asking him to suggest some possible reversible cholinesterase inhibitor structures that might be appropriate candidates for oral or parenteral therapy in Alzheimer's disease. He offers the following four structures shown:

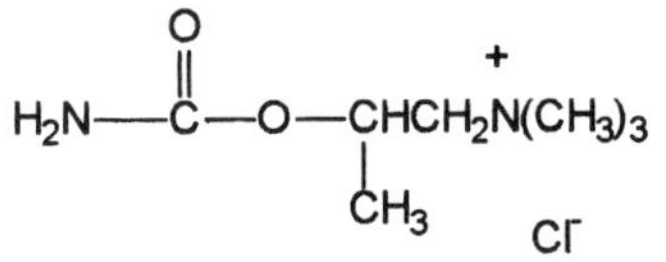

69.1

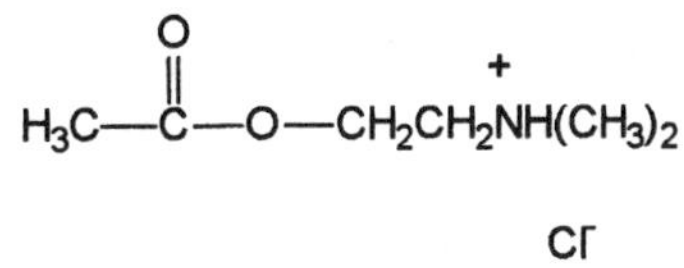

69.2

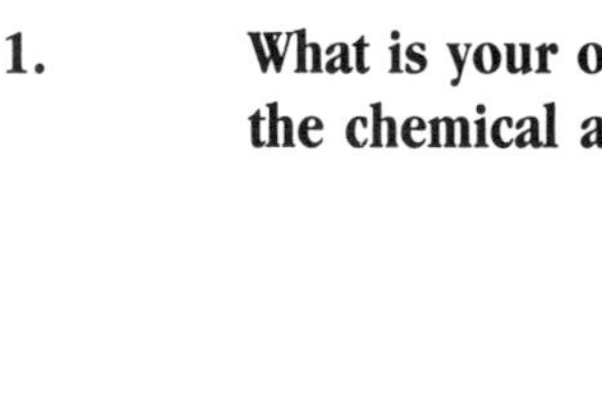

69.3

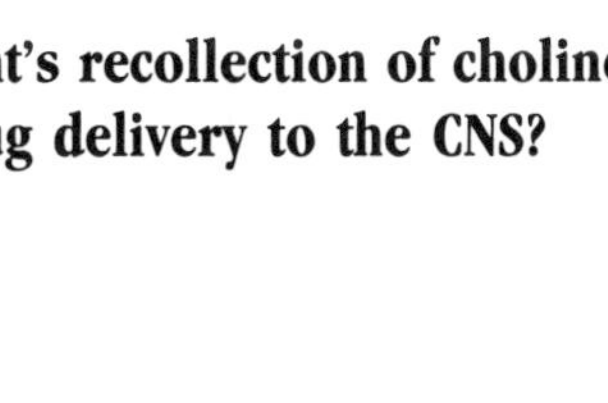

69.4

1. **What is your opinion of this student's recollection of cholinesterase chemistry and the chemical aspects critical to drug delivery to the CNS?**

2. How many of the student's structures (69.1–69.4) do you think would be viable candidates for therapy in Alzheimer's disease? What is wrong with the structures with which you disagree?

Case 70

Submitted by Matthias C. Lu and George Udeani

Droxidopa (70.1), an antiparkinsonism agent recently marketed in Japan (1989), is a synthetic amino acid precursor of (-) norepinephrine, which is absorbed from the gut and transported into the brain.

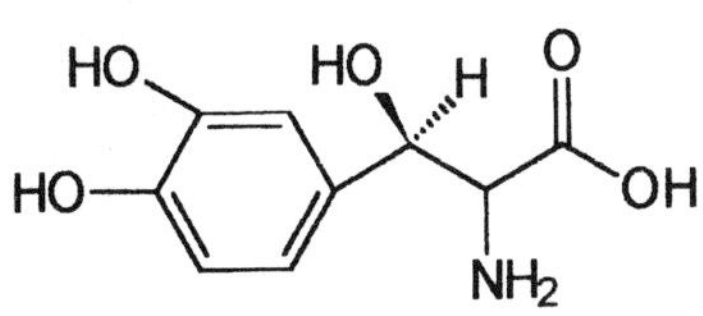

70.1

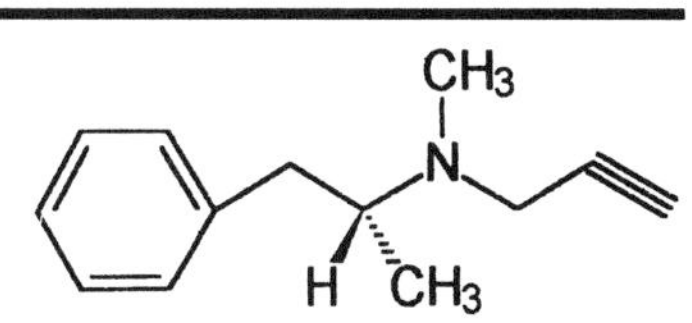

70.2

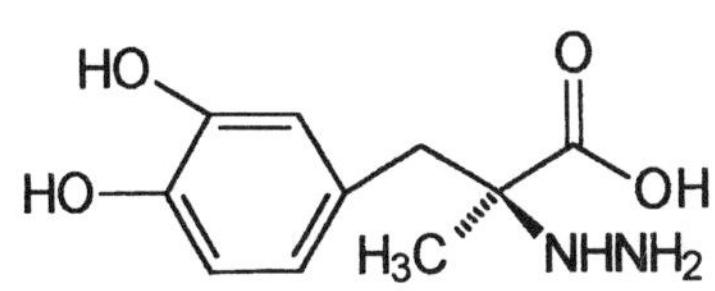

70.3

1. Explain how droxidopa (70.1) works in Parkinson's disease?

2. When (-) deprenyl (70.2) is added to a therapeutic regimen of droxidopa (70.1), the dose of droxidopa can be greatly reduced without the loss of its therapeutic efficacy. Why?

3. **With your knowledge of how (-) deprenyl (70.2) metabolizes in vivo, provide a reason why the (+)-isomer should not be used?**

4. **Since (-) deprenyl (70.2) is not yet approved for the treatment of Parkinson's disease in the U.S., can carbidopa (70.3) be substituted for (-) deprenyl (70.2)?**

Case 71

Submitted by Victoria F. Roche

JJ, a 37-year-old woman, consults you concerning severe depression of 6 months' duration. JJ is employed at a neighborhood day care center, has recently divorced, and is now raising two adolescent children alone. She is currently on no medications, but complains of occassional "fainting feelings." Her depressive state is beginning to interfere with her ability to do her job, which she fears losing. In consultation with the physician, you request a laboratory determination of urinary 3-methoxy-4-hydroxyphenylglycol (MHPG), and find the level to be within the normal range. Assume you have only the following antidepressant structures (71.1–71.3) available.

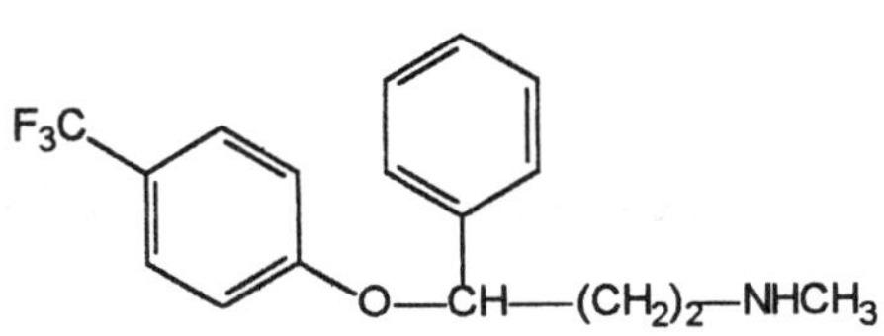

71.1

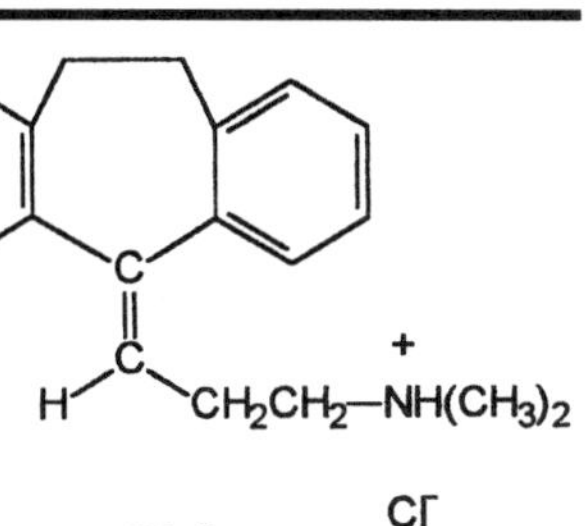

71.2

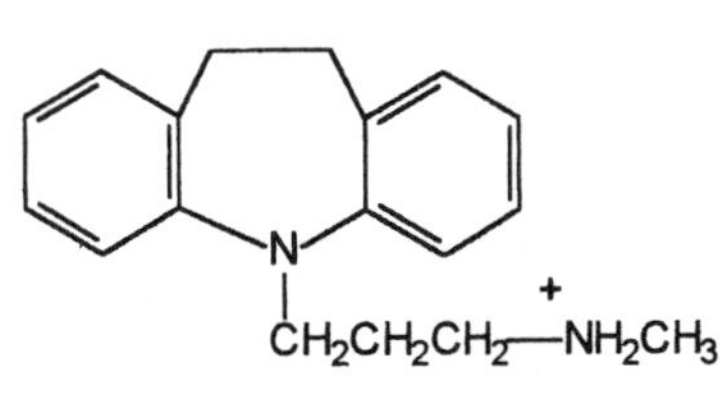

71.3

1. **What are the possible biochemical etiologies of endogenous depression?**

2. **What is the significance of the MHPG laboratory finding (please draw the structure of this metabolite).**

3. **What are the specific mechanisms of antidepressant action of each agent (71.1–71.3)?**

4. **Which of the three antidepressants (71.1–71.3) would you recommend for this patient?**

Case 72

Submitted by Victoria F. Roche

RD, a 31-year-old male construction worker, comes into your pharmacy with a prescription for Dalmane (flurazepam) (72.1). Upon consulting with this patient, you find he is experiencing moderate anxiety over an upcoming surgical procedure and, although he is able to function adequately during the day, is unable to sleep at night. You also discover that he has a significant pollen allergy, and is currently taking OTC chlorpheniramine maleate (72.2). In addition to Dalmane (72.1), your pharmacy carries three other anxiolytic products (72.3–72.5).

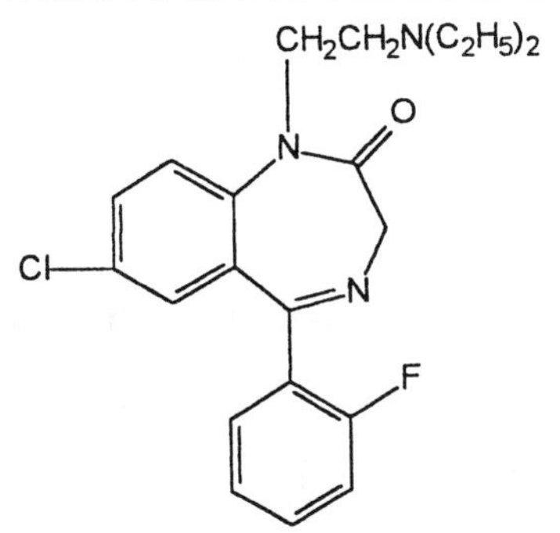

72.1

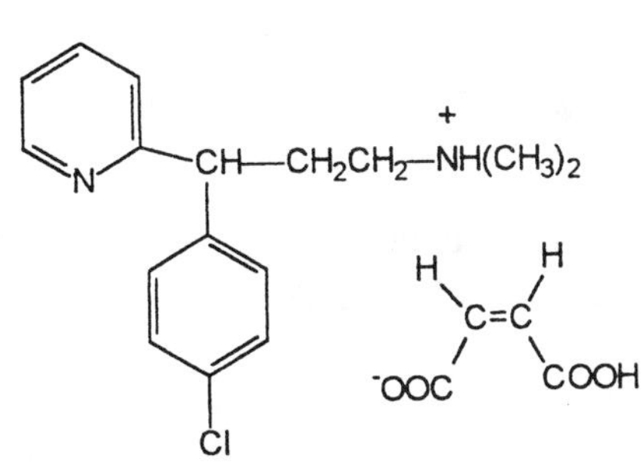

72.2

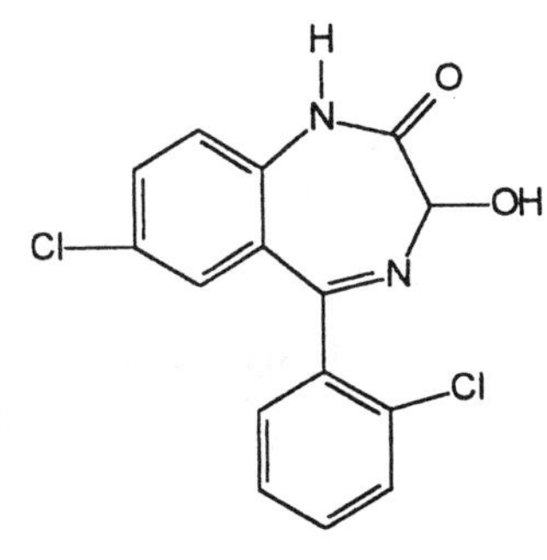

72.3

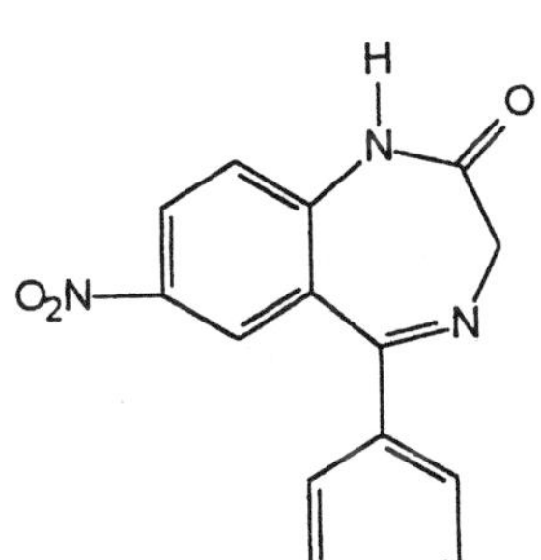

72.4

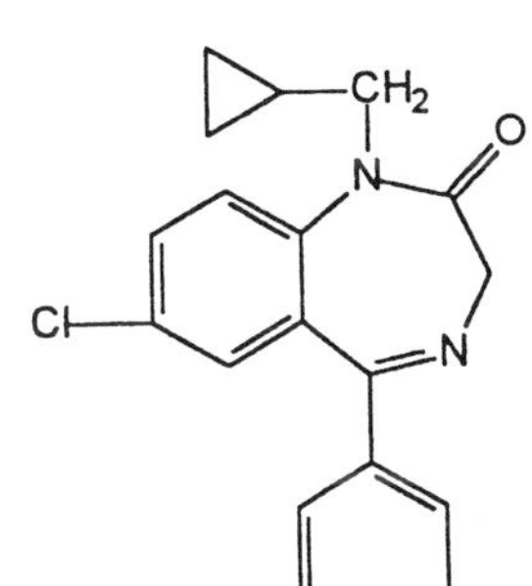

72.5

1. **What are RD's most pressing therapeutic needs?**

2. **What are the pros and cons of each of the drug structures provided (72.3–72.5) in meeting these needs?**

3. **Which of the four anxiolytic agents, Dalmane (72.1) or the three alternative choices (72.3–72.5), would you recommend?**

Case 73

Submitted by Arnold R. Martin and Richard N. Herrier

DK is a 32-year-old female part-time nursery school teacher/housewife and mother of two, who suffers occasional bouts of extreme anxiety (panic) when she is in crowded places (i.e., the subway, grocery stores, symphony, basketball games, in her car at rush hour, etc.). These attacks typically occur with little warning, and seldom last more than 15 to 20 minutes. However, DK is worried that during an attack she may do something terribly embarrassing or even dangerous. She very much desires to maintain her occupation and her family life, which requires that she drive to work, to the grocery store, and to her children's activities.

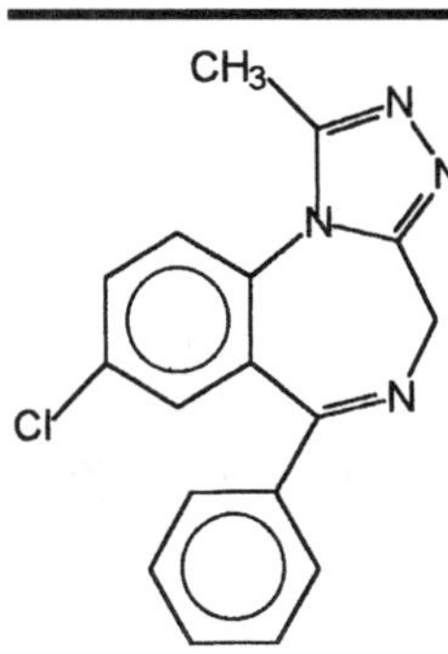

73.1

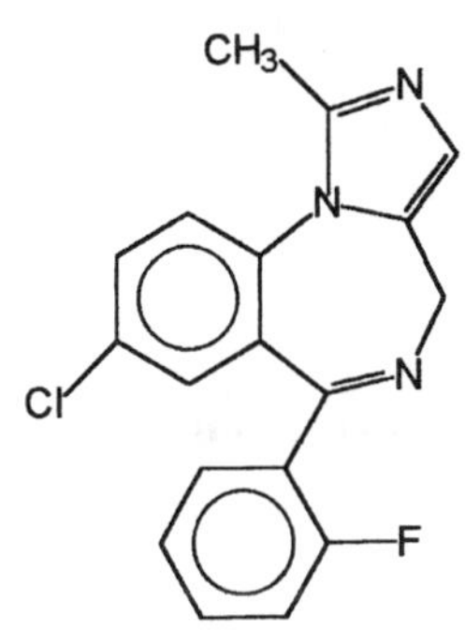

73.2

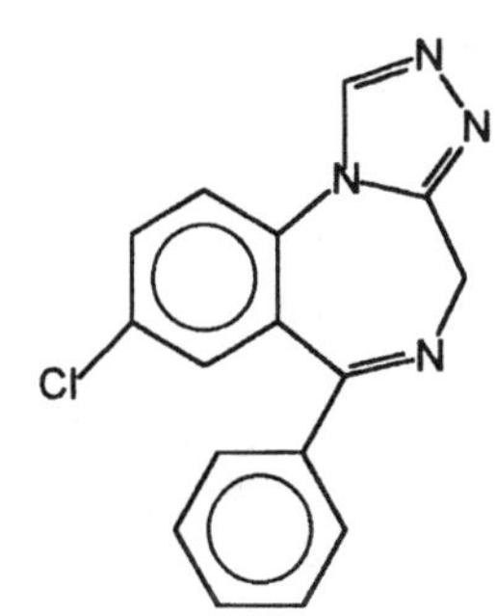

73.3

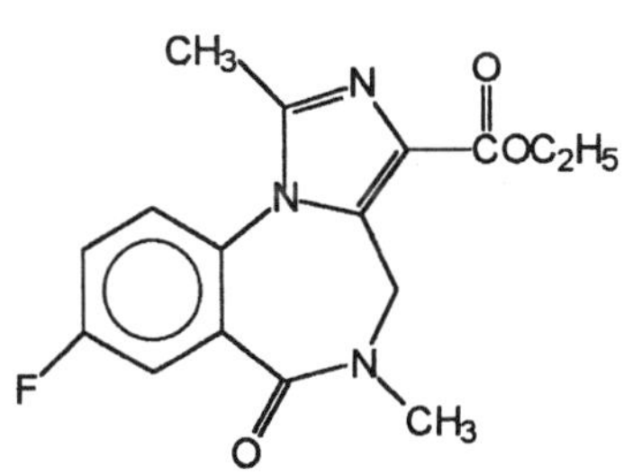

73.4

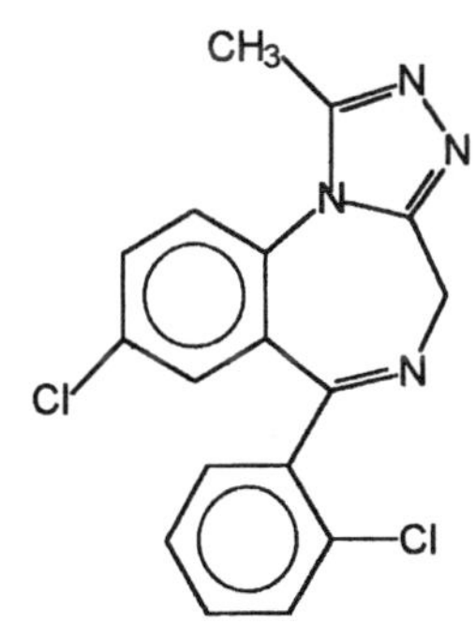

73.5

1. **Select the most appropriate anxiolytic benzodiazepine from the structures (73.1–73.5) given. Justify your choice on the basis of comparative potencies and pharmacokinetics of the structures.**

2. **Rank the benzodiazepines given in order of increasing potency (least to most potent).**

3. **Rank the benzodiazepines given in order of increasing duration of action (shortest to longest).**

4. **One of the benzodiazepine structures is an antagonist. Indicate the functional group that confers this property.**

Case 74

Submitted by Arnold R. Martin and Richard N. Herrier

A 36-year-old southern Arizona cotton farmer, DL, checks into his HMO complaining of flulike symptoms that have persisted for more than 2 weeks. He has alternating fever (low grade) and chills, muscle aches, joint pain, and bouts of coughing. A coccidioidomycosis skin test was positive at 15-mm induration. Serum antibody titers confirmed the diagnosis of active coccidioidomycosis (valley fever). His internist gives him a prescription for ketoconazole (74.1) 200 mg, two tablets daily. While the technician is filling the prescription, the HMO pharmacist checks DL's patient record only to discover that 2 months ago the pharmacy had dispensed 40 pentobarbital (74.2) capsules (50 mg), with the directions to take one capsule at bedtime as needed for insomnia. An interview with DL reveals that he occasionally has difficulty getting to sleep, "especially during harvest time when he has a lot on his mind." He therefore persuaded his doctor to prescribe "sleeping pills." DL further states that he has at least half of the capsules remaining.

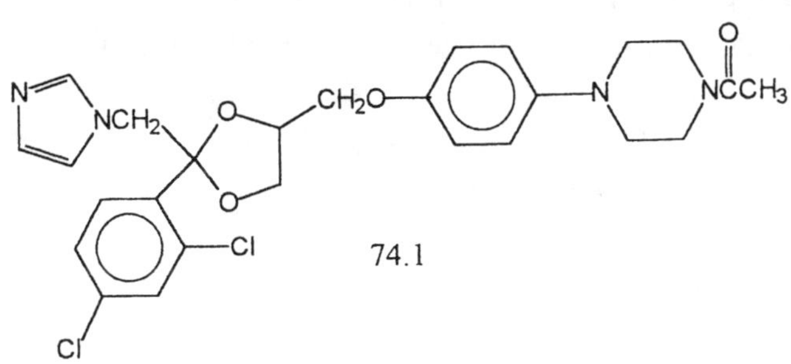

74.1

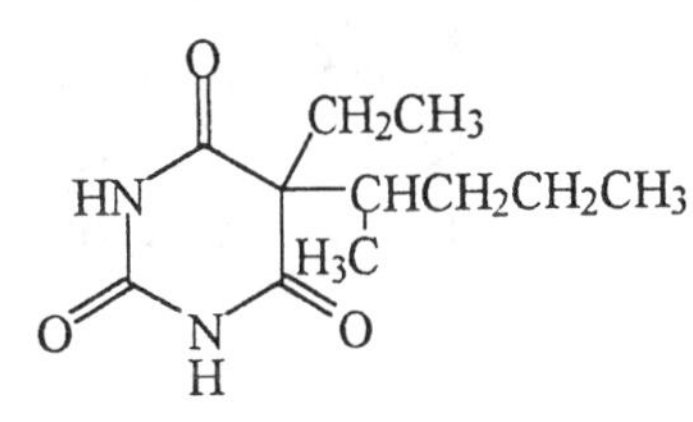

74.2

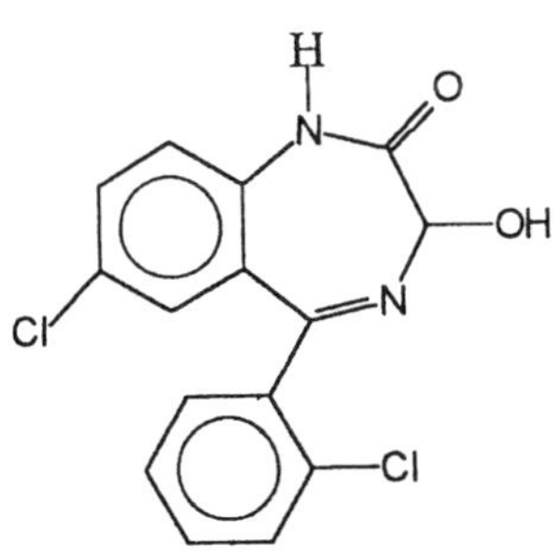

74.3

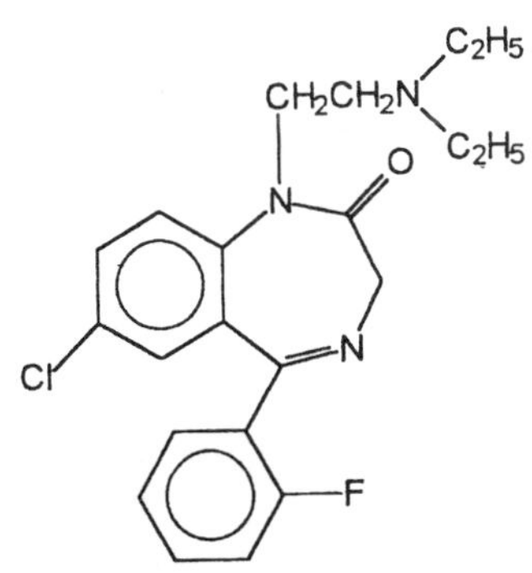

74.4

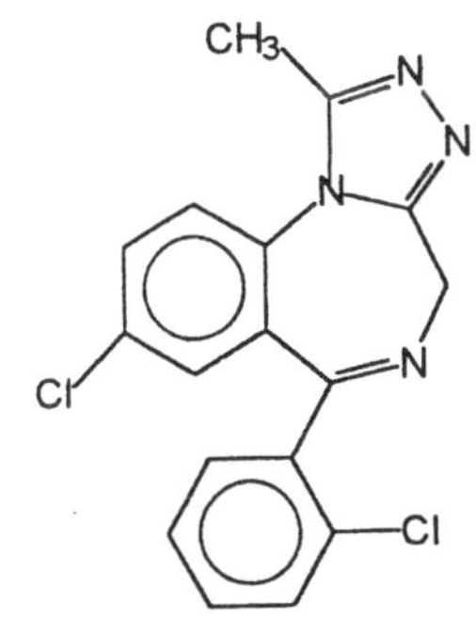

74.5

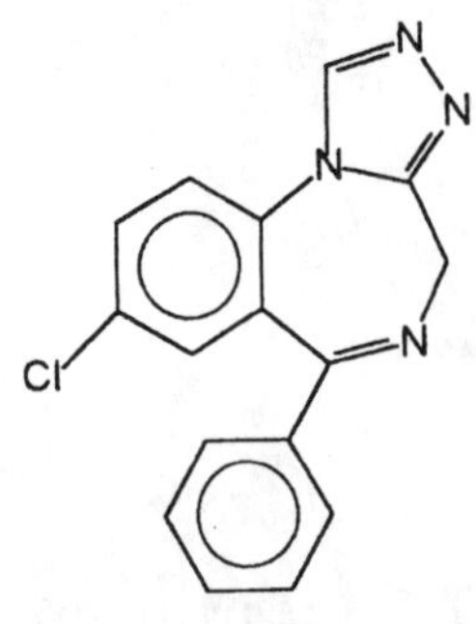

74.6

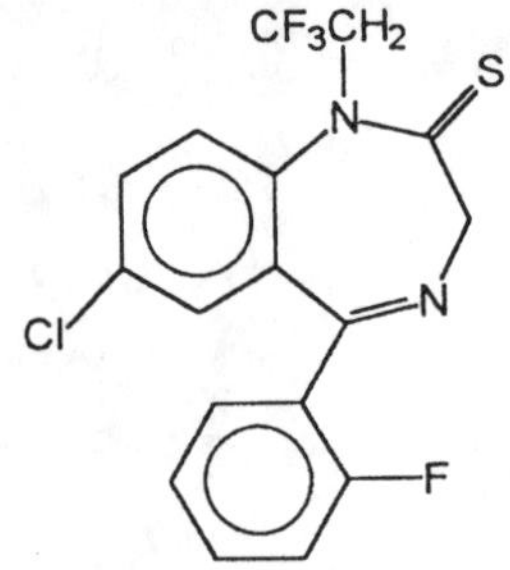

74.7

1. **Should the HMO pharmacist contact the doctor concerning the medications that DL is taking? Explain fully.**

2. **Assume that the doctor wishes to continue the prn sedative/hypnotic, but given the advice from the pharmacist, realizes that pentobarbital (74.2) can no longer be used. He is aware that a number of benzodiazepines are employed as hypnotics, and asks the pharmacist to recommend the best choice. Given the structures of the benzodiazepines (74.3–74.7) that are recommended for sedative/hypnotic use, select one that is appropriate for this particular patient. Justify your choice and explain in detail why the remaining structures were rejected.**

Case 75

Submitted by Jeffrey J. Christoff and Mary L. Chavez

GJ is a 58-year-old grandmother who has sought treatment for depression. She has received counseling for over 2 years, but her depression has become progressively worse. She has a history of narcolepsy (recurrent attacks of sleep), isolated systolic hypertension (which is under control with the diuretic chlorthalidone 12.5 mg qd), and glaucoma (treatment with epinephrine dipivaloate).

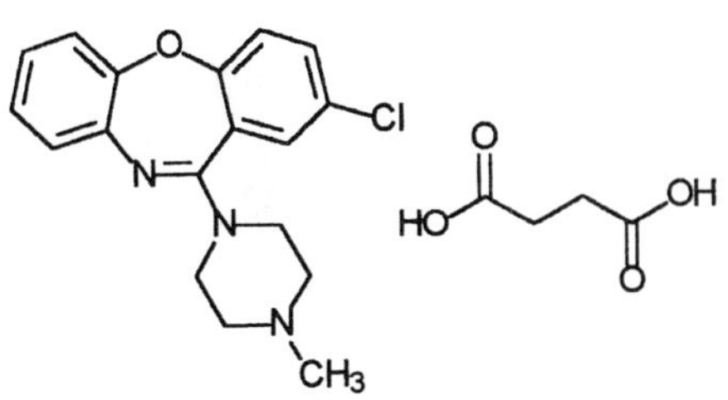

75.1

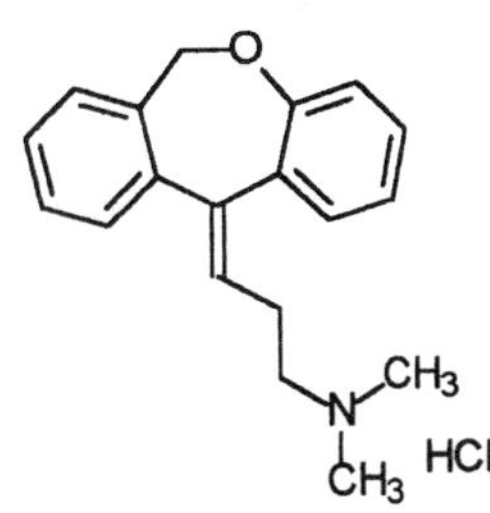

75.2

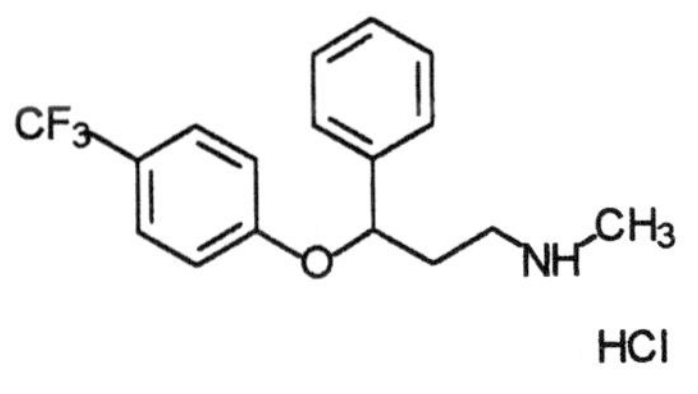

75.3

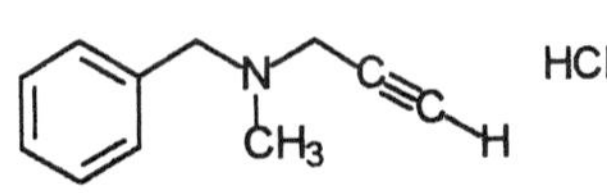

75.4

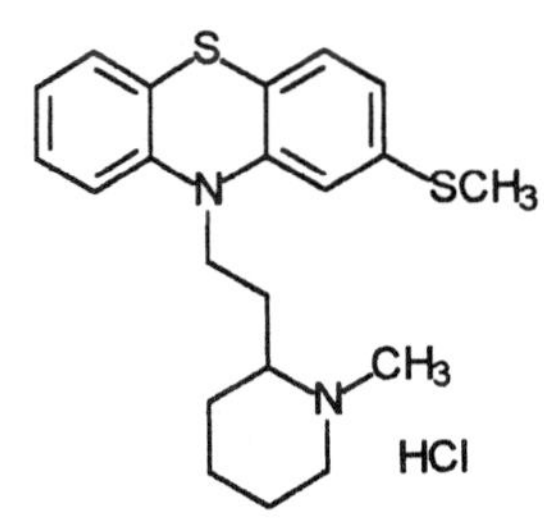

75.5

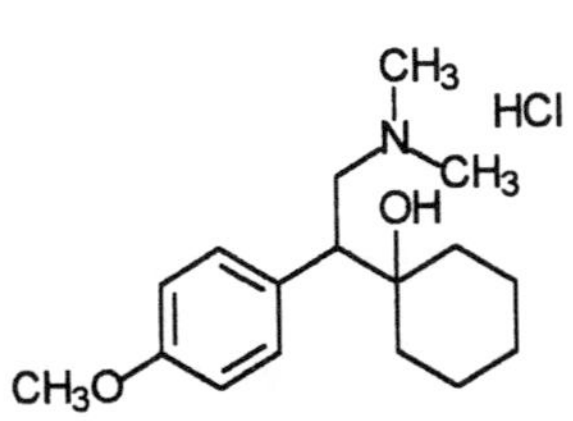

75.6

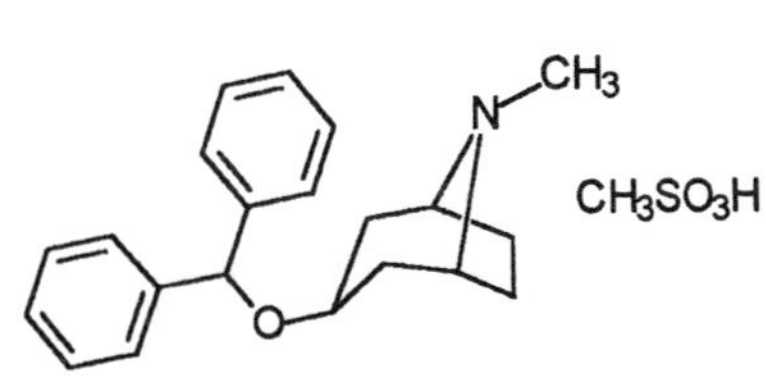

75.7

1. **Which of the following drugs (75.1–75.7) is the appropriate choice for treating this patient's endogenous depression? Justify your reasons for selecting the appropriate drug as well as any criteria to eliminate alternative choices.**
   a. **Loxapine succinate (75.1)**
   b. **Doxepin hydrochloride (75.2)**
   c. **Fluoxetine hydrochloride (75.3)**
   d. **Pargyline hydrochloride (75.4)**
   e. **Thioridazine hydrochloride (75.5)**
   f. **Venlafaxine hydrochloride (75.6)**
   g. **Benztropine mesylate (75.7)**

Case 76

Submitted by Victoria F. Roche

A physician consults you concerning PS, a 32-year-old woman recently admitted to the psychiatric hospital with emotional, agitated daytime anxiety coupled with insomnia. As PS is non-compliant, the physician wishes to minimize therapeutic problems by prescribing only one medication to be taken hs. You have three benzodiazepine anxiolytic agents (76.1–76.3) on hand.

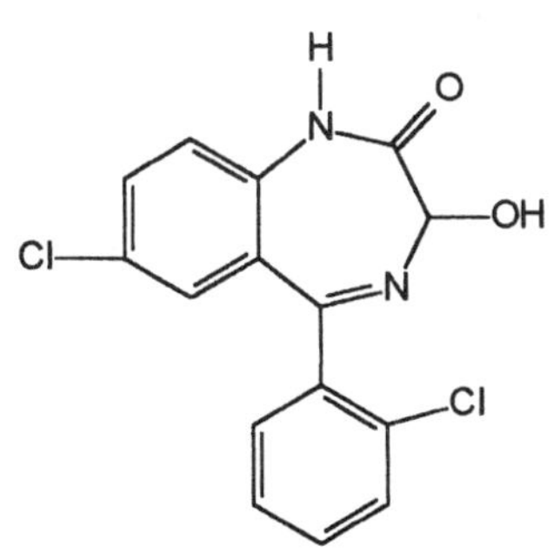

76.1

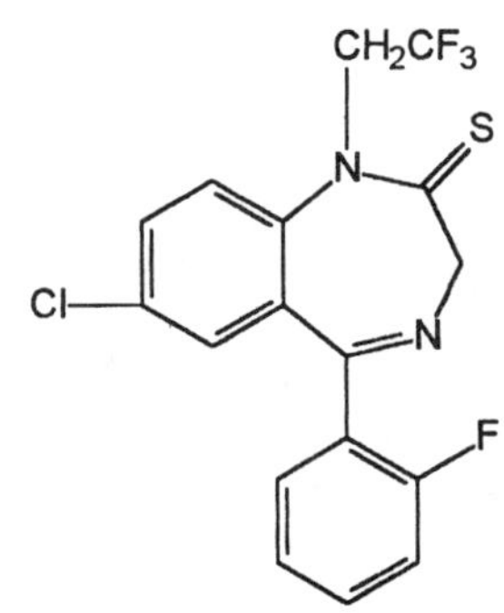

76.2

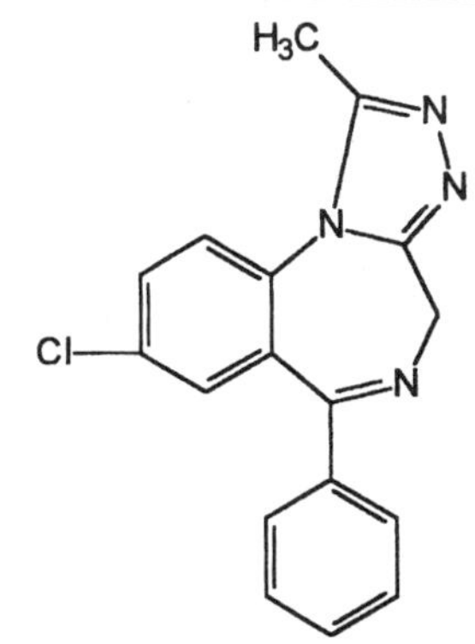

76.3

1. **What therapeutic considerations should be taken into account when selecting appropriate drug therapy for PS?**

2. **What anticipated pharmacologic actions would be expected from each of these agents based upon their chemical structures?**

3. Which of the three anxiolytic agents (76.1–76.3) would you recommend for PS?

Case 77

Submitted by Arnold R. Martin and Richard N. Herrier

CJ is a 37-year-old man who is brought to the emergency psychiatric unit by the police. He was found naked, huddled in a yoga position atop a Ford Bronco parked in the middle of the university quadrangle and screaming profanities at bypassers. CJ's Medic-Alert bracelet revealed that he is suffering from chronic schizophrenia. A call was made to the caregiver who stated that several hours ago, CJ had apparently wandered away from the psychiatric care home where he has resided for the past 16 years.

CJ has been a troublesome patient to control on antipsychotic medications. Either his response has not been satisfactory, or he has experienced intolerable side effects. Although high-dose thioridazine (77.1) was reasonably effective initially, it proved to be inadequate to control CJ's symptoms over time. Two years ago, he was switched to haloperidol (77.2). He now takes 20 mg bid po, which gave good control until a few weeks ago when his behavior became increasingly inappropriate, culminating in his "escape." CJ's caregiver expressed the concern that CJ had not been taking his medication as given. Recently, CJ had complained of insomnia, and appeared to be nervous. He had also been observed pacing about his room and restlessly tapping his foot when seated on his bed. During the initial evaluation of CJ, the attending physician noted irregular movements of his mouth (sucking, smacking, and chewing) that appeared to be involuntary. After stabilization of CJ with haloperidol (77.2), the physician decided that a change in medication was indicated.

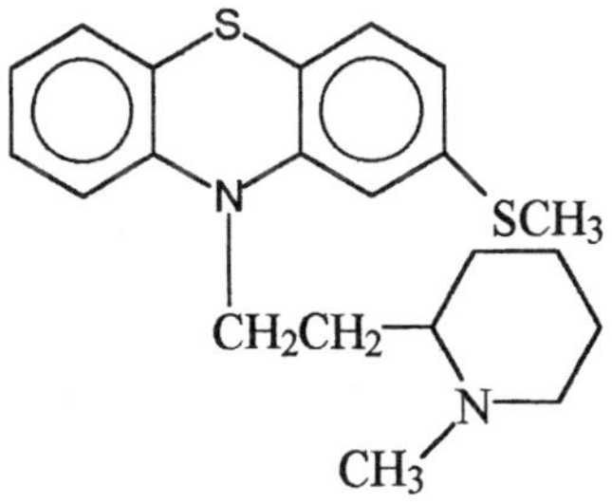

77.1

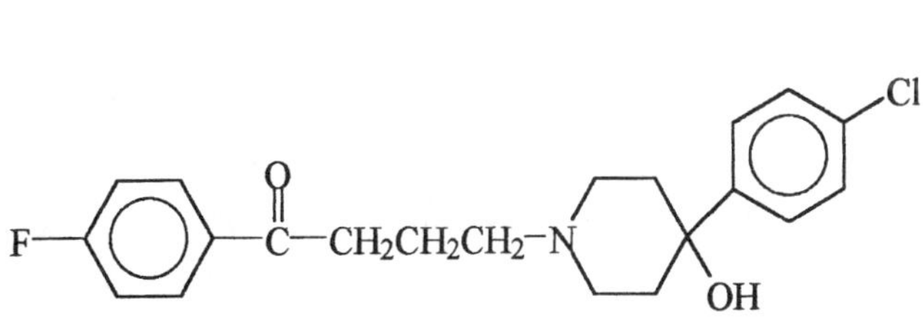

77.2

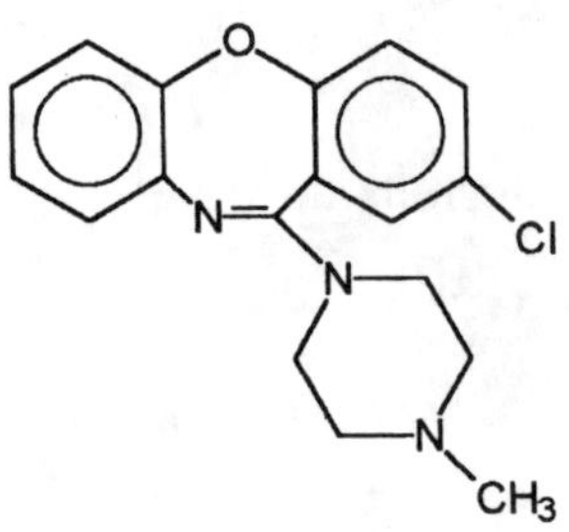

77.3

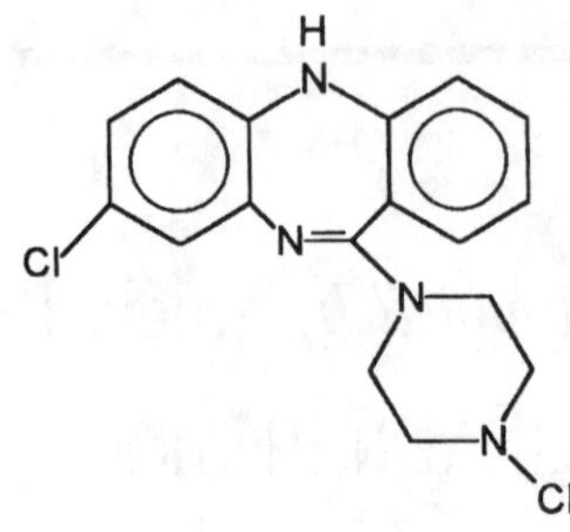

77.4

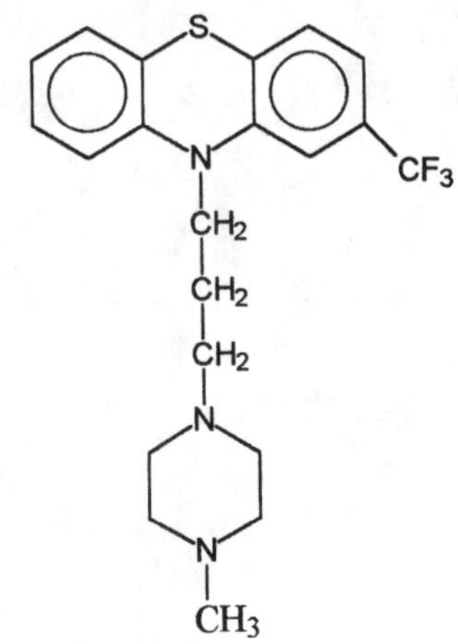

77.5

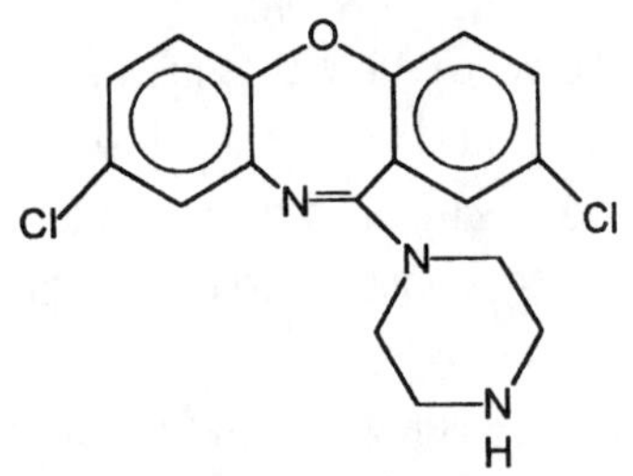

77.6

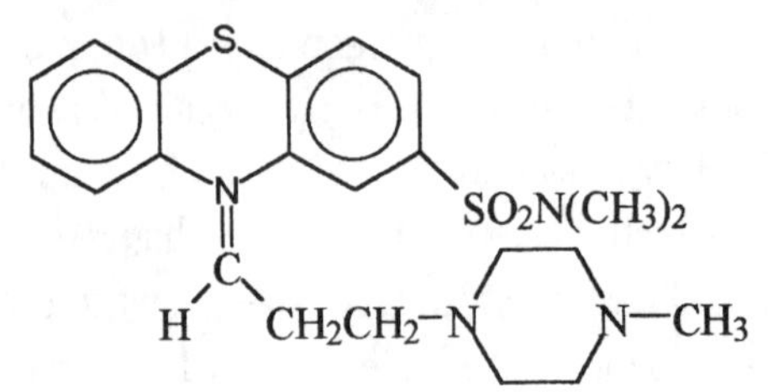

77.7

1. **What are the nature and cause of CJ's involuntary symptoms? How can they be prevented and CJ's illness controlled at the same time?**

2. **From the antipsychotic structures (77.3–77.7) provided, select an agent that would minimize the side effects caused by haloperidol (77.2) and yet control CJ's schizophrenia. Justify your choice on a structure-activity basis and explain why each of the other structures were eliminated as choices.**

3. For the medication that you have chosen, indicate the most serious adverse effects that might be anticipated.

Case 78

Submitted by Jeffrey J. Christoff

SY is a 33-year-old woman with clinically diagnosed endogenous depression. She is a pilot for a package delivery service, a job which frequently requires her to work at night for next day deliveries. Her physician has prescribed trimipramine (78.1) to treat her depression.

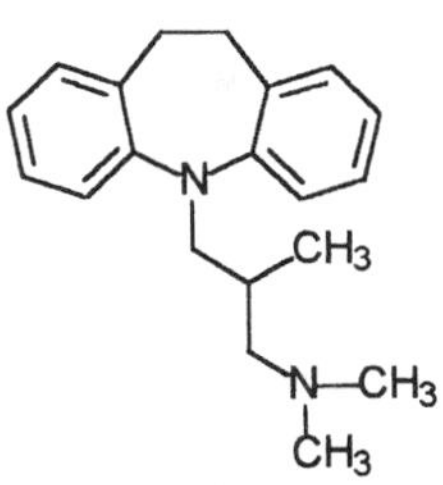

78.1

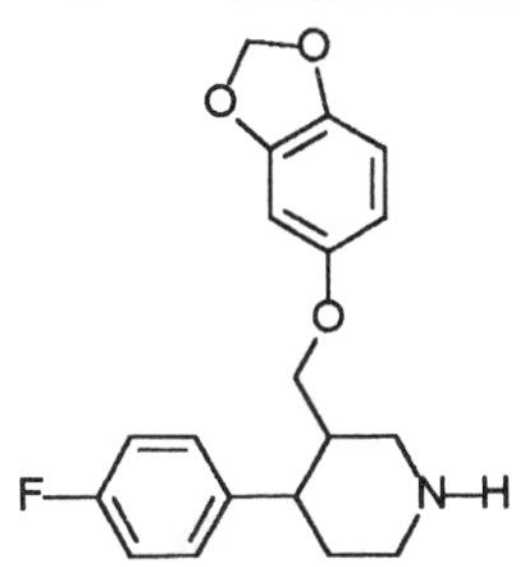

78.2

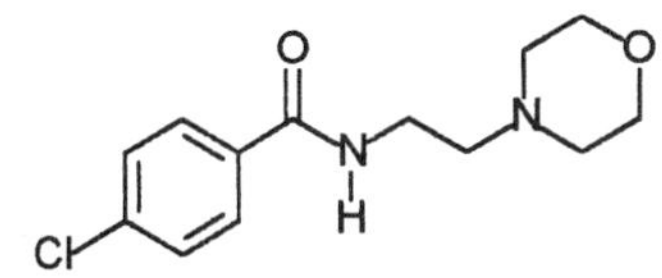

78.3

1. **Briefly describe the mechanism of action for trimipramine (78.1) (include brief structure-activity relationship [SAR] information).**

2. **At first glance, the tricyclic nature of trimipramine (78.1) could lead a student to believe (incorrectly) that this drug is an antipsychotic agent. Describe the key conformational property of trimipramine (78.1) that provides for its antidepressant activity.**

3. **What side effects could be associated with using trimipramine for SY? Briefly explain your chemical rationale (include brief SAR for the side effects).**

4. **Each of the other drugs can be an appropriate alternative to trimipramine (78.1). Select ONE of these drugs for SY and explain your rationale (using mechanism of action and brief SAR of each drug) for its use as an antidepressant.**

Case 79

Submitted by Victoria F. Roche

A physican telephones you regarding GN, a second-year pharmacy student suffering from debilitating anxiety over approaching final examinations. The pharmacy student is desperate to get into medical school and deems optimal performance on finals as essential to fulfilling her career goal. The student has no history of mental disturbance, and the physician believes she would not abuse an anxiolytic if prescribed short term.

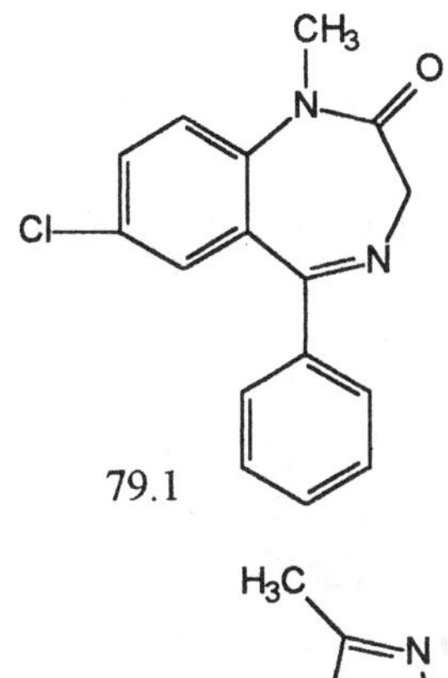

79.1

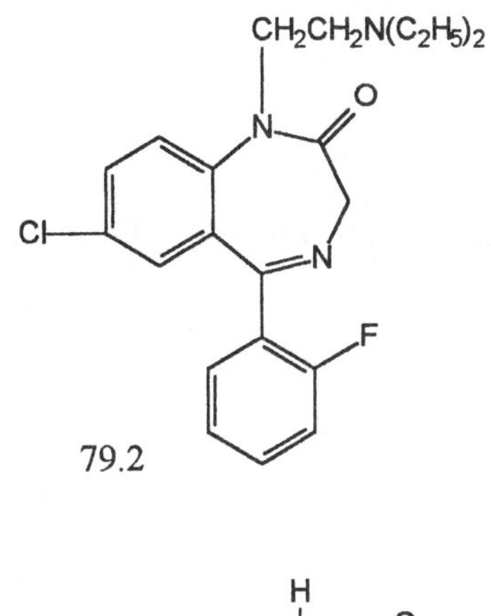

79.2

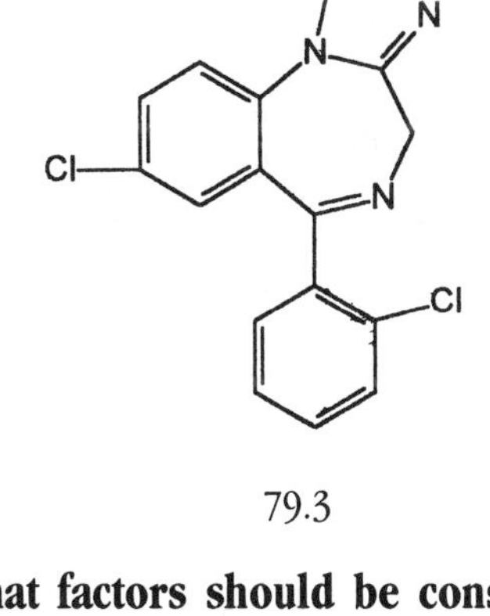

79.3

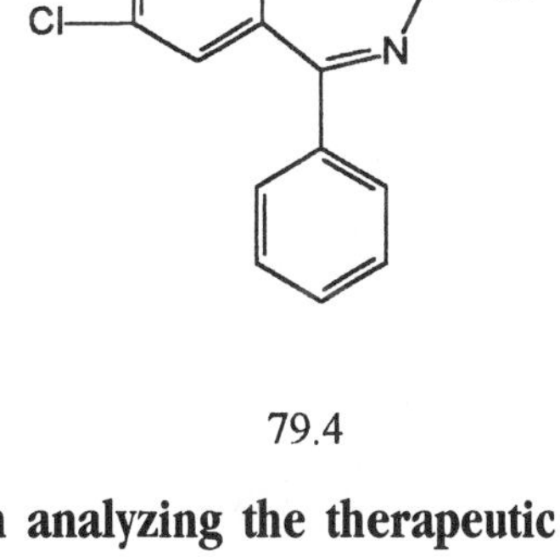

79.4

1. **What factors should be considered when analyzing the therapeutic needs of this patient?**

2. **What are the pros and cons of each anxiolytic agent (79.1–79.4) provided in this case?**

3. **Which of the four antianxiety agents (79.1–79.4) would you recommend?**

Case **80**

Submitted by Arnold R. Martin and Richard N. Herrier

JP is a 68-year-old male former auto mechanic who presents to the geriatric outpatient clinic for a routine follow-up for his medical problems. JP was diagnosed with Parkinson's disease 9 years ago. As the symptoms gradually intensified, JP was no longer able to work consistently and was forced to retire 6 years ago. For the first 2 years, JP's symptoms were largely controlled with low to moderate doses of the anticholinergic drug benztropine mesylate (80.8) (2 mg/d, initially, increased to 4 mg/d). Then, therapy with levodopa (80.6)/carbidopa (80.7) (10/100 mg bid) was initiated and benztropine maintained at 4 mg/d. Currently, JP is taking 25/250 mg levodopa/carbidopa twice daily, and his benztropine has been raised to 6 mg/d. Meanwhile, JP's parkinsonian symptoms (ie, pillrolling, cogwheel rigidity, hypokinesis) have intensified slightly. In addition, he is experiencing blurred vision, constipation, and difficulty in urinating. His BP is 130/80, and his HR is 125. An eye examination reveals that JP is suffering from narrow angle glaucoma and probably requires surgery. The attending neurologist decides to discontinue the anticholinergic therapy to prevent further worsening of narrow angle glaucoma and institute therapy with a monoamine oxidase inhibitor (MAO). She plans to continue the levodopa/carbidopa at the present dosage level.

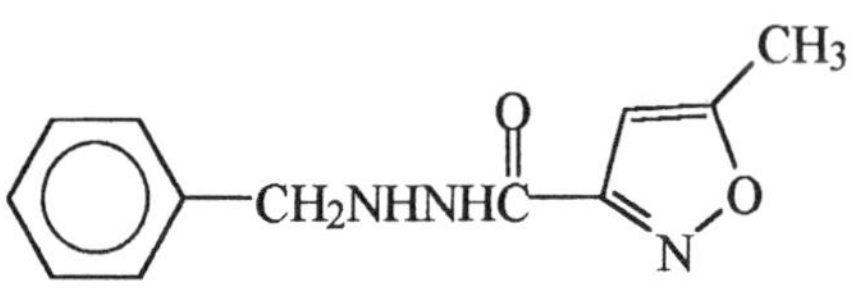

80.1

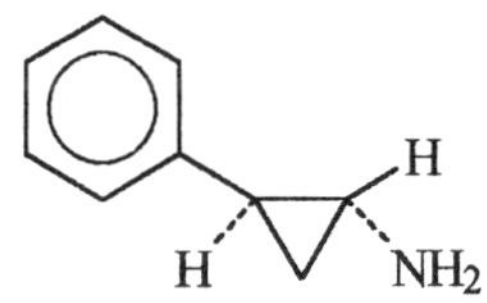

80.2

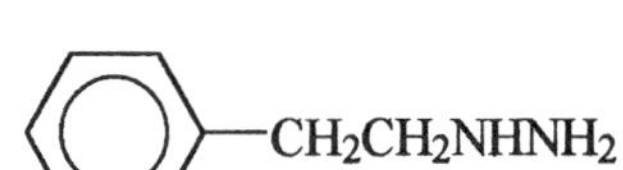

80.3

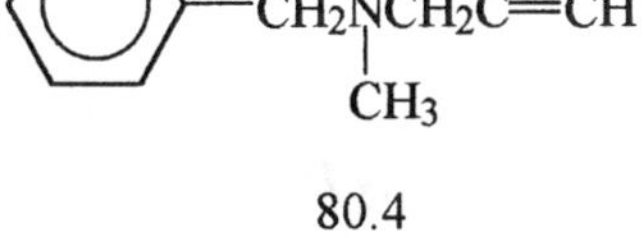

80.4

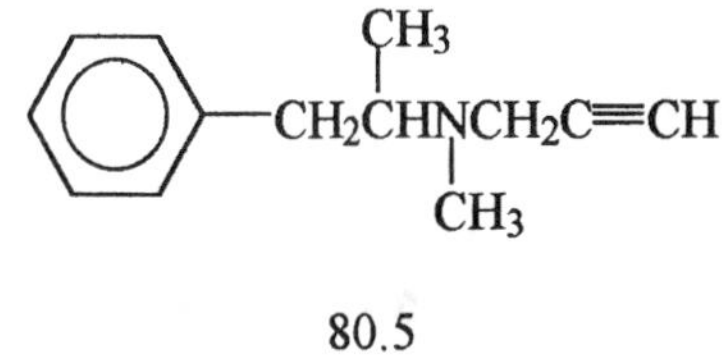

80.5

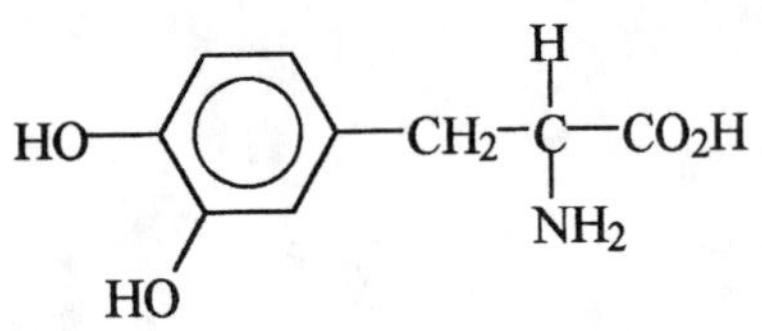

80.6

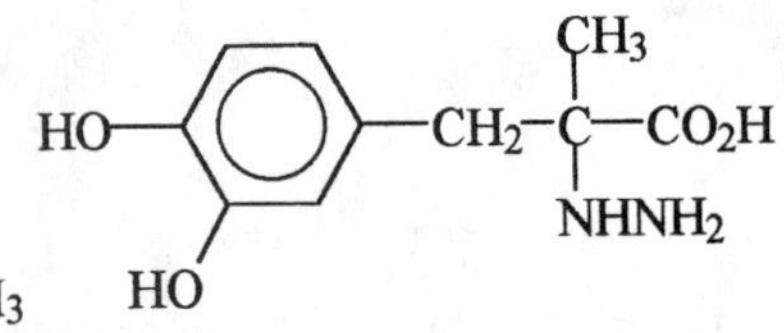

80.7

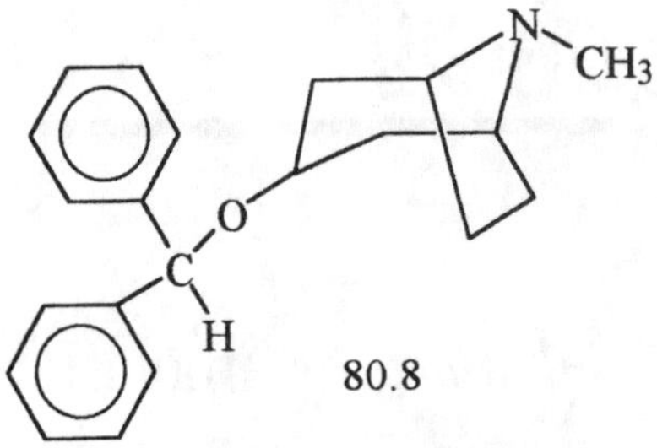

80.8

1. Explain fully the reasons for the discontinuation of anticholinergic treatment.

2. Select a monoamine oxidase inhibitor (MAOI) from the structures (80.1–80.5) provided that is the most appropriate as an adjunct for treating Parkinson's disease. Justify your choice fully. Explain why each of the remaining structures were rejected.

3. Why is carbidopa (80.7) given in combination with levodopa (80.6)?

4. Hypertensive reactions are reported to occur when MAOIs are employed in combination with levodopa (80.6) therapy. Are such reactions likely to occur when selegiline (80.1) and levodopa/carbidopa are administered to JP? Explain fully.

Case 81

Submitted by Arnold R. Martin and Richard N. Herrier

LC is a 34-year-old man who is admitted to the burn center after being rescued from a house fire by the fire department. LC has suffered 65% total body surface area burns (primarily second-degree with 10% eschar-free third-degree). Before being admitted to the burn center, LC received the following treatments in the ER: (1) an intravenous catheter inserted and lactated Ringer's solution infused at 750 mL/h (1.5 L administered in the ER); (2) an endotracheal tube inserted during elective laryngoscopy/bronchoscopy that demonstrated inflammation and edema; (3) intravenous injections of 0.5 mL diphtheria/tetanus toxoids; (4) insertion of an indwelling Foley catheter; (5) insertion of a nasogastric tube; and (6) application of mafenide acetate (81.1) cream to all burned areas.

LC's past medical history reveals that an appendectomy was performed following an acute appendicitis 2 years ago. At that time, LC experienced an urticarial anaphylactoid reaction to the skeletal muscle relaxant tubocurarine (81.3). LC has a history of alcohol abuse and has been homeless since he was 22.

Physical examination of LC is as follows: BP 94/60; HR 145; RR 26, rales and rhonchi heard in all lung fields, inspiratory and expiratory wheezes. The total volume of urine collected in the ER stay was 38 mL. The urine is reddish brown. The patient does not respond to verbal stimuli, but responds to and localizes pain. Laboratory tests reveal a blood alcohol level of 48 mmol/L, metabolic acidosis, and hyperkalemia.

Anticipating the likelihood that as LC becomes more alert he may attempt to remove his endotracheal tube or attempt to override the ventilator, the attending physician contemplates the administration of both a neuromuscular blocking agent and a sedative. The doctor also decides to replace the topical mafenide with silver sulfadiazine (81.2).

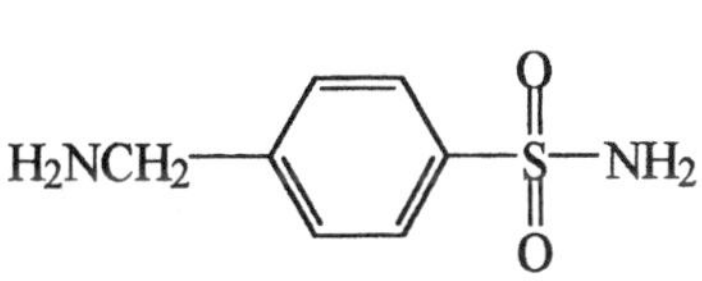

81.1

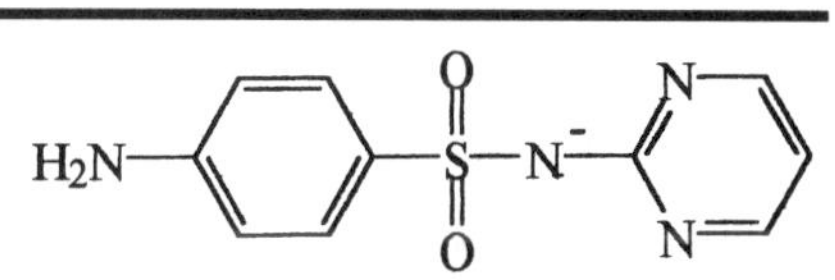

81.2 Ag$^+$

1. Explain why the physician decided to change the topical antibacterial medication.

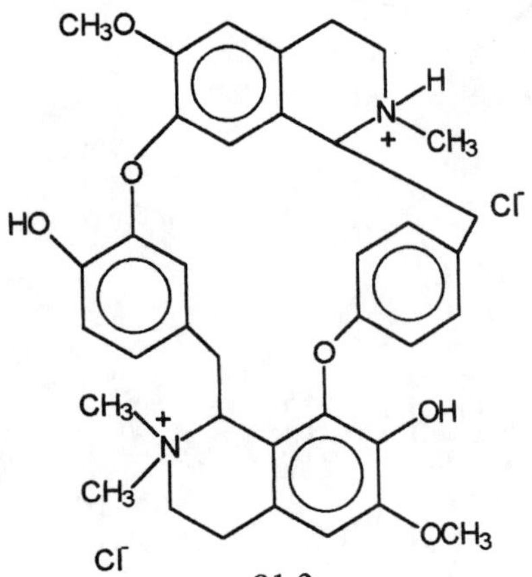

81.3

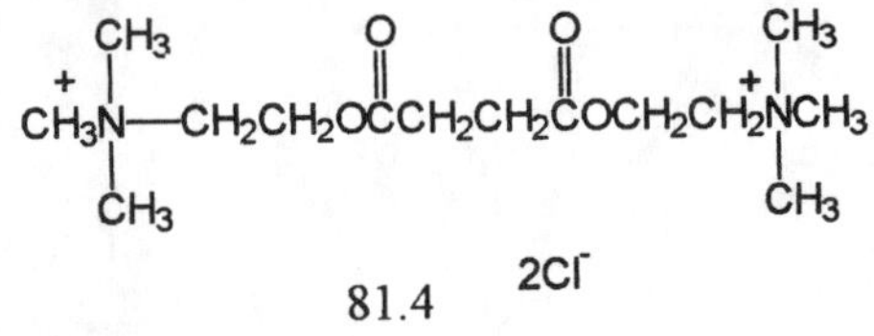

81.4

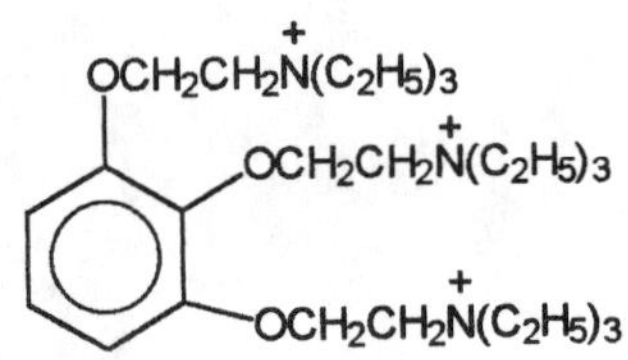

81.5

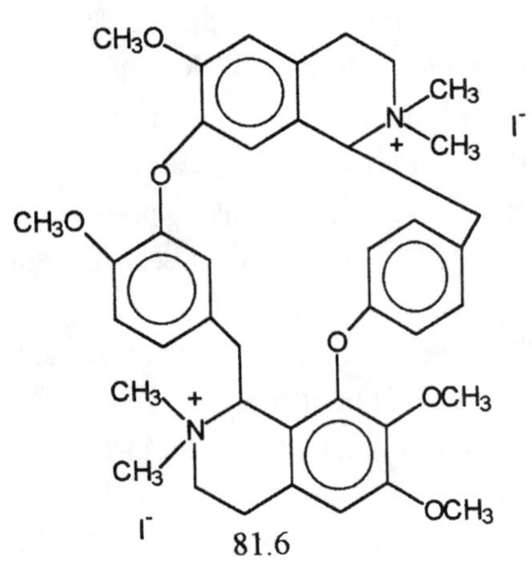

81.6

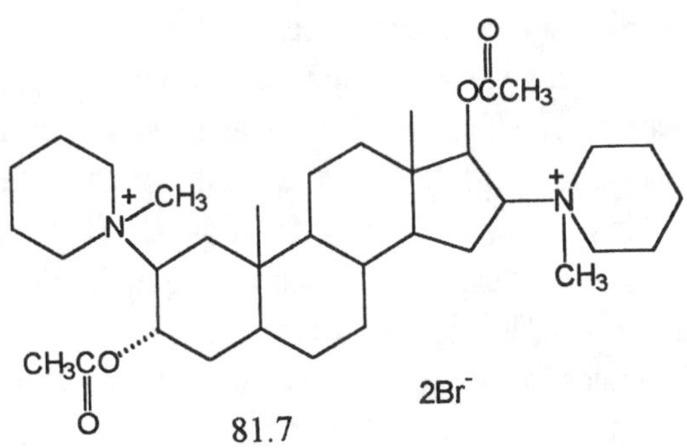

81.7

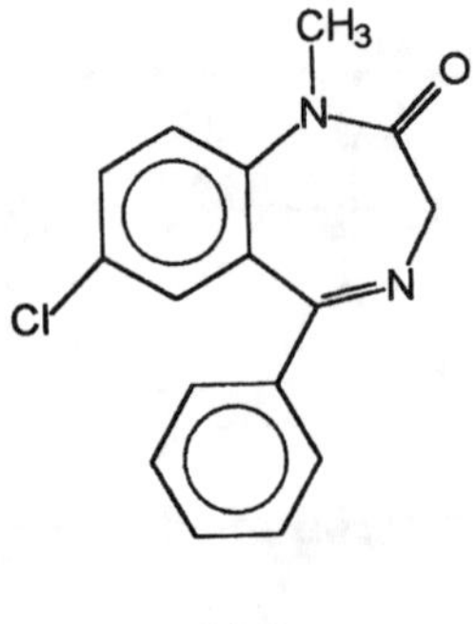

81.8

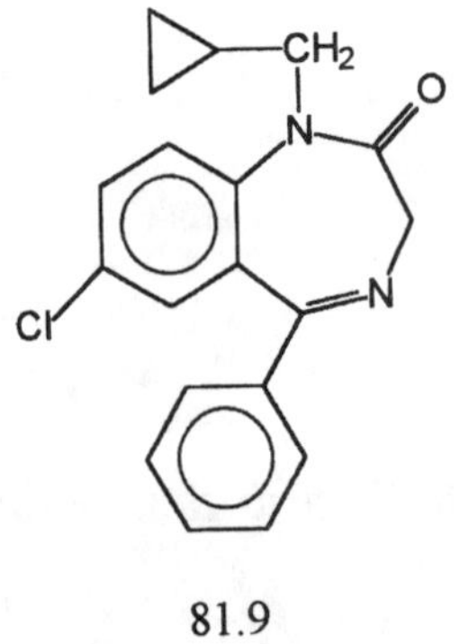

81.9

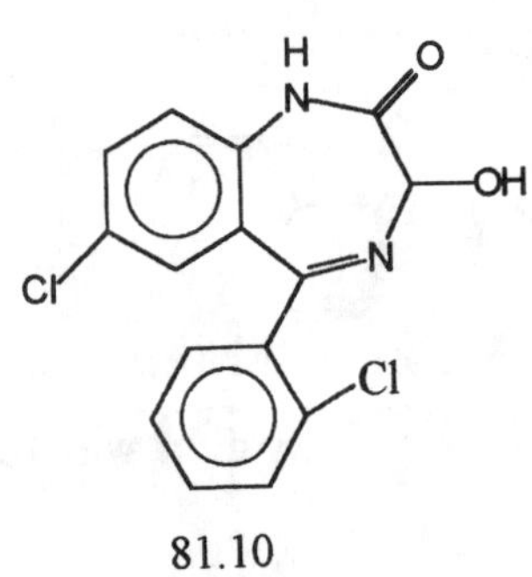

81.10

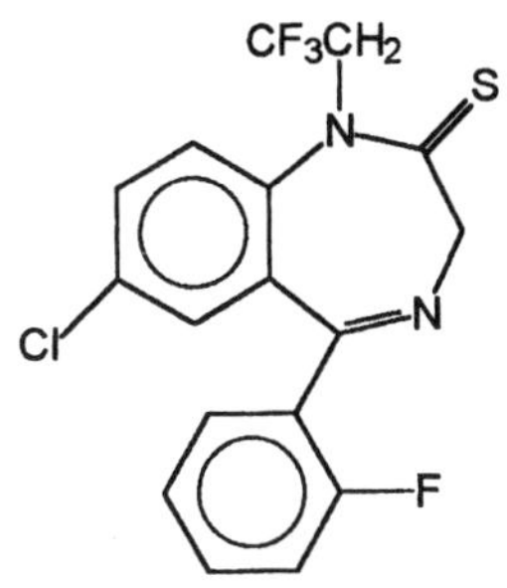

81.11

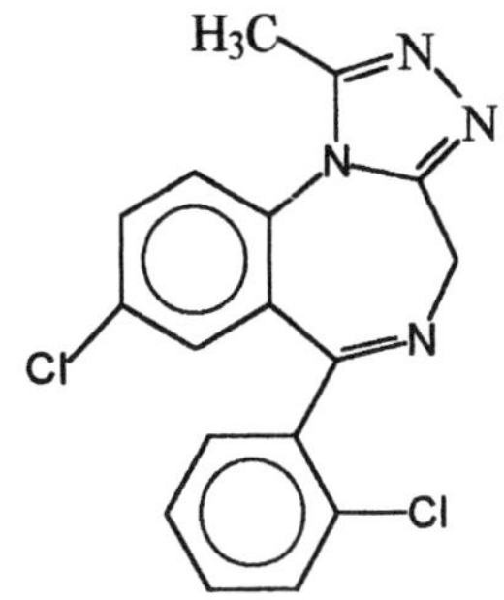

81.12

2. **Select the most appropriate neuromuscular blocking agent from the structures (81.3–81.7) provided to prevent LC from thrashing about. Justify your choice and explain why other possible choices were eliminated.**

3. **Select the most appropriate sedative from the benzodiazepine structures (81.8–81.12). Explain the basis for your selection and why the other structures were not chosen.**

Case **82**

Submitted by Victoria F. Roche

You are the medicinal chemist at the state crime laboratory. The police have confiscated unmarked capsules from a high school student suspected of dealing in "uppers" (centrally active α agonists). After analysis, you find that the capsules contain 2-amino-1-m-hydroxyphenylpropanol bitartrate (82.1).

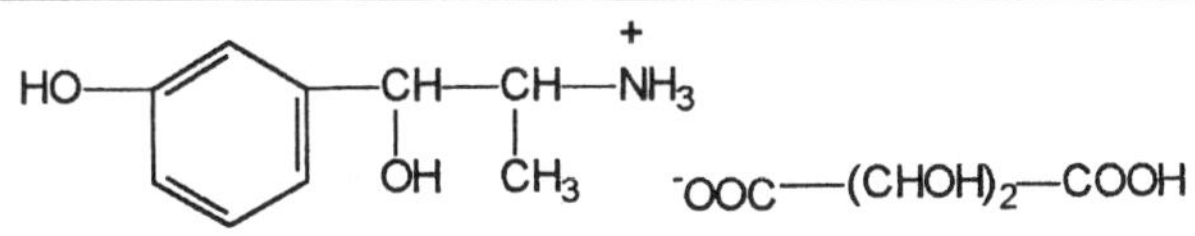

82.1

1. **Do you tell the police to book the student, or let him go?**

2. **What chemical features of this confiscated agent (82.1) support your answer?**

Case **83**

Submitted by Arnold R. Martin and Richard N. Herrier

FP is a 62-year-old male vintner with hypertension that is not well controlled on hydrochlorothiazide 50 mg daily and captopril 50 mg tid. Blood pressures on treatment ranges from 160 to 170 systolic and 95 to 105 diastolic. Ten days ago, FP developed a cold that initially got somewhat better, but then worsened over the last 3 days, with the development of facial pain below the eyes, which is made worse when he bends over; a purulent nasal discharge that "tastes and smells foul"; and a fever of 101F. The doctor plans to treat the sinusitis with amoxicillin 500 mg tid for 3 weeks and wants decongestant treatment for the first 3 to 5 days, but is concerned about its effect on FP's blood pressure. FP used Dristan nasal spray for his cold, but stopped because "it made me jittery." The doctor calls asking for a recommendation for a decongestant that will exert minimal effect on blood pressure and won't make FP feel jittery.

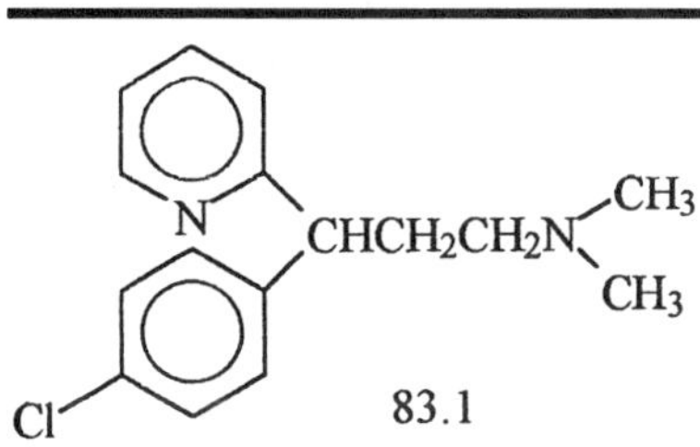

83.1

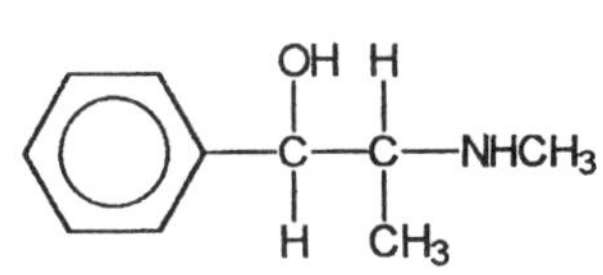

83.2

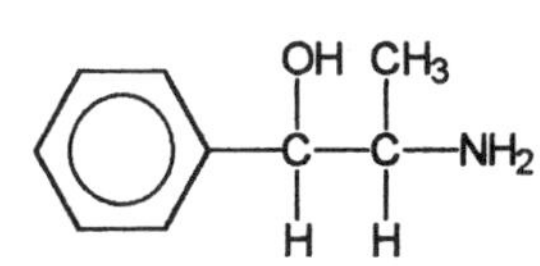

83.3

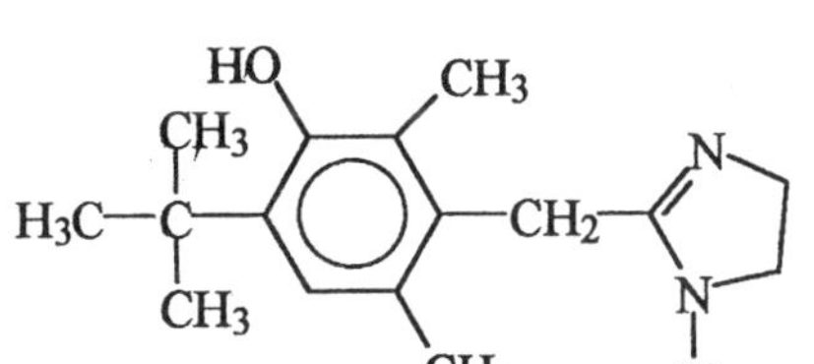

83.4

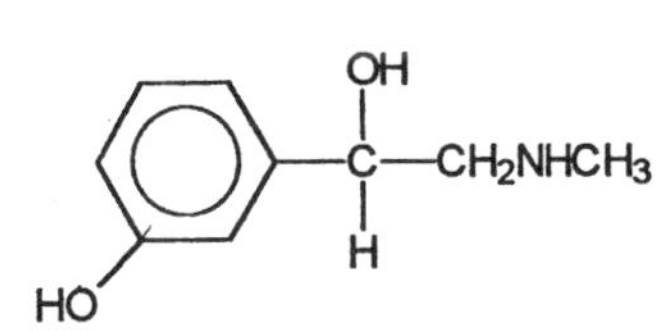

83.5

1. **Select a decongestant from the structures (83.1–83.5) provided that you would recommend for FP. Explain your rationale on a structure/activity basis.**

Answer Phenylephrine (83.5), as a nasal spray solution, would be the best choice for the following reasons: it has the least CNS activity of the available drugs (due to more hydroxyl substitution); nasal administration limits the extent of systemic absorption that can occur; what little that is absorbed will be rapidly metabolized (O-glucuronidation and sulfation of the *m*-hydroxy group and oxidative deamination) in the liver; and this drug has only direct α-adrenergic vasoconstrictive effects (ie, it has little or no β-adrenergic actions through indirect effects *via* release of norepinephrine).

Pseudoephedrine (83.2) and phenylpropanolamine (83.3) have greater CNS stimulant effects than 83.5 because they possess fewer hydroxyl groups and have α-carbon branching that increases lipid solubility and thus CNS activity. These properties also cause significant amounts of 83.2 and 83.3 to be absorbed by oral or nasal routes. Finally, systemic effects of 83.2 and 83.3 are both direct (α) and indirect (β). Oxymetazoline (83.4), although available for topical (intranasal) use, has both CNS stimulant and indirect adrenergic effects. It is also resistant to metabolism.

Case 84

Submitted by Jeffrey J. Christoff

As the inpatient pharmacist at Bigshots University Hospital, you've received a call from a medical student for some drug information. This medical student is working on a case presentation for a patient who has been ordered Aldomet 500 mg IV q6h. His question for you is: Why would an IV solution be prepared with α-methyldopate hydrochloride (Aldomet IV [84.1]) rather than α-methyldopa (Aldomet [84.2])?

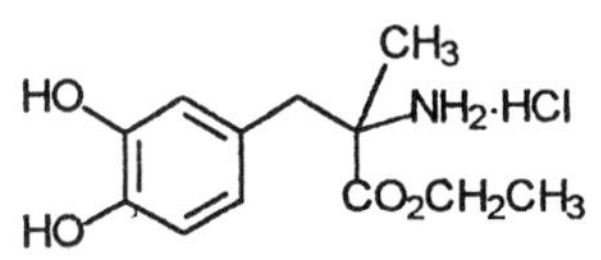

84.1

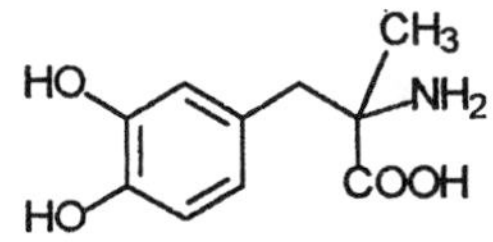

84.2

1. **Why would an IV solution be prepared with α-methyldopate hydrochloride (Aldomet IV [84.1]) rather than α-methyldopa (Aldomet [84.2])?**

Answer α-Methyldopate hydrochloride (Aldomet IV [84.1]) is an ester and salt derivative of α-methyldopa (Aldomet [84.2]). α-Methyldopa is a zwitterion and is considerably less water soluble than its ester salt. The water solubility of methyldopa is approximately 10 mg/mL (pH dependent), whereas the solubility of methyldopate hydrochloride is approximately 10 to 300 mg/mL (pH dependent). Thus, the increased water solubility of α-methyldopate hydrochloride limits the amount of intravenous fluid that must be administered to a hypertensive patient.

Case 85

Submitted by Victoria F. Roche

You have been hired as a medicinal chemist at Currie Pharmaceuticals. Dr. Currie is concerned about falling behind in the ophthalmics market, and wants to produce a new reversible cholinesterase inhibitor, zitostigmine bromide, that would have a faster onset of action than Roche's neostigmine bromide (85.1), and a longer duration of action than Alcon's carbachol chloride (85.2).

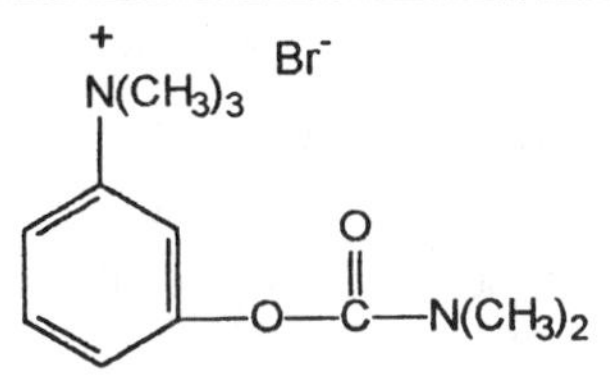

85.1

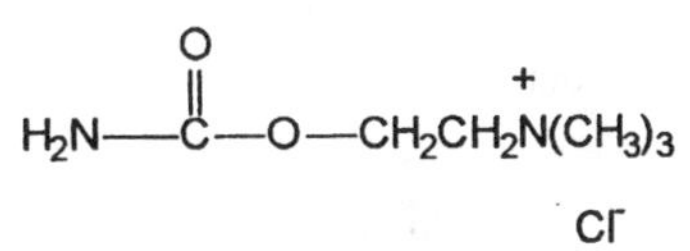

85.2

1. How do reversible cholinesterase inhibitors work? What structural features of reversible cholinesterase inhibitors promote high acetylcholinesterase (AChE) affinity? How is AChE regenerated after inactivation by reversible cholinesterase inhibitors?

Answer Reversible cholinesterase inhibitors work by carbamylating the nucleophilic serine residue of the enzyme AChE. This important serine residue is responsible for hydrolyzing the endogenous neurotransmitter acetylcholine. When this serine is carbamylated by a reversible cholinesterase inhibitor, the serine oxygen is no longer nucleophilic since its electron density is delocalized through resonance with the attached carbamoyl unit. To carbamylate AChE, reversible inhibitors must first anchor to the enzyme's anionic site through an ion-ion bond. This requires a cationic amino nitrogen on the drug structure. All commercially available carbamate cholinesterase inhibitors, but one, are quaternary amines. Quaternary amines are ionized, neutral species, and retain their cationic charge at all pHs. Whether quaternary or tertiary, the amino nitrogen of reversible cholinesterase inhibitors has small alkyl substituents (eg, CH_3) in order to minimize hindrance to ion-ion anchoring. The carbamate carbonyl group (which is the site of attack by AChE serine) must be kept separated from the anchoring cationic amino nitrogen by the same distance that the enzyme anionic site and nucleophilic serine residue are separated (approximately 3–5Å). The appropriate amount of separation can be achieved if there is a three or four atom chain between the amino and carbonyl groups.

2. **Based on this inhibition mechanism, what important structural considerations must be taken into account when designing Zitostigmine?**

Answer Since zitostigmine must act faster than neostigmine bromide, it must show a higher affinity for AChE serine. Neostigmine bromide is an aminophenolcarbamate, and extended resonance between the carbamate group and conjugated aromatic ring promotes an electrophilic carbonyl carbon that is attractive to the serine. Neostigmine is a tertiary carbamate, and the two CH_3 groups on the carbamate nitrogen do provide some steric hindrance to serine's approach. In addition, their weak positive inductive effect diminishes the electrophilic character of the carbonyl carbon over what would be seen with a secondary or primary carbamate. To provide an onset of cholinesterase inhibition faster than neostigmine bromide, zitostigmine would have to retain the aminophenolcarbamate structure to keep the resonance advantage, but have fewer CH_3 groups (ie, one or none) on the carbamate nitrogen so as to decrease hindrance to the approach of the AChE serine.

Zitostigmine must also provide a longer duration of action than carbachol chloride, which is a primary carbamate. Once the serine residue of AChE is carbamylated, the enzyme is inactive until a molecule of activated water can hydrolyze the carbamylated residue, releasing the free, active AChE serine. The greater the number of CH_3 groups on the carbamate nitrogen of the carbamylated serine, the slower the regeneration will be. When on the serine residue, the carbamate CH_3 groups provide the same steric and electronic hindrance to the approach of activated water as they did to the approach of the AChE serine when they were part of the drug structure. To provide a longer duration of indirect cholinergic action than carbachol chloride, zitostigmine must have one CH_3 group attached to the carbamate nitrogen. Although two carbamate methyl groups would also provide an even longer duration of AChE inhibiting action, they would also provide an onset equivalent to (rather than faster than) neostigmine bromide.

3. **Design your own molecule of zitostigmine and discuss how your molecule would meet Dr. Currie's activity requirements.**

Answer A zitostigmine structure (85.3) that would meet Dr. Currie's expectations is provided. Note that the quaternary amine structure has been retained for high affinity binding to the AChE anionic site.

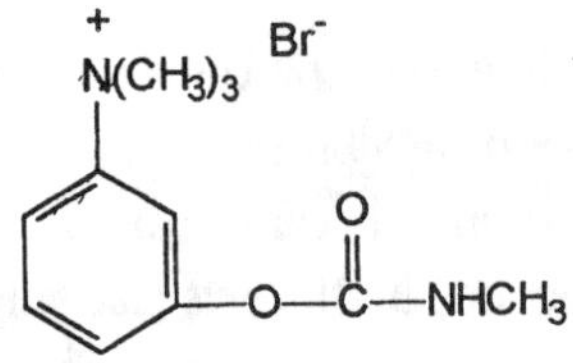

85.3

Case 86

Submitted by Victoria F. Roche

HS, a western Nebraska farmer, is brought by his wife to the emergency room of your local hospital with a broken leg. He is also complaining of weakness and dizziness, which he is attributing to the pain of his break. You also note that he is salivating and sweating excessively, has a low heart rate, and is breathing with some difficulty. You discover that, in the farm accident which caused the broken leg, he came in contact with large quantities of the insecticide Parathion (86.1).

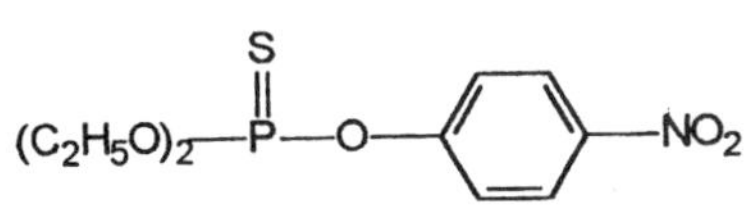

86.1

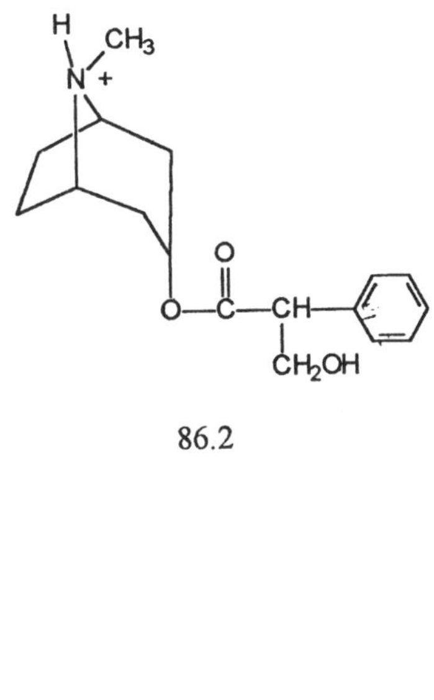

86.2

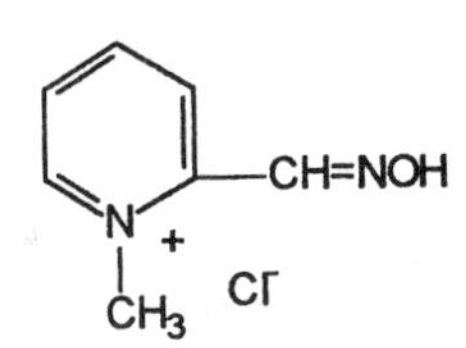

86.3

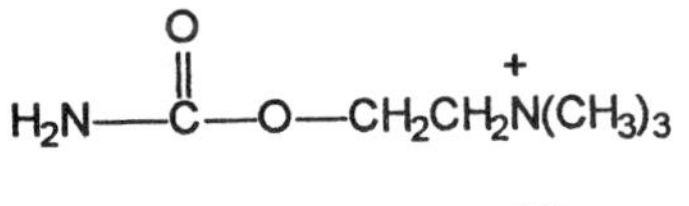

86.4

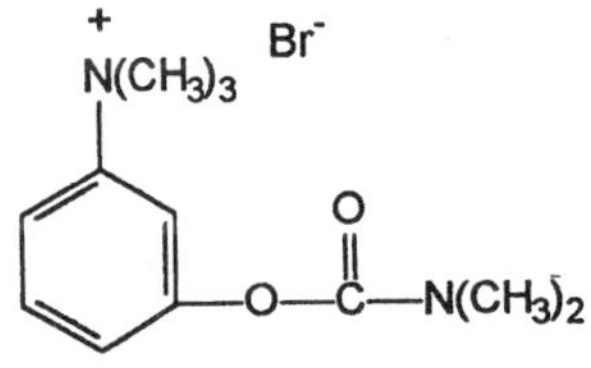

86.5

1. Explain the cause of the farmer's symptoms.

2. **Which of the structures (86.2-86.5) should be immediately administered? Why?**

Case 87

Submitted by Victoria F. Roche

You have volunteered your services as a pharmacist in the Dominican Republic through your university's international outreach program. You will be assisting a physician in performing eye surgery on a 74-year-old man. The surgery will be taking place in a makeshift hospital in a remote village "campo," and drugs are in short supply. The physician asks you to select an appropriate agent from among those donated to maintain miosis during the operation. The drug will be administered topically, and she wants it to act rapidly and provide a relatively short duration of action so that the extent of miosis can be easily controlled.

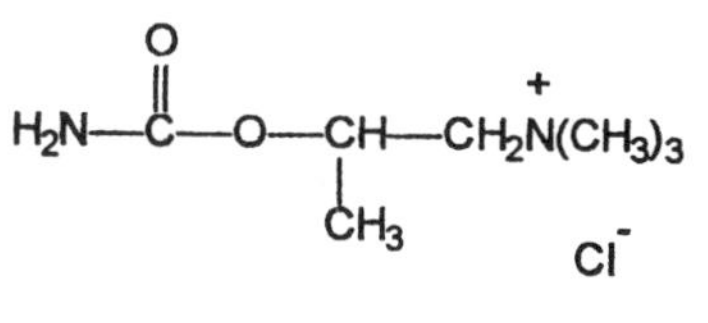

87.1

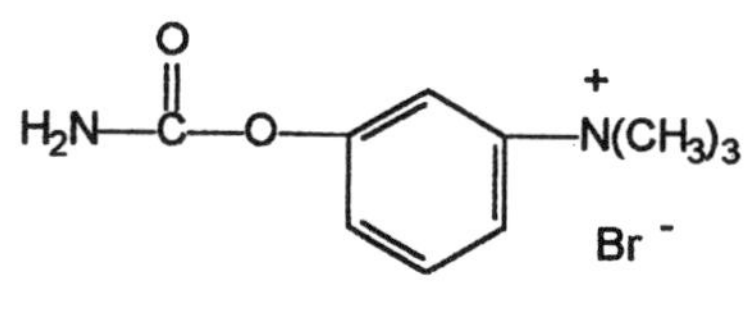

87.2

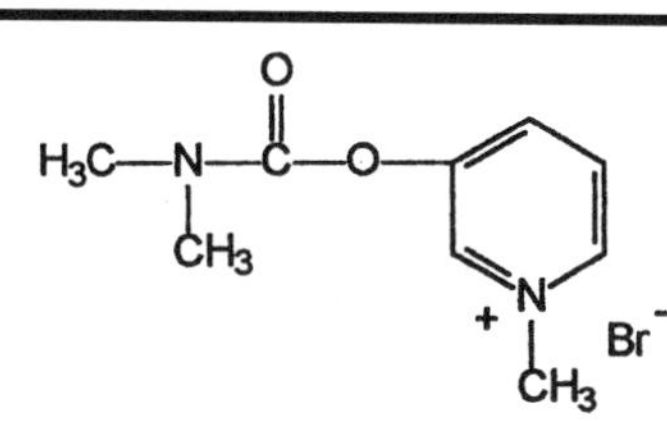

87.3

1. **Which of the three donated drug agents (87.1–87.3) would you recommend? Defend your choice, based on the activities anticipated from an analysis of chemical structure.**

2. **Your patient is predisposed to bradycardia and, during the surgery, the physician notices a slowing of the heart rate. Could the drug you selected for miosis be at least partially responsible for this effect? Explain.**

Case **88**

Submitted by Arnold R. Martin and Richard N. Herrier

LN is a 74-year-old woman who presents to her HMO with malaise, shortness of breath, and fatigue. She also complains of weight gain and swollen hands and feet. LN has a 6-month history of congestive heart failure (CHF), a 3-year history of hypercholesterolemia, and a 6-year history of adult onset (type II) diabetes mellitus. Two weeks ago, LN suffered a mild stroke and was hospitalized for 3 days. She was then discharged with a prescription for oral warfarin (5 mg daily) for 3 weeks. Her other current medications include: digoxin (0.125 mg daily), lovastatin (20 mg at night), and tolazamide (250 mg daily). The patient's creatinine clearance is 35 mL/min. The cardiologist determines that the digoxin alone is insufficient to control LN's congestive heart failure. Because previous attempts to use angiotensin-converting enzyme inhibitors caused intolerable hypotension, it is decided to employ diuretic therapy.

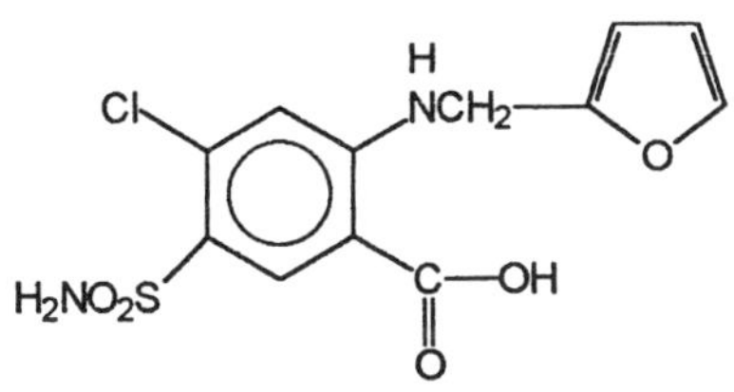

88.1

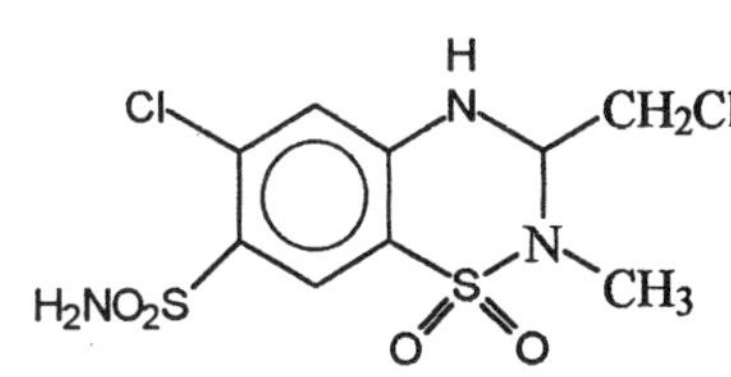

88.2

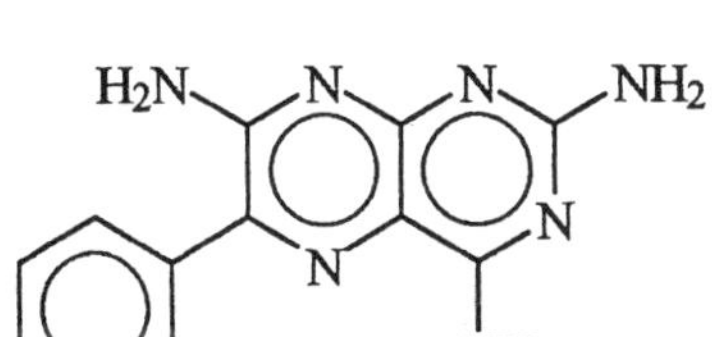

88.3

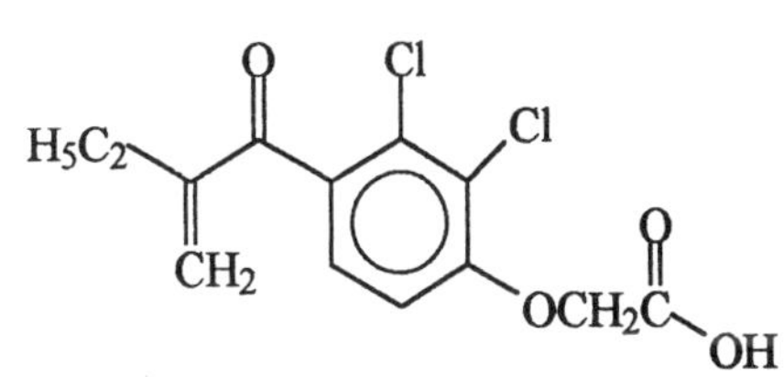

88.4

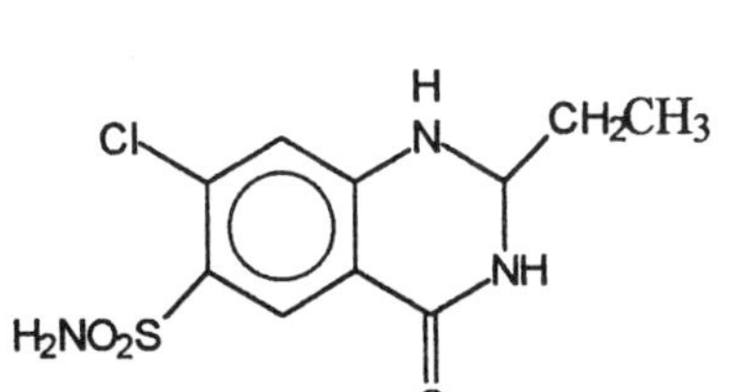

88.5

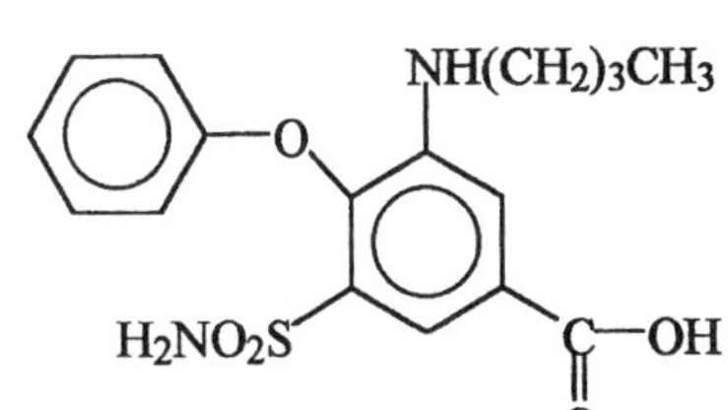

88.6

1. **Which diuretic(s), of the structures (88.1–88.6) shown, would provide the most benefit? Justify your choice(s).**

2. **Indicate agents that would be inappropriate and/or less effective than your first choice. Explain why.**

Case 89

Submitted by Arnold R. Martin and Richard N. Herrier

KC is a 59-year-old man who is admitted to the hospital with severe headaches, nausea, malaise, dizziness, blurred vision, and shortness of breath. He also complains of difficulty in sleeping due to nightime breathing problems (paroxysmal nocturnal dyspnea) and must prop himself up in bed with pillows to sleep comfortably. On admission, his physical signs are: BP, 215/140; HR, 98; and RR, 15. Laboratory values include potassium 3.2 and creatinine 1.3. KC currently takes medication for high blood pressure (lisinopril 5 mg qd); congestive heart failure (furosemide [89.1] 20 mg qd); deep vein thrombosis, 2 incidents (warfarin 5 mg qd); and arthritis (ibuprofen 800 mg bid). KC admits that he religiously takes his blood thinning and arthritis medicines, but uses the others only when he needs them.

The hypertensive crisis was controlled with sodium nitroprusside (35 mg/min IV) until the diastolic BP fell to 110 mm. IV furosemide (89.1) was then administered to reduce the edema. IV potassium chloride was given to maintain serum potassium. Before his discharge, KC was counseled regarding the importance of taking his blood pressure and diuretic medications. At this time, KC complained that the potassium chloride tablets upset his stomach. KC's cardiologist thus contemplates alternative diuretic therapy.

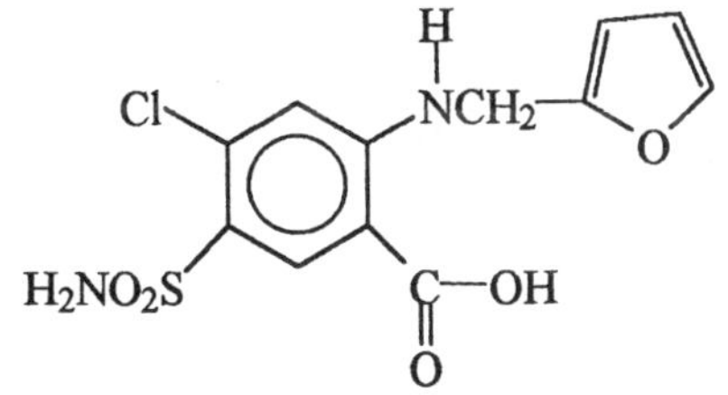

89.1

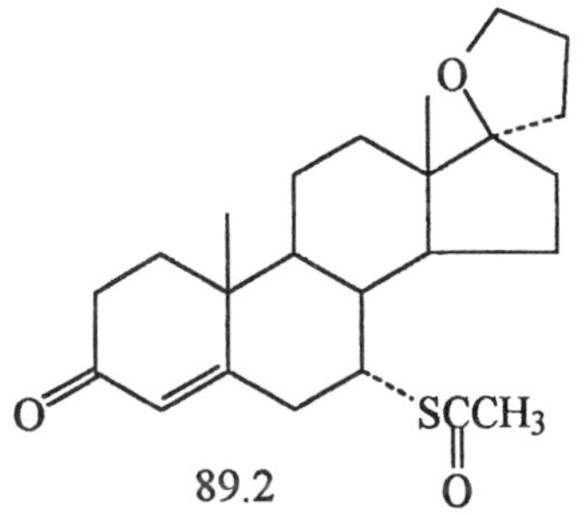

89.2

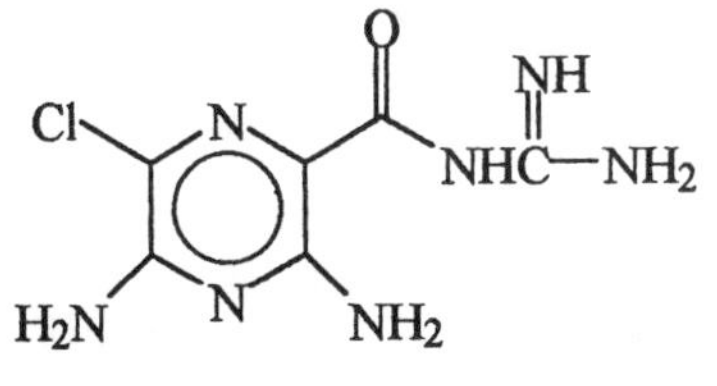

89.3

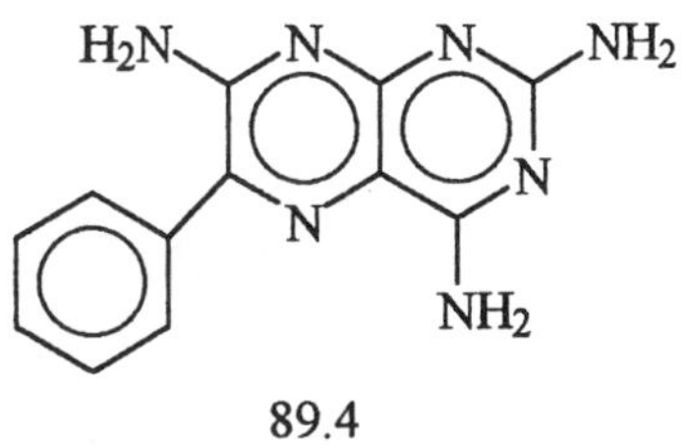

89.4

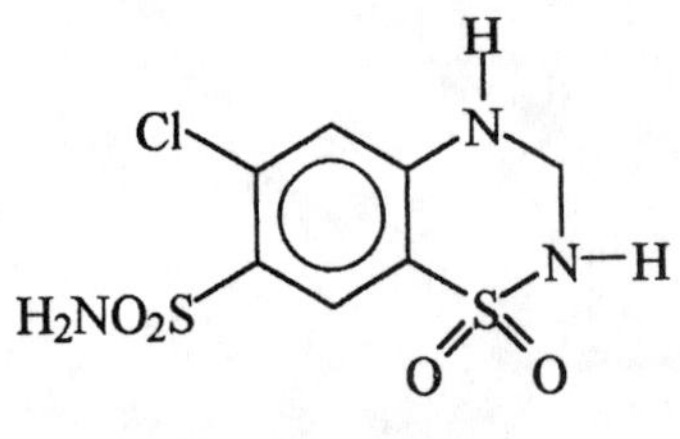

89.5

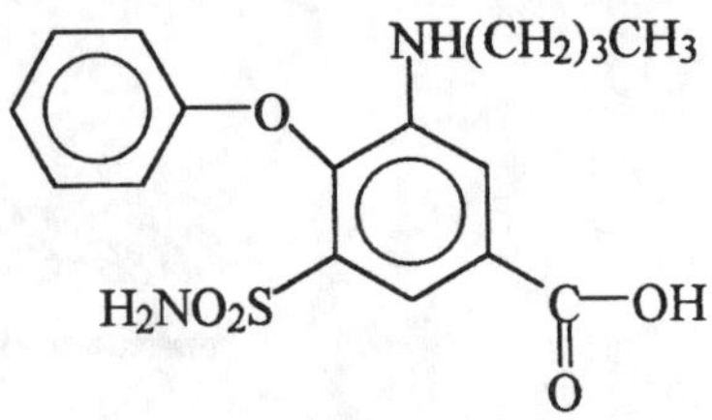

89.6

1. **From the diuretic structures (89.1–89.6) provided, suggest alternative diuretic therapy (which includes furosemide [89.1]) that will help control KC's hypertension, prevent congestive heart failure, and avoid hypokalemia. Justify your choice on the basis of the properties of the drug structure(s) selected. Explain the basis for rejecting any structures and discuss alternative choices.**

Case 90

Submitted by Arnold R. Martin and Richard N. Herrier

AG is a 56-year-old construction worker with a 15-year history of hypertension and ischemic heart disease that was well controlled until 6 months ago, when he began having anginal chest pain with increasing frequency and severity. Over the past 2 months, he has been seen twice in the ER with chest pain unrelieved by sublingual nitroglycerin. On a third occasion, he was hospitalized with the same symptoms and was given a complete workup. He has severe two vessel disease, but refuses angioplasty or bypass surgery at this time. His current medications include nifedipine XL, captopril, nitroglycerin, diltiazem, and aspirin. During the past week, he experienced several anginal attacks each day that were relieved by nitroglycerin; finally, AG has decided to seek more definitive treatment. Tonight he presents to the ER with chest pain unrelieved by nitroglycerin and is sent to the critical care unit with the diagnosis of unstable angina pectoris. The attending physician orders IV nitroglycerin and wants to start an IV β-adrenergic blocker to counteract any tachycardia, but is concerned about its long-term effects on overall cardiac function.

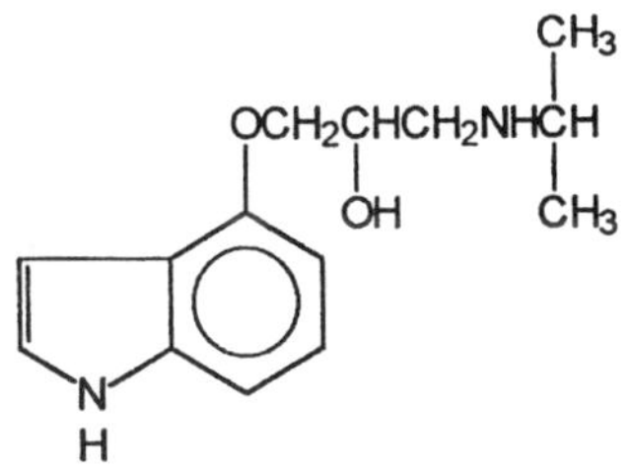

90.1

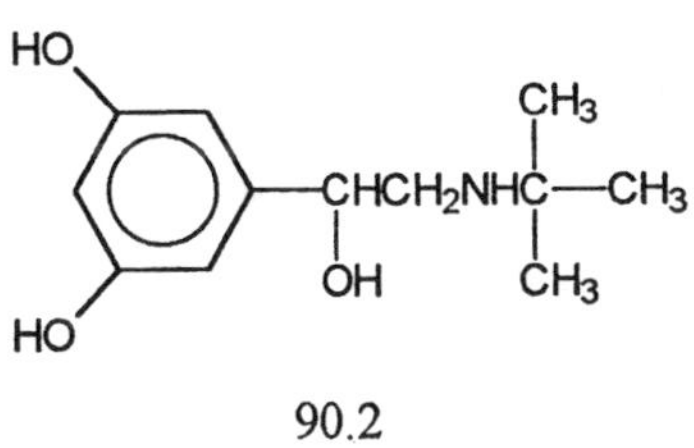

90.2

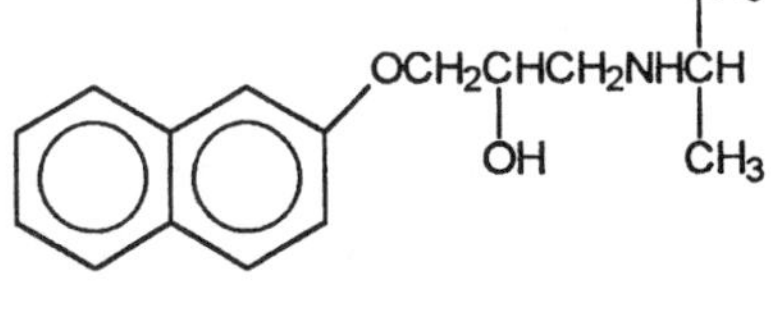

90.3

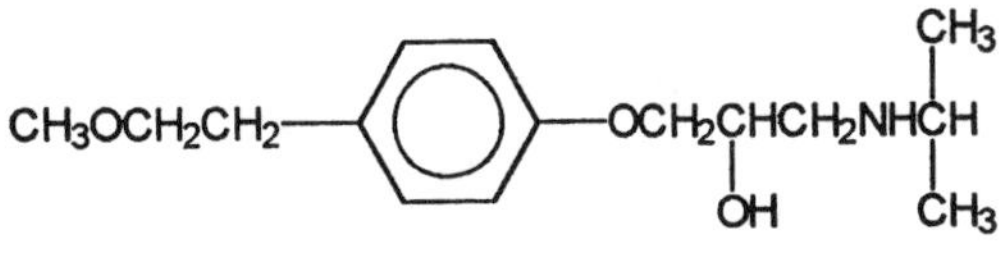

90.4

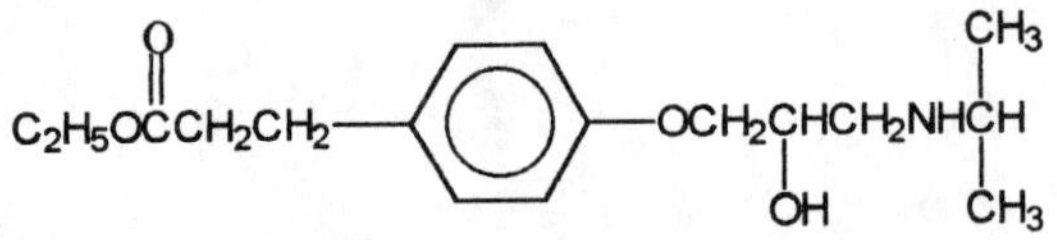

90.5

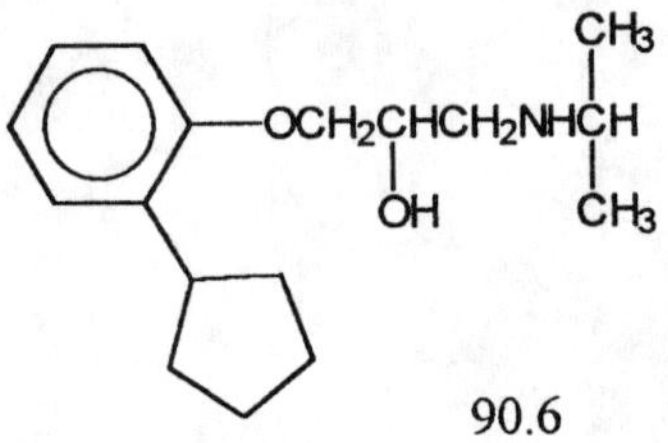

90.6

1. **Select the most appropriate β-adrenergic antagonist from the structures (90.1–90.6) provided. Justify your choice on the basis of the pharmacology and the comparative pharmacokinetics of the available structures.**

2. **One of the structures is contraindicated in this patient. Which one is it, and why is it contraindicated?**

Case 91

Submitted by Robin M. Zavod

BJ, a 35-year-old mother of two, was recently admitted to the ER with severe leg pain. She was diagnosed with deep vein thrombosis, treated (no oral anti-coagulant therapy), and was discharged with a prescription for warfarin sodium (91.1). As soon as she arrived home, she decided to "ensure" that this would never happen again and took a 5-day supply at once.

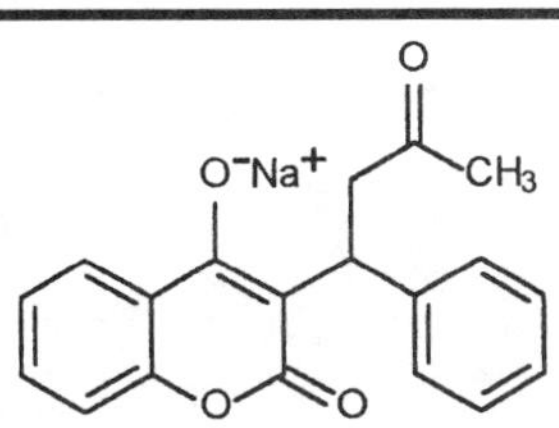

91.1

1. **Could the patient have been given tissue plasminogen activator (tPA) as part of the deep vein thrombosis treatment? Why, or why not? Base your answer on the mechanism of action of tPA.**

2. **Since the patient was not given any warfarin (91.1) in the emergency room, did she change the rate at which warfarin (91.1) would be effective by taking a 5-day supply? Please explain why, or why not based on the mechanism of action of warfarin (91.1).**

Case 92

Submitted by Arnold R. Martin and Richard N. Herrier

BD is a 36-year-old woman who presents to her primary care physician with the following symptoms: headache, lethargy, fatigue, muscle weakness, palpitations, thirst, and urinary incontinence. BD's BP was 130/95. Laboratory tests reveal hypokalemia, hypernatremia, hyperaldosteronemia (but normal plasma cortisone levels), low plasma renin, and a urinary pH of 7.0. A tentative diagnosis was made of primary aldosteronemia due to an adenoma of the adrenal cortex. The diagnosis was confirmed by CT scan. Before the surgical removal of the adenoma, the consulting surgeon wishes to correct the hypokalemia using a potassium sparing diuretic.

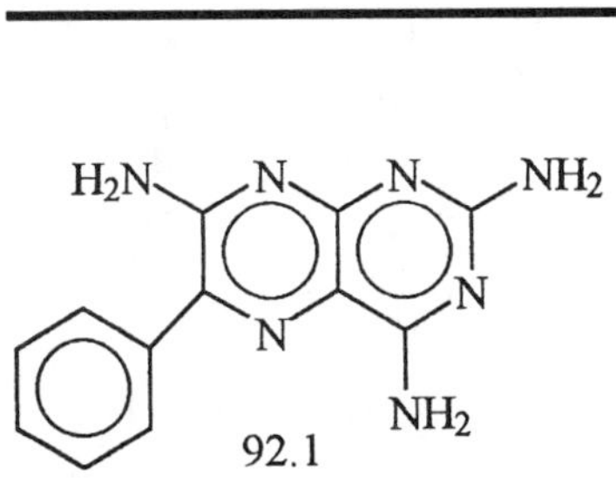

92.1

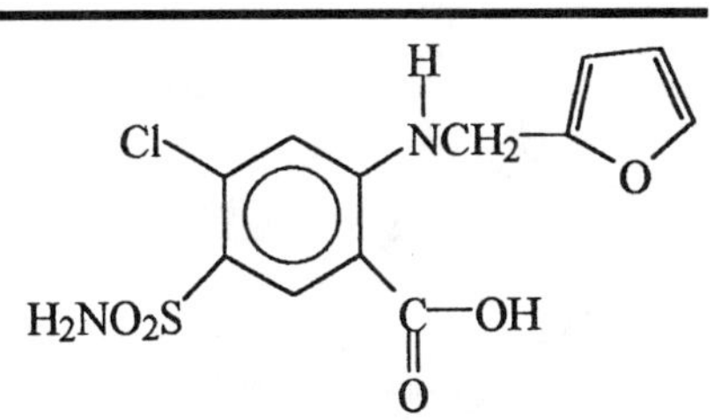

92.2

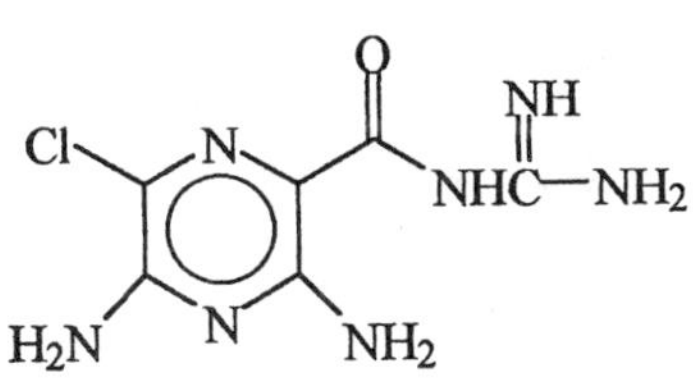

92.3

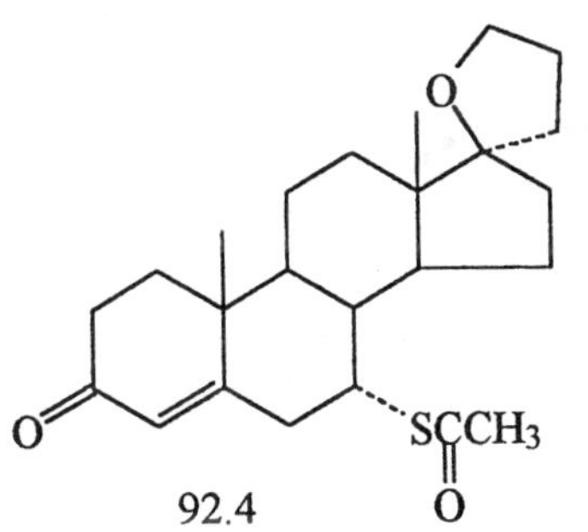

92.4

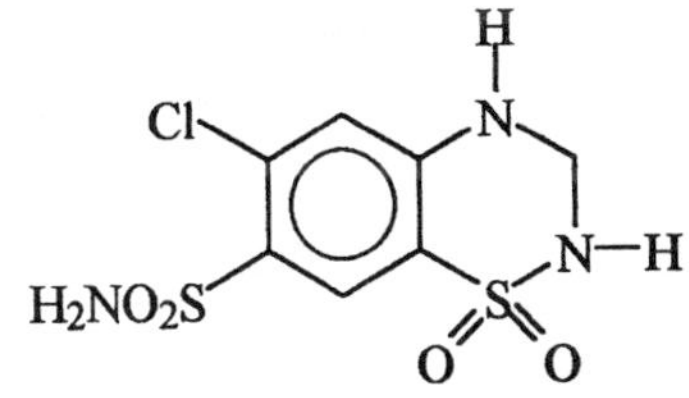

92.5

1. **Select an appropriate diuretic from the structures (92.1–92.5) shown. Provide a rationale for your choice and explain why the remaining structures were not chosen.**

Case 93

Submitted by Arnold R. Martin and Richard N. Herrier

MP is a 43-year-old man who is admitted to a health service for complications arising from his chronic alcohol use and liver cirrhosis. MP has a history of alcohol abuse beginning as a teenager. Until 2 months ago, his alcohol abuse consisted of 4 or 5 beers or a bottle of wine each day, with heavy binges involving up to a case of beer on the weekends. Despite the alcohol abuse, MP had no apparent symptoms until 3 years ago when he had an attack of pancreatitis that required hospitalization. Since then, he has been hospitalized three times: once for pancreatitis, once for alcoholic hepatitis, and on the last occasion 2 months ago, for a tonic-clonic seizure coincident with an attempt to quit drinking. After his discharge from the clinic following the seizure, MP decided to abstain from alcohol. He has been taking Antabuse for the past 2 months and attends Alcoholics Anonymous meetings on a regular basis.

On admission to the clinic, MP complains of several recent episodes of vomiting (some of the vomitus was blood tinged), abdominal tenderness, malaise, and weight gain. MP's liver was tender and palpable, and esophageal varices were present on the upper GI. MP presented with the following measurements: BP, 140/90; HR, 75; height, 172 cm; weight, 96 kg; and girth, 140 cm. Liver enzymes, bilirubin, and albumin were all above normal levels. Cirrhosis of the liver with ascites was diagnosed and confirmed by biopsy. Diuretic therapy was initiated.

On the first day of hospitalization, MP was started on a 20-mg IV dose of furosemide (93.1), followed by oral doses of 40 mg on the second day and 80 mg on the third day. His weight and girth steadily dropped (to 88 kg and 125 cm, respectively, by the end of the 2nd day) for the first 2 days of furosemide (93.1) treatment, but did not change on the 3rd and 4th days. A check of blood electrolytes revealed normonatremia and hypokalemia. Concluding that furosemide (93.1) had reached the limit of its effectiveness in correcting MP's hepatic edema, the attending physician contemplates changing diuretic medication.

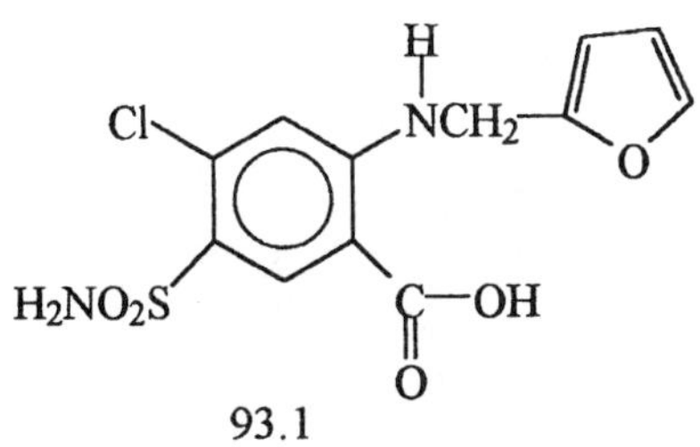

93.1

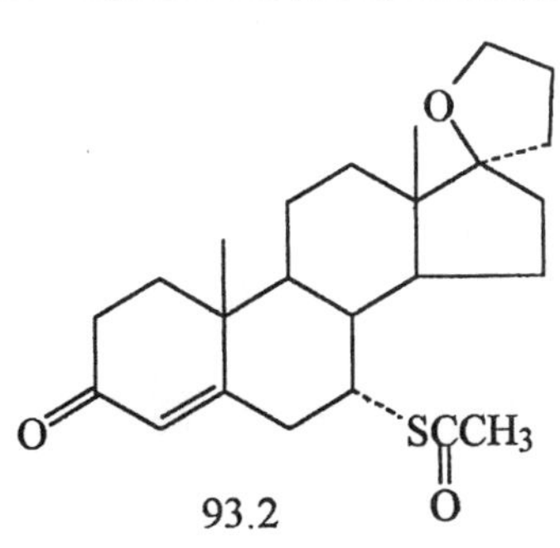

93.2

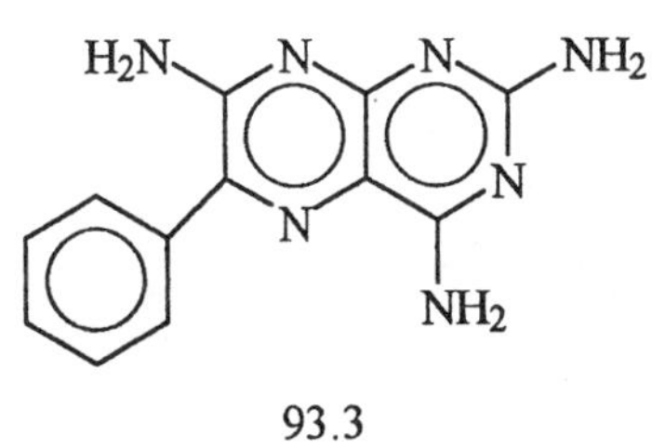

93.3

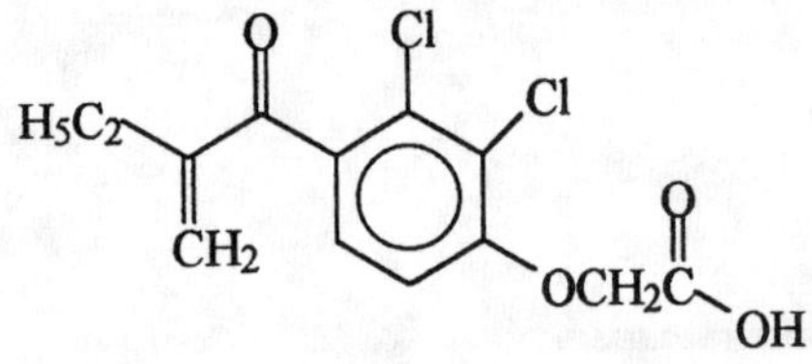

93.4

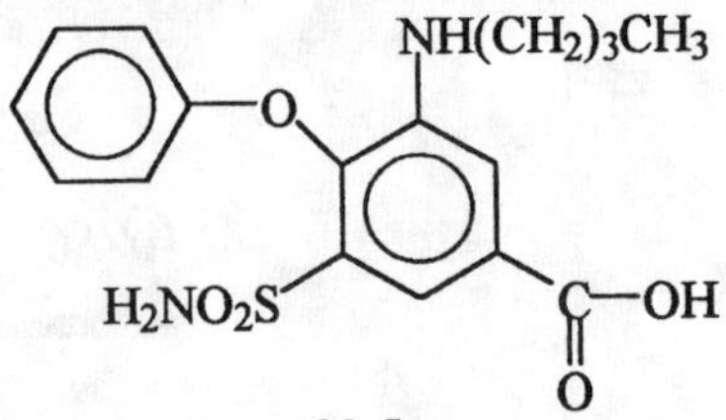

93.5

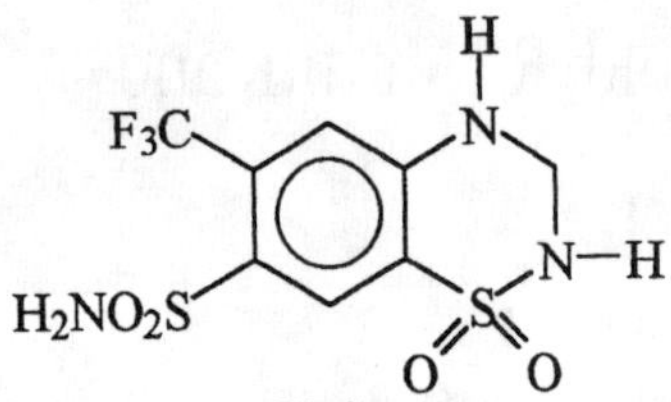

93.6

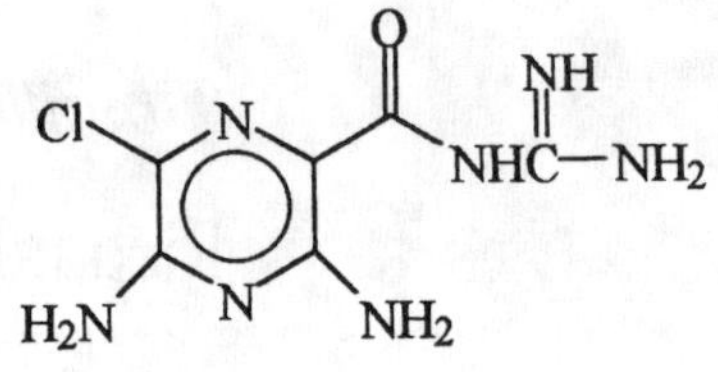

93.7

1. **From the diuretic structures (93.2–93.7) provided, select an agent that would be effective in further reducing MP's hepatic edema and also in correcting the electrolyte imbalance. Justify your choice on the basis of the pharmacologic properties of the drug chosen. Explain the basis for eliminating other choices.**

Case 94

Submitted by Arnold R. Martin and Richard N. Herrier

BL is a 71-year-old male retired copper miner, who presents to the clinic for follow-up of chronic medical problems. His chief complaints are shortness of breath, palpitations, muscle weakness, and chronic fatigue. BL worked in the mines since he was 18. He was laid off in 1978 when the price of copper dropped. He contracted tuberculosis when he was in his early 40s and was successfully treated with a course of streptomycin (1 year), isoniazid (2 years), and p-aminosalicylic acid (intermittent over a 2-year period). He suffered significant hearing loss, attributed to streptomycin, that did not fully reverse. To relieve the chronic pain associated with his work in the mines, BL has abused acetaminophen over a 10-year period. The acetaminophen abuse is believed to be a contributor to BL's chronic renal insufficiency (current creatine clearance, 25 mL/min). He has been treated for the past 5 years for congestive heart failure with digoxin (0.125 mg qd). BL has had hypertension for more than 20 years and has been treated with a variety of medications to control his blood pressure. The current medications, enalapril and hydrochlorothiazide (94.1)/triamterene (94.2), appear to be reasonably effective (current BP 120/95) in controlling BL's blood pressure, except for the edema. Plasma electrolytes reveal hyperkalemia (moderate), hyperphosphatemia, and normal sodium levels. BL's cardiologist contemplates a change in diuretic therapy.

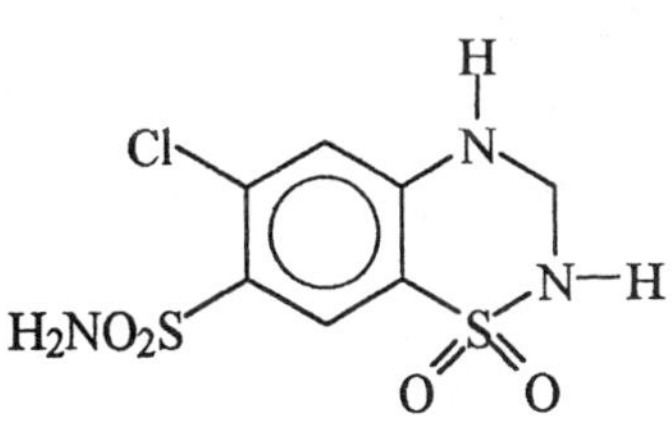

94.1

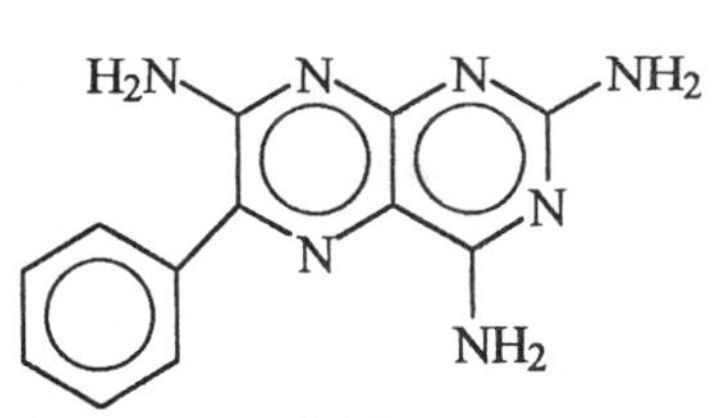

94.2

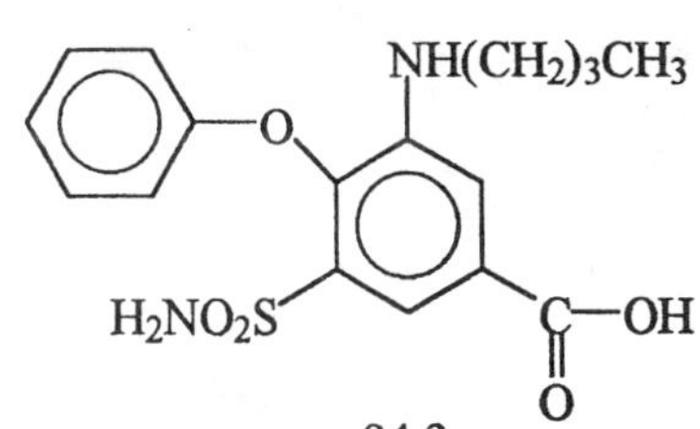

94.3

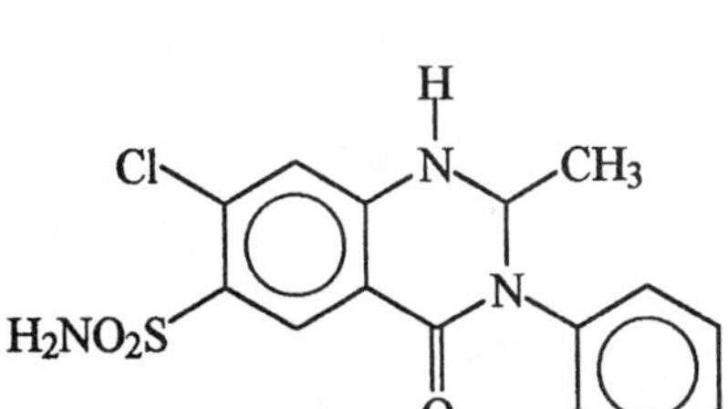

94.4

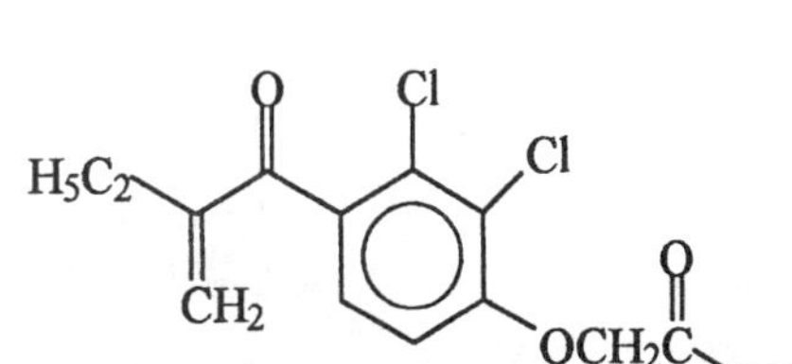

94.5

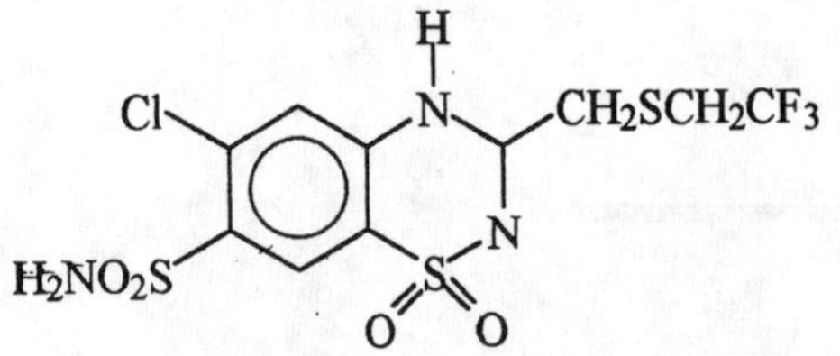

94.6

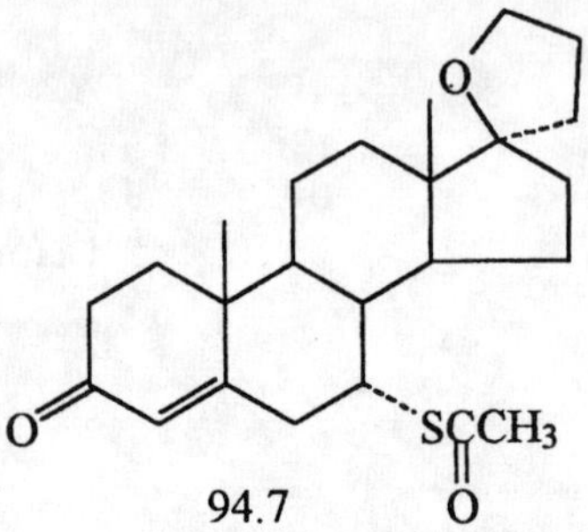

94.7

1. **Select an appropriate diuretic from the structures (94.1–94.7) provided that will: (1) correct the edema; (2) restore the electrolyte balance; and (3) minimize the risk to the patient. Provide a detailed rationale for your choice and explain why each of the other possible choices were rejected.**

Case 95

Submitted by Marc W. Harrold

FF comes into the ER last evening complaining of sharp, severe pains in his side and lower back. Following examination and testing, he was admitted with a diagnosis of urate kidney stones. At the time of admission, his uric acid level was 8.5 mg/dL. He is currently being treated with parenteral codeine as needed for the pain, and $NaHCO_3$ in D_5W, 1 L q6h. A medical history obtained from his family doctor reveals that FF has been hyperuricemic for more than 8 years. He has suffered periodic bouts of gout in the past; however, his last occurrence was more than 4 years ago. The periodic attacks were resolved with short-term combination therapy of colchicine and probenecid. His last uric acid level before hospitalization was taken 2 weeks ago during a routine checkup and was 8.3 mg/dL. In addition to hyperuricemia, FF was diagnosed with glaucoma 4 years ago. This condition had been treated with acetazolamide (95.1) up until 6 weeks ago. At that time, FF saw a new ophthalmologist (his old one retired), who switched his medication from acetazolamide (95.1) to timolol (95.2).

95.1

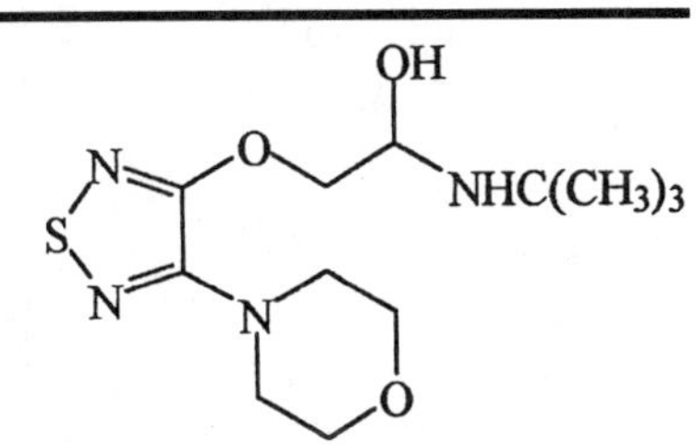

95.2

1. **What is the purpose of using $NaHCO_3$ in D_5W in this patient?**

2. **What are the mechanisms of action of timolol (95.2) and acetazolamide (95.1) in the treatment of glaucoma? What advantage could potentially be gained by switching a patient from acetazolamide (95.1) to timolol (95.2)?**

3. **Explain how the change in the patient's glaucoma medication could have contributed to the formation of urate kidney stones.**

Case 96

Submitted by Victoria F. Roche

GS, a 53-year-old hospitalized man with cardiovascular disease who was stabilized on 80 mg bid of the agent shown, began to hallucinate, exhibit confusion and agitation, and became emotionally labile and verbally abusive to hospital staff.

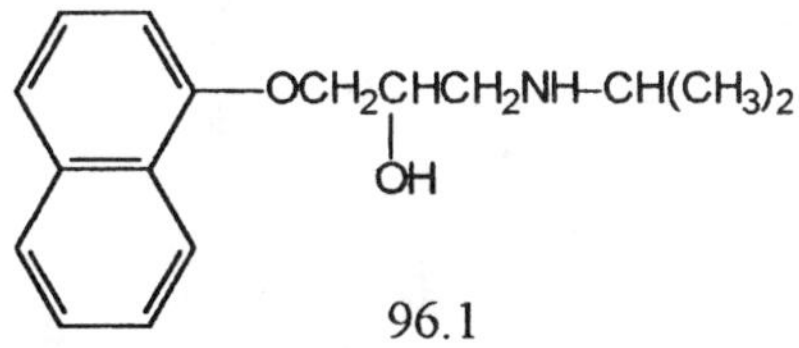

96.1

1. How could this drug structure have been responsible for this man's symptoms?

2. How could you modify the structure to attenuate these undesired side effects?

3. **How would the duration of action of your new agent compare with the original agent after po administration?**

Case 97

Submitted by Victoria F. Roche

A gardener comes into your community pharmacy with red, irritated hands. He had been handling the insecticide Sarin (97.1) with wet hands when they began burning.

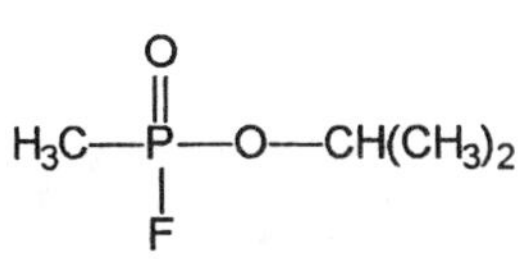

97.1

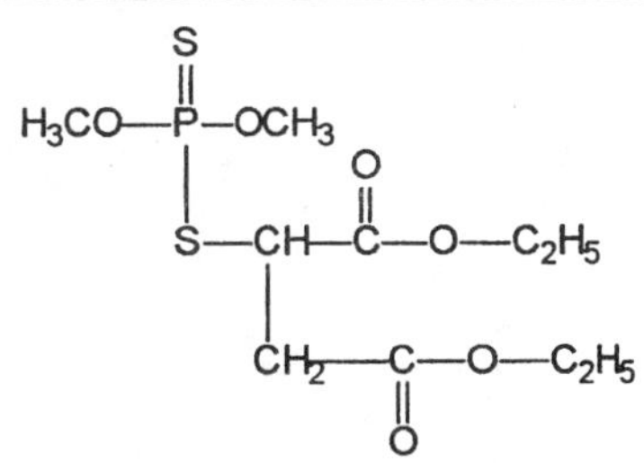

97.2

1. **What reaction can occur between sarin (97.1) and water?**

2. **How do you explain the irritation experienced by this gardener?**

3. **The gardener just purchased a different insecticide (malathion [97.2]) from a nursery located in the same strip mall as your pharmacy, and he shows you the container. He wants to know if this product will be as dangerous as sarin (97.1), the remains of which are now resting peacefully in a dumpster. You check the nomenclature of malathion and deduce the structure to be as shown (97.2). How do you respond to this question?**

Case 98

Submitted by Victoria F. Roche

As the respected town pharmacist in a small Nebraska community, you are giving a drug information talk at the local PTA meeting, and you are asked a question about medication advertising by the local curmudgeon, IM. IM is questioning whether or not drug companies can really back up their advertising claims with scientific proof, or if the whole business is all Wall Street fantasy. He quotes an ad for the OTC drug Sudafed, which claims to provide "sinus relief without putting you to sleep." IM said he took this drug during allergy season and felt drowsy. Sudafed contains **only** threo-(+)-2-methylamino-1-phenylpropanol (98.1).

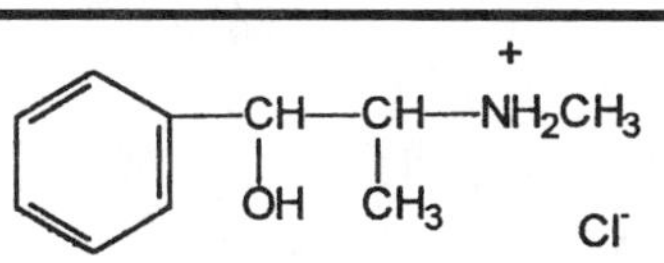

98.1

1. **How do you answer this conscientious but skeptical consumer? (By the way, IM teaches chemistry at the local high school, so go ahead and give him a chemically based explanation.)**

2. **What other questions might you ask IM in order to explain why he experienced drowsiness when taking Sudafed?**

Case 99

Submitted by Victoria F. Roche

You are the senior pharmacist at a community pharmacy that offers extemporaneous compounding as a service to its patients. An ophthalmologist telephones to discuss cholinesterase inhibitor therapy for a patient with glaucoma. To enhance compliance, he wants a long-acting compound, and the patient has requested that an aqueous solution dosage form be dispensed because the idea of putting an ointment in his eye grosses him out. You have the following cholinesterase inhibitors readily available.

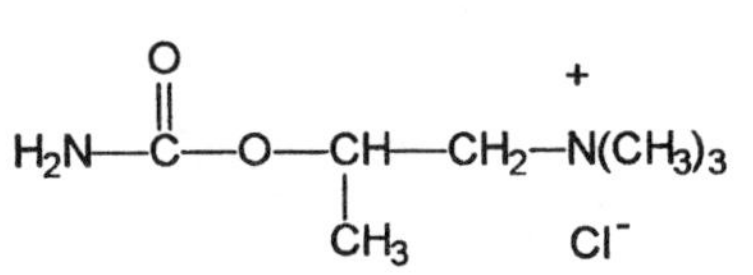

99.1

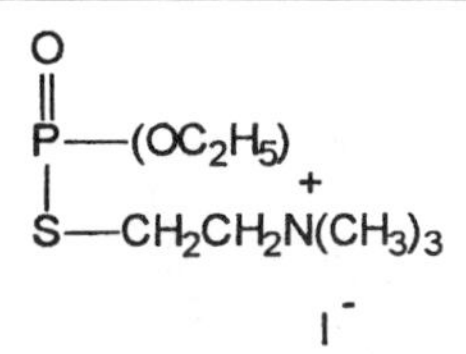

99.2

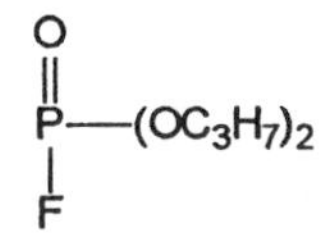

99.3

1. Which of the cholinesterase inhibitors (99.1–99.3) would be water soluble?

2. What is the relationship between chemical structure and duration of cholinesterase inhibition for the compounds shown in this case?

3. **Which structure (99.1–99.3) would you select to fill this prescription?**

Case 100

Submitted by Victoria F. Roche

RV, a student from the local community college, is standing in the OTC section of your pharmacy looking miserable and reading the backs of three product packages. He is trying to select something that will help him get over the cold of the century, and expresses anxiety about being ill with autumn midterm examinations coming up in a few days. Normally quite healthy, he appears very congested and feels "achy," but his lungs are still clear. He is taking Nodoz prn (which contains caffeine) to help him stay up nights to study, and admits to no allergies. The products he has pulled off the shelf contain the following chemical compounds.

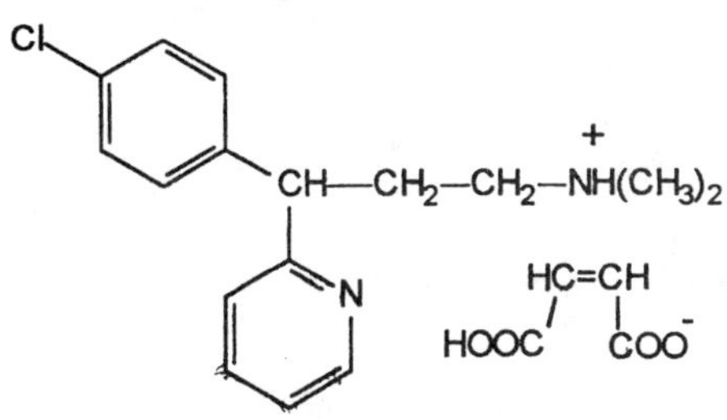

100.1

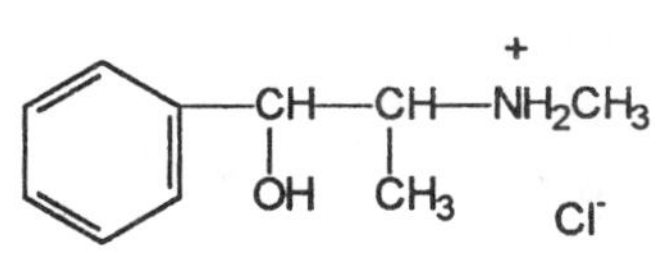

100.2

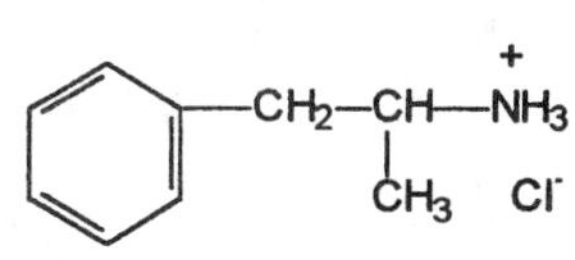

100.3

1. **What pharmacologic class of molecule does RV require? Which of the structures provided (100.1–100.3) fall into this pharmacologic class?**

2. **Of these agents (100.1–100.3), which would you recommend? How does the chemistry of your agent of choice support your recommendation?**

3. How would you counsel RV?

Case 101

Submitted by Victoria F. Roche

OR, a hypertensive 58-year-old man, is brought to the ER suffering from status asthmaticus. After injecting terbutaline sulfate, a β_2 selective agonist, the physician calls for corticosteroid therapy. You find the following compounds on hand (101.1–101.4).

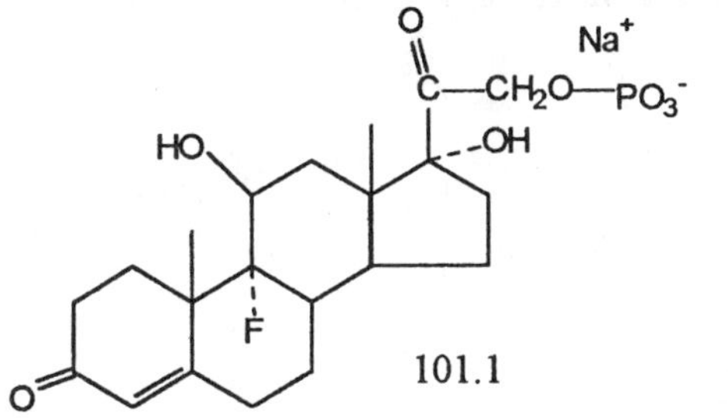

101.1

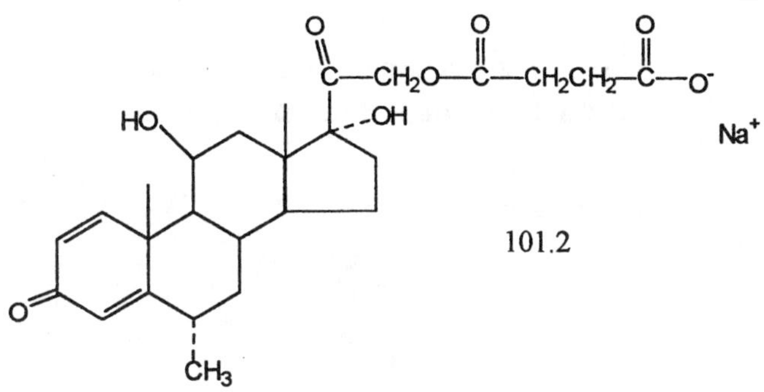

101.2

101.3

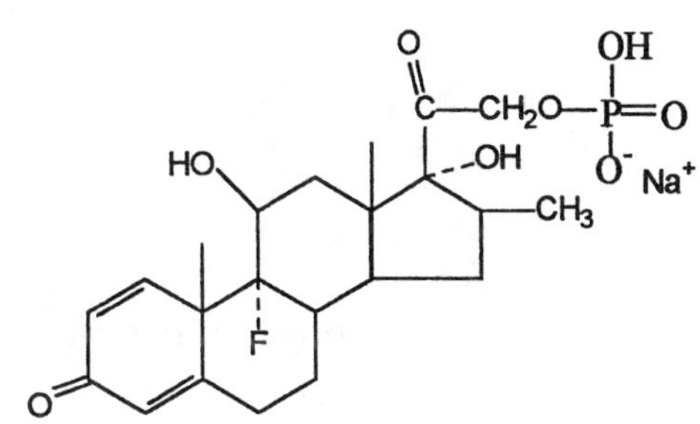

101.4

1. Why was the β-adrenergic agonist administered? Was it necessary to use a β_2 selective agonist? Why, or why not?

2. **What route of administration is required for the corticosteroid about to be dispensed? Why is this route necessary?**

3. **What pharmacologic and chemical properties of the steroid molecules under consideration are important in this case?**

4. **Which agent(s), if any, would you recommend?**

Case 102

Submitted by Jeffrey J. Christoff,
Robert L. Chapman, and Robin M. Zavod

MV is a 64-year-old woman admitted to the hospital with an exacerbation of her congestive heart failure (CHF). She has been on full therapeutic doses of furosemide (102.1) and digoxin (102.4) before her hospital admission.

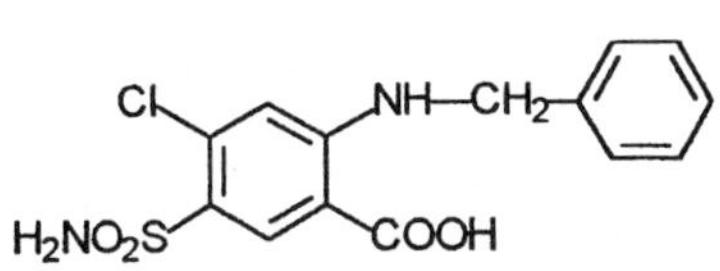

102.1

$Na_2[Fe(CN)_5NO]$

102.2

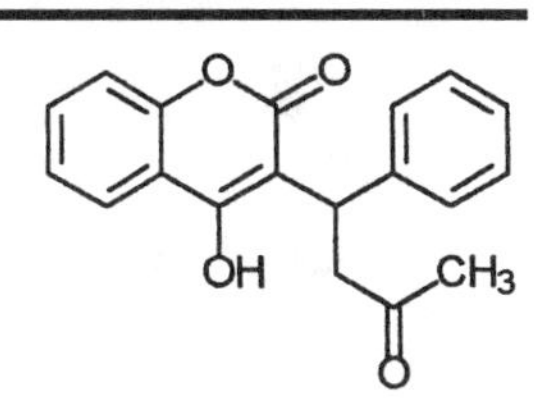

102.3

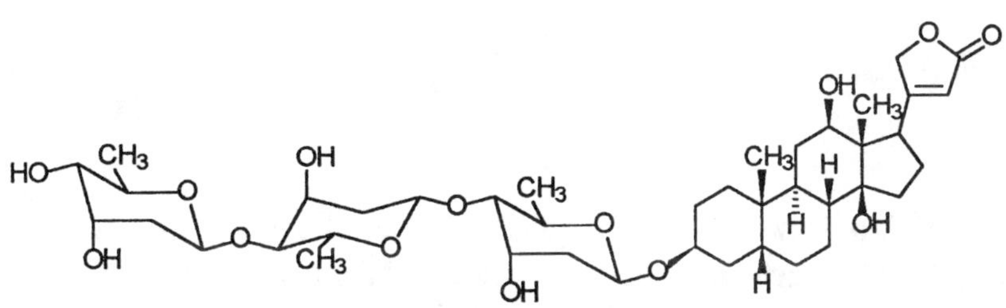

102.4

1. **Explain the biochemical basis for the effects of furosemide (102.1) on serum K^+.**

2. **Discuss the importance of three major structural features needed for digoxin (102.4) to inhibit cardiac Na^+/K^+ adenosine triphosphatase (ATPase).**

3. **For treatment of an exacerbating CHF, sodium nitroprusside (102.2) was started. Which portion of sodium nitroprusside (102.2) is responsible for its vasodilatory effects? Which portion is responsible for its toxic effects?**

4. **The physician is worried about thromboembolic events in MV. Would a 50-mg loading dose of warfarin (102.3) be indicated for this patient? Discuss the biochemical reasons to support your decision (normal daily dose, 2–10 mg).**

Case 103

Submitted by Jeffrey J. Christoff

KC is a 49-year-old woman who has a 15-year history of hypertension. She is currently being treated with Moduretic 5/50 qAM (amiloride hydrochloride [103.1] 5 mg + hydrochlorothiazide [HCTZ][103.2] 50 mg), Cardura (103.3) 1 mg qd (doxazosin mesylate), and Procardia XL (103.4) 30 mg qd (extended release nifedipine). KC is not feeling very well today. She describes her symptoms as dizziness, headache, an uncomfortable hot sensation, and a racing heart. When she took her medication this morning, she accidentally dropped an empty cereal bowl on her pills. The crushed medications were brushed into her glass of juice, and she drank the mixture.

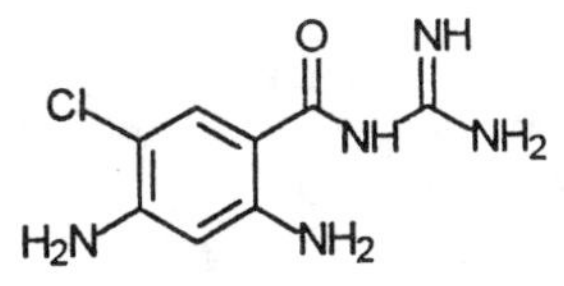

103.1

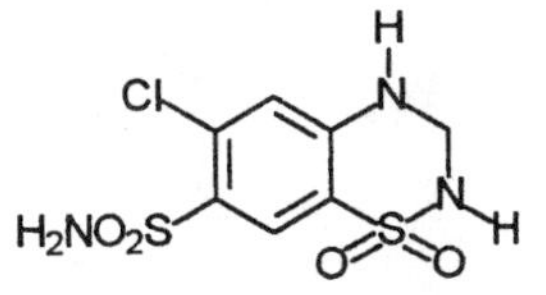

103.2

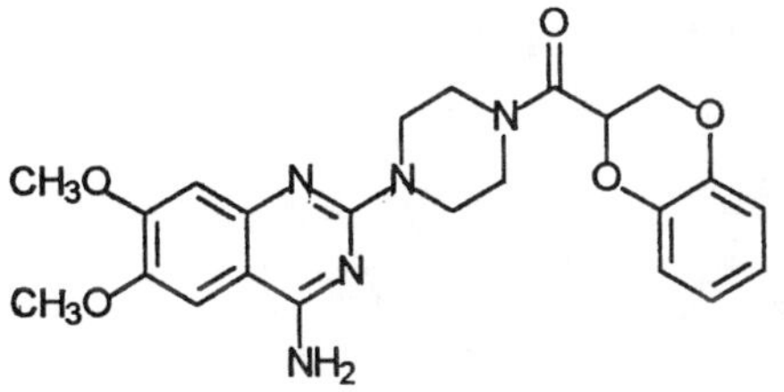

103.3

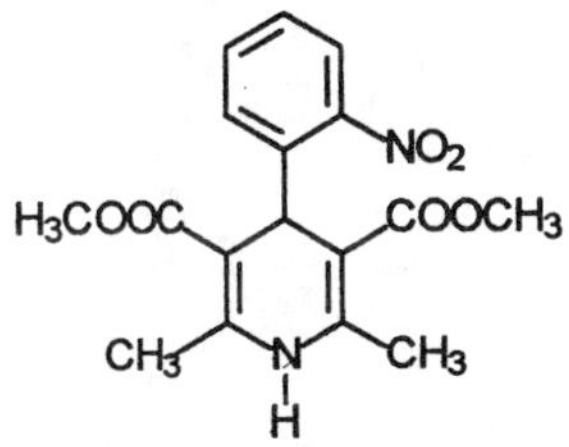

103.4

1. Describe the mechanism of action for each component of the diuretic (103.1 and 103.2). Why are these drugs used in combination?

2. **What is the mechanism of action for doxazosin mesylate (103.3)?**

3. **Explain why KC is suffering from dizziness, headache, flushing, and tachycardia.**

Case 104

Submitted by Robert L. Chapman and Kathleen A. Snella

MB is a 62-year-old woman whose mother passed away 2 years ago from a stroke at age 82, and whose 66-year-old brother had a coronary artery bypass graft that same year. She is without a history of smoking, diabetes mellitus, hypothyroidism, nephrotic syndrome, or obstructive liver disease. Her serum lipid profile has shown elevated LDL and serum cholesterol concentrations on two successive visits during the past 2 months:

Total Cholesterol = 280 mg/dL
LDL = 170 mg/dL
HDL = 75 mg/dL
Triglycerides = 175 mg/dL

MB has hypertension that is controlled with captopril, 50 mg tid. She also has a recent history of angina, for which the physician has now prescribed nitroglycerin transdermal patches (release rate, 0.4 mg/h) to be used daily. She has been compliant on her low-sodium, low-fat diet, and has lost 15 pounds over the last 2 years. However, her physician believes that at this point, dietary therapy alone will not bring about the desired decrease in serum LDL concentration below 100 mg/dL, and decides to begin antihyperlipidemic drug therapy. The physician consults with you for suggestions on using an antihyperlipidemic agent.

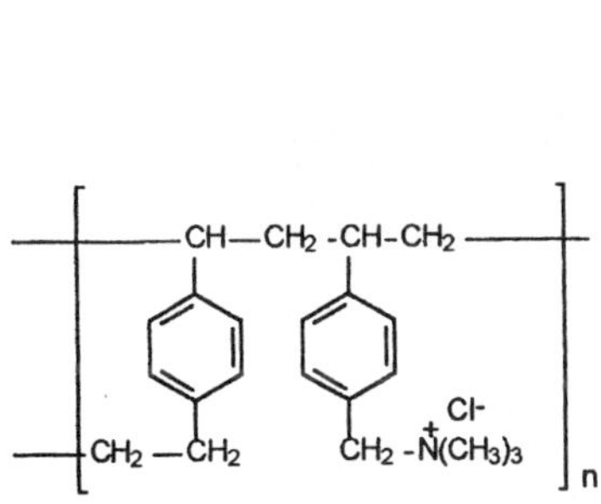

104.1

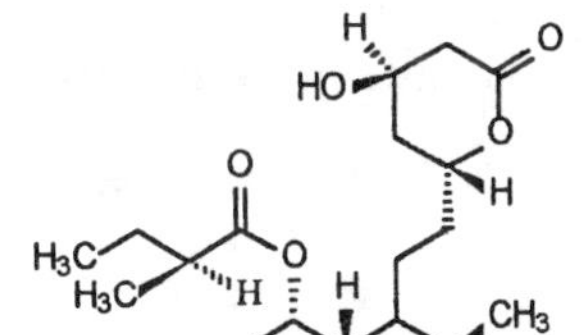

104.2

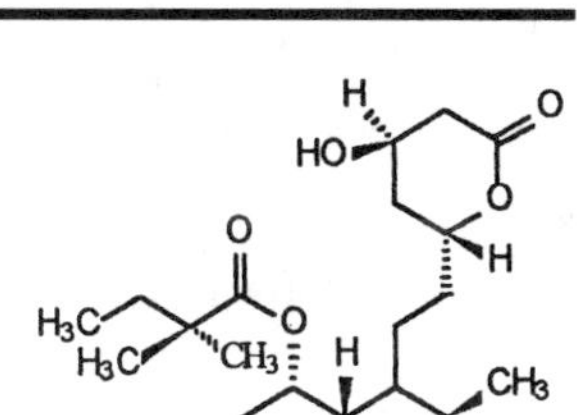

104.3

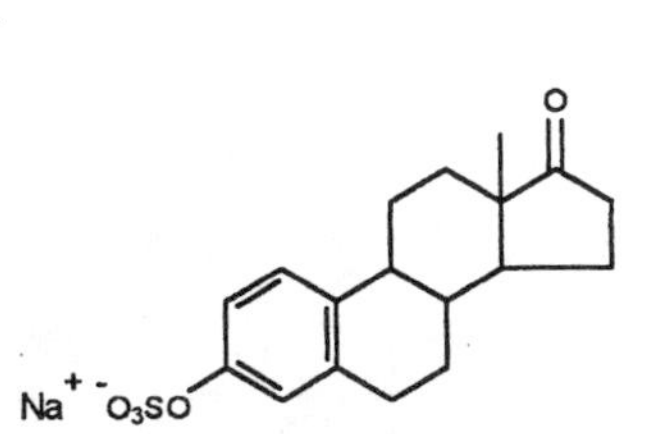

104.4

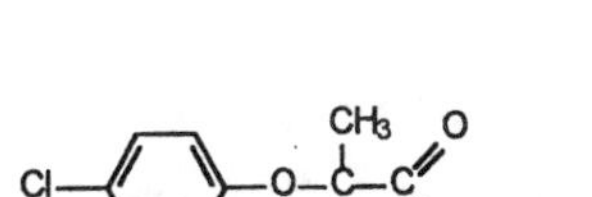

104.5

1. **Discuss the mechanisms of action for the following therapeutic options that you could suggest to MB's physician:**
   a. Cholestyramine (104.1):

   b. Lovastatin (104.2) and Simvastatin (104.3):

   c. Clofibrate (104.5):

   d. Conjugated estrogens (104.4):

2. **Considering MB's other disease states and medication profile, why might high-dose niacin therapy not be an option for MB?**

3. **Do bile acid sequestrants take part in any significant drug interactions? Why, or why not?**

4. **Which of the agents in Question 1 are prodrugs? Draw structures to explain how they are bioactivated.**

Case **105**

Submitted by Robin M. Zavod,
Jeffrey J. Christoff, and Robert L. Chapman

KV, a 51-year-old man, was recently hospitalized and underwent surgery to replace a defective heart valve. He has a history of elevated serum cholesterol levels and grand mal seizures. The patient was taking warfarin (105.5) before hospital admission and was switched from warfarin (105.5) to heparin therapy several days before the surgical procedure. Several days after his discharge, KV visited his pharmacy and requested a complete listing of his medications to be sent to an additional cardiologist.
This is what the cardiologist received:

Warfarin (105.5) 7.5 mg qd
Dipyridamole (105.6) 75 mg qid
Clofibrate (105.2) 500 mg qid
Phenytoin (105.1) 100 mg tid

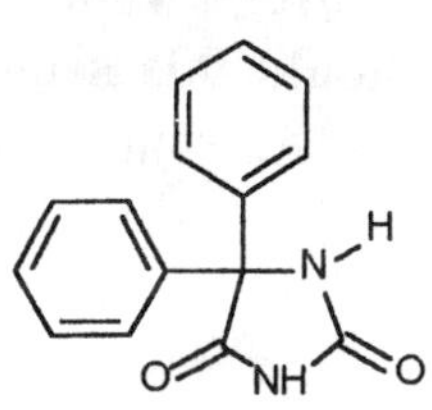

105.1

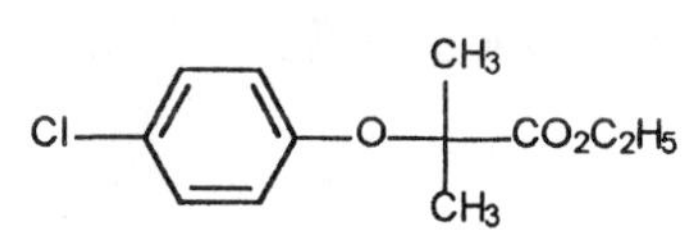

105.2

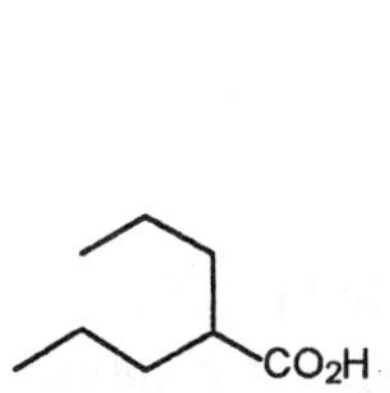

105.3

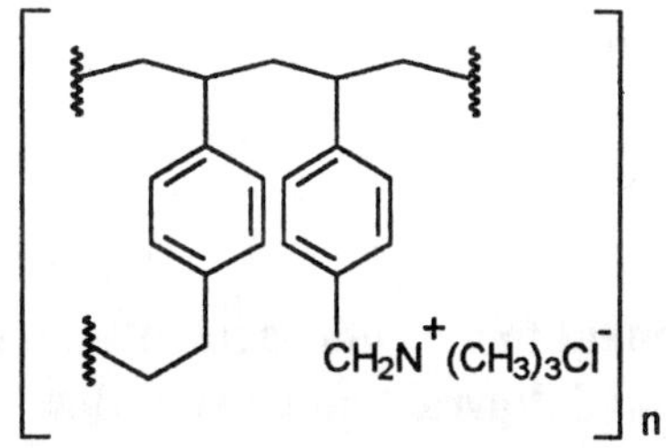

105.4

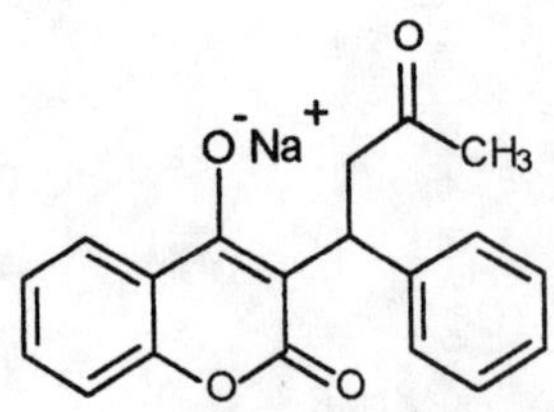

105.5

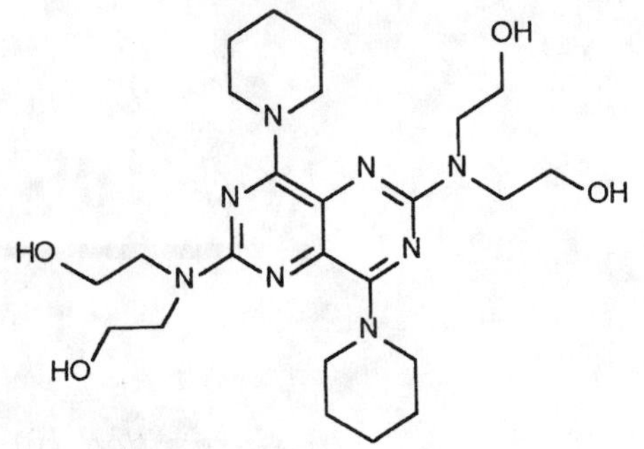

105.6

1. **Why was the patient switched from warfarin (105.5) to heparin therapy before surgery? Describe the mechanisms of action of warfarin and heparin, including relevant pharmacokinetic information.**

2. **Following surgery, what sequence of anticoagulant drugs would be required to return the patient to warfarin (105.5) therapy? Should this sequence include a warfarin (105.5) loading dose? Why, or why not? Explain your rationale based on the mechansim of action of each drug.**

3. **Subsequent to his anticoagulant stabilization (warfarin [105.5]), a medical resident prescribed dipyridamole (105.6) as adjunct therapy. Is the combination of warfarin (105.5) and dipyridamole (105.6) appropriate for this patient? Why, or why not?**

4. **Assess KV's medication profile. Are there any potential drug interactions? If so, what are they? If KV's anticonvulsant therapy was changed to valproic acid (105.3), how would his warfarin (105.5) therapy be affected? If the clofibrate (105.2) was changed to cholestyramine (105.4), how might this affect his warfarin (105.5) therapy?**

Case 106

Submitted by Victoria F. Roche

A physician telephones your pharmacy to consult about the treatment of a hypertensive patient, SS, who was recently diagnosed with chronic obstructive pulmonary disease (COPD). SS's hypertension is currently being managed with propranolol (106.1). SS reports that, since starting propranolol (106.1), he has been having more trouble breathing than he used to, which is what brought him to his physician's office. He reluctantly admitted to having experienced bouts of fatigue and impotence while taking propranolol (106.1), and he really wants nothing more to do with the drug. You stock four additional β-adrenergic blocking agents (106.2–106.5) in your pharmacy.

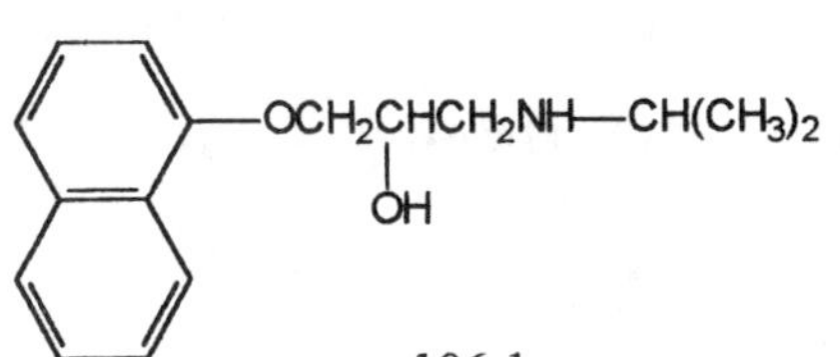

106.1

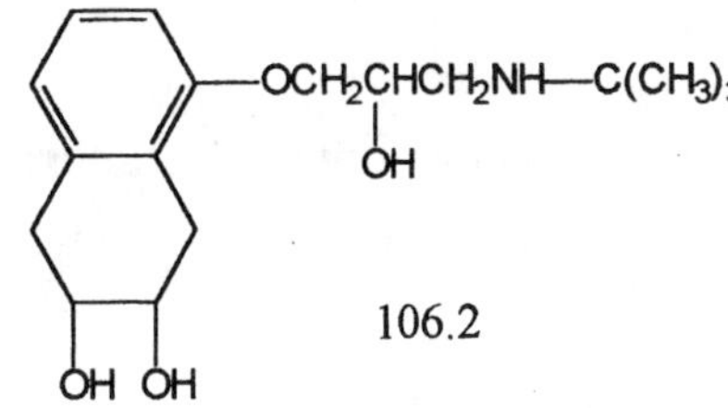

106.2

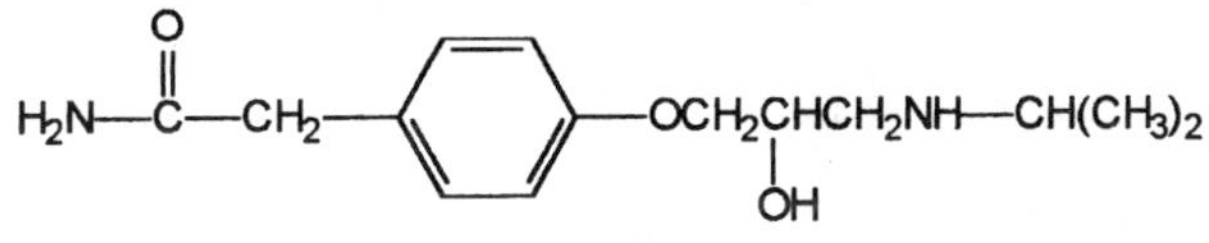

106.3

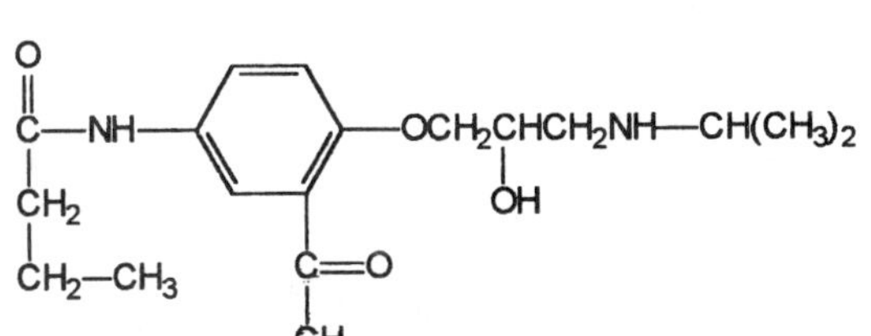

106.4

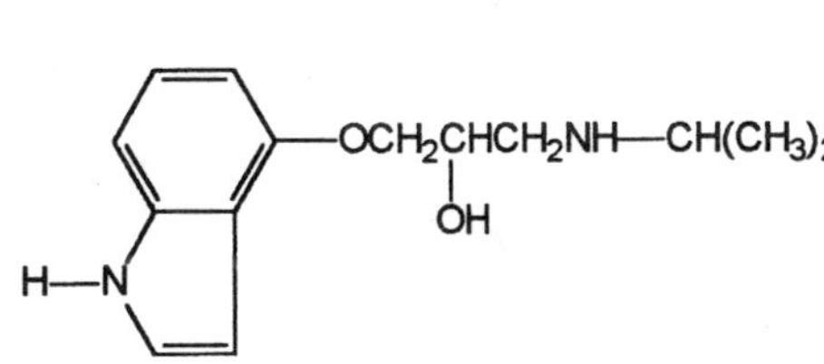

106.5

1. **Could SS's fatigue and impotence have been drug induced, or was he just tired and anxious about his hypertension? How does the chemistry of propranolol (106.1) influence your answer?**

2. What therapeutic aims must you keep in mind when selecting a β-blocking agent for SS?

3. Which of the four alternative β-blocking agents (106.2–106.5) drawn, if any, would you recommend?

Case **107**

Submitted by Victoria F. Roche

AF, a 65-year-old retiree who jogs regularly, collapsed in V-fib while undergoing an annual stress test. He was successfully defibrillated and, after a brief hospitalization, was discharged on a maintenance dose of carvedilol (107.1). After 2 weeks of therapy, he presents at your pharmacy complaining of difficulty in breathing. Upon taking his medical history, you find he has been in excellent health except for a mild, but chronic, bronchitis. Upon further counseling, you discover that he is experiencing some symptoms of depression (fatigue, lethargy), and has been having bizarre dreams.

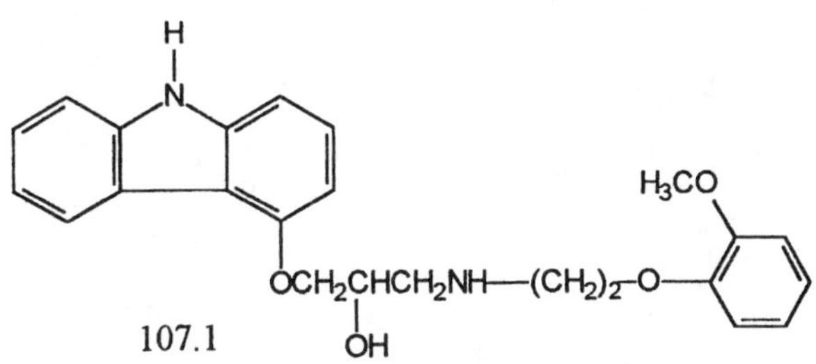

107.1

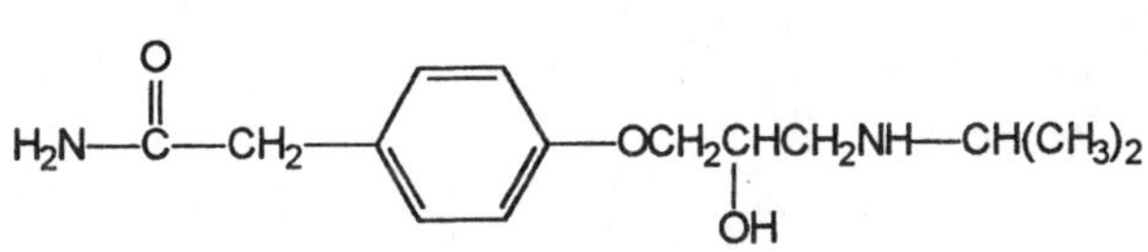

107.2

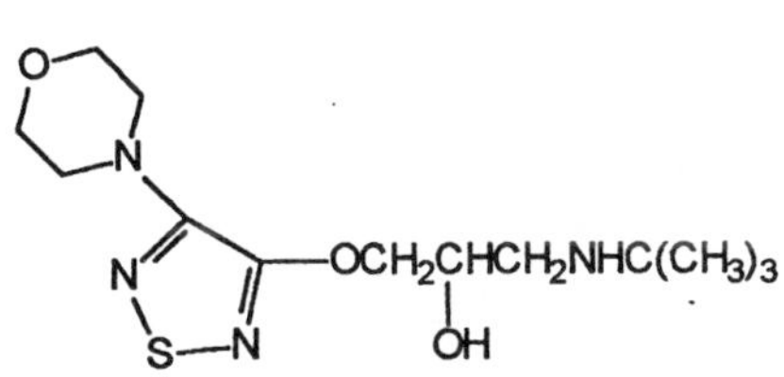

107.3

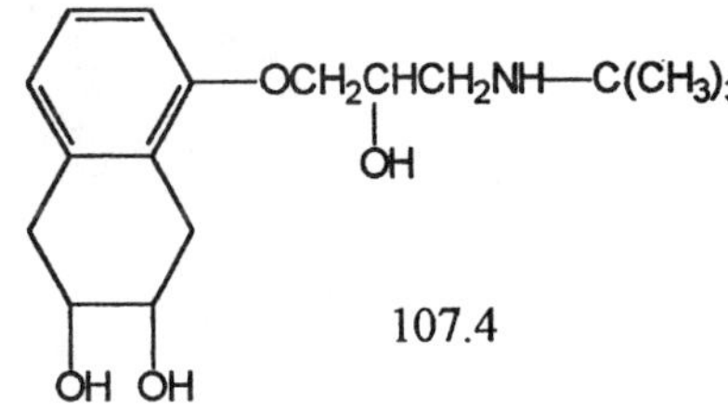

107.4

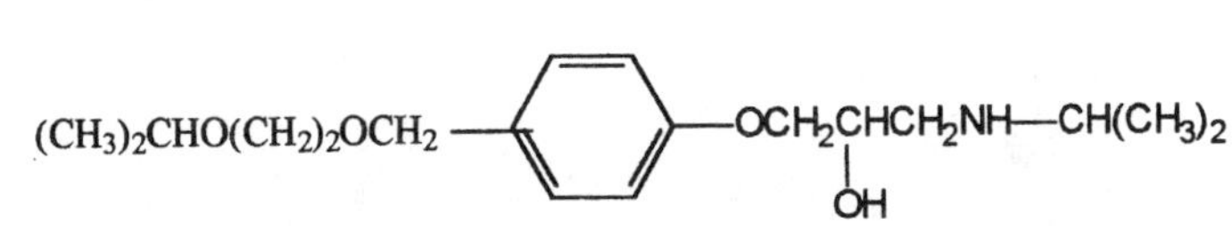

107.5

1. **Explain how AF's new complaints might be related to his cardiac therapy.**

2. **Is a change in therapy warranted? If so, which of the four alternative β-blocker structures (107.2–107.5) would you recommend?**

Case 108

Submitted by Victoria F. Roche

AK, a 55-year-old man with a 30-year history of cigarette smoking, has developed emphysema. Upon taking his medical history, you find that he is predisposed to mild tachycardia, and is currently being treated with low-dose acebutolol (108.1). Your first order of business is to recommend appropriate therapy for the pulmonary disorder, and consult your hospital's formulary. You find the three following β-adrenergic agonists available (structures 108.2–108.4).

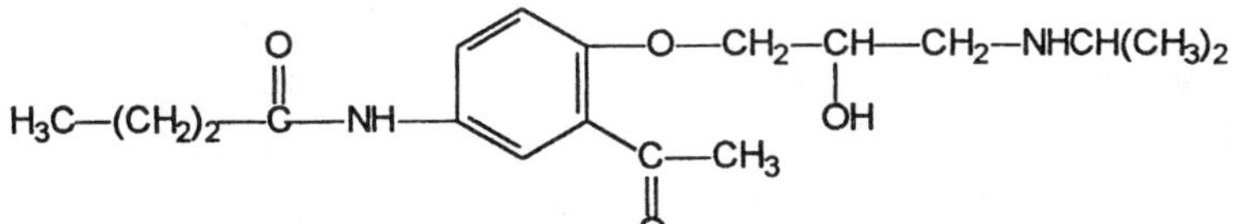

108.1

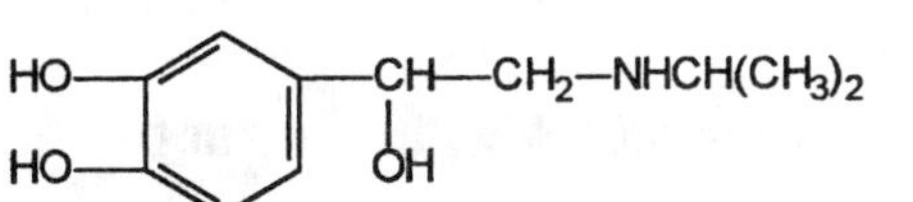

108.2

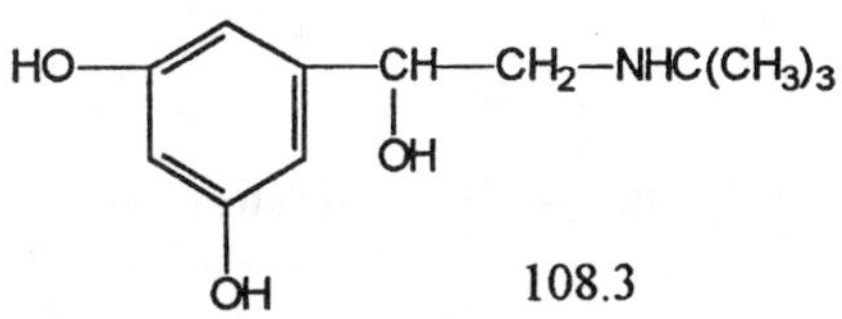

108.3

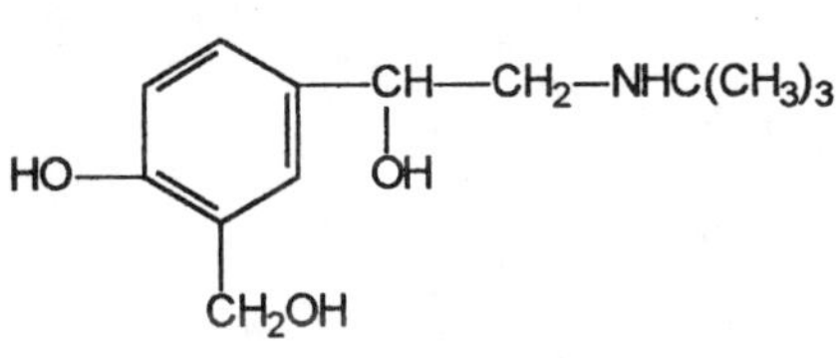

108.4

1. **What are the therapeutic needs of this patient?**

2. **Which of the three β-adrenergic agonist agents (108.2–108.4) might be acceptable in this patient?**

3. **What is the chemical basis for accepting or rejecting any of the β-agonist structures (108.2–108.4) in this case?**

4. **Is a change in β-blocker therapy indicated for AK? Why, or why not?**

Case 109

Submitted by Victoria F. Roche

RO is an 82-year-old male resident of a skilled care facility. A smoker his entire life, he suffers from severe emphysema and requires continuous oxygen. His routinely high blood pressure is being successfully controlled with enalapril (an angiotensin converting enzyme inhibitor), but he has now developed ocular hypertension that his physician wishes to treat with a topical β_1-selective adrenergic blocking agent. The physician had graduated from pharmacy school before entering medical school 25 years ago, and he swears he can still hear his old medicinal chemistry teacher harping that β_1-selective blockers require a *p*-substituted phenyl ring attached to an aryl oxypropanolamine side chain. He suggests trying a relatively new agent, metipranolol hydrochloride (109.1), to treat RO's ocular hypertension.

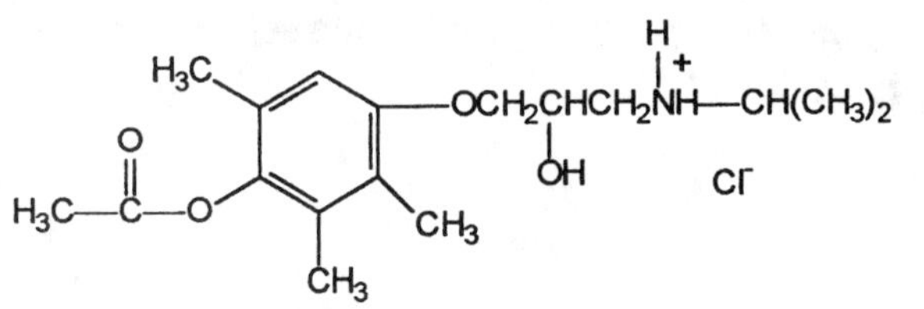

109.1

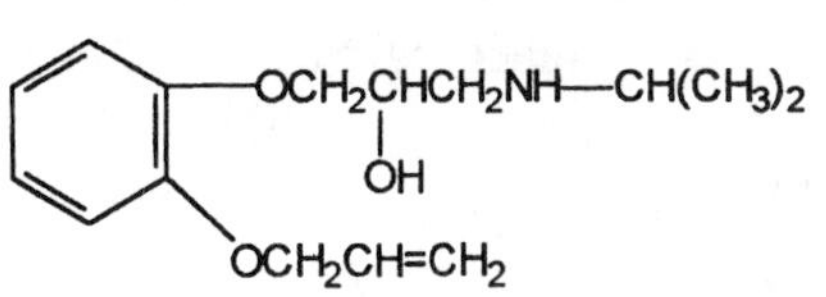

109.2

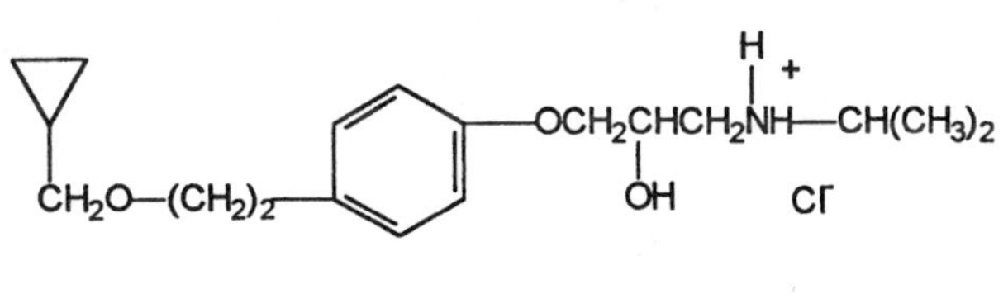

109.3

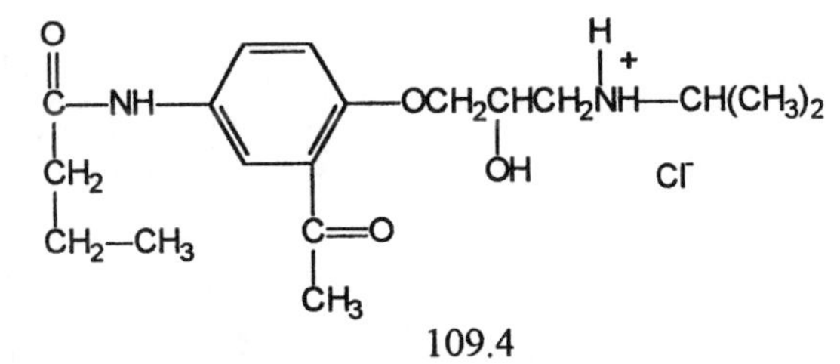

109.4

1. What is your opinion of this physician's medicinal chemistry reminiscences?

2. **Would metipranolol hydrochloride (109.1) be an appropriate agent for RO?**

3. **Which of the alternative structures (109.2–109.4) would provide the selective pharmacologic action that this physician is seeking?**

Case 110

Submitted by Victoria F. Roche

BR, a single working mother, has a 5-year-old child with glaucoma. Her pediatrician has prescribed treatment with a cholinesterase inhibitor and seeks your therapeutic recommendation. The day care center that cares for the child will not administer medication, and the mother's workplace is 15 miles away from the day care center. You have the following cholinergic structures (110.1–110.3) in your pharmacy.

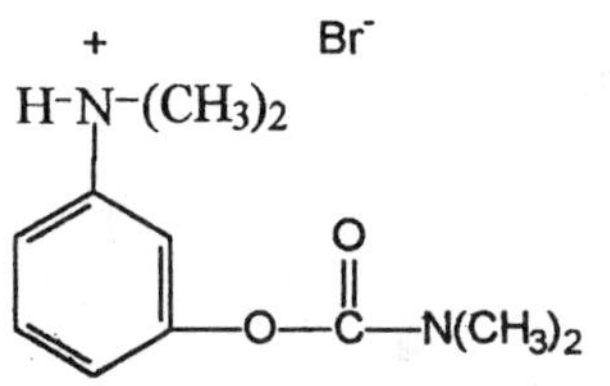

110.1

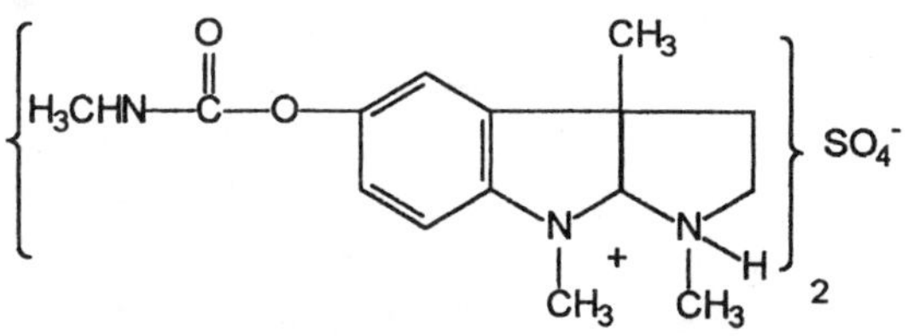

110.2

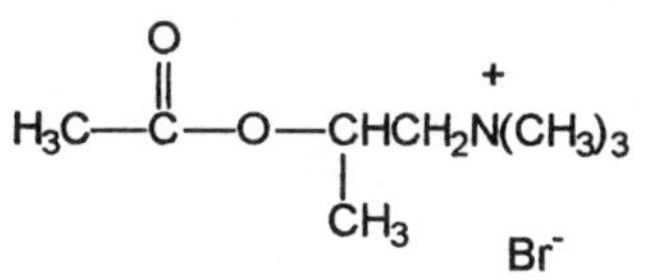

110.3

1. **What is the mechanism of cholinergic action of each of the three candidates (110.1–110.3)?**

2. **How does the chemical structure of the cholinesterase inhibitors (110.1–110.3) influence onset and duration of action?**

3. **Which of the three structures (110.1–110.3) shown would best meet the therapeutic and personal needs of this family?**

Case 111

Submitted by Victoria F. Roche

KK, an 85-year-old woman with chronic bradycardia and limited mental functioning, develops glaucoma. This patient is incapable of self-medication, and relies on her son (a CEO for a major insurance company) to administer her medications. You have been asked to recommend the most appropriate topical therapeutic agent to the attending physician, and ponder the merits of the three cholinesterase inhibitors (111.1–111.3) shown.

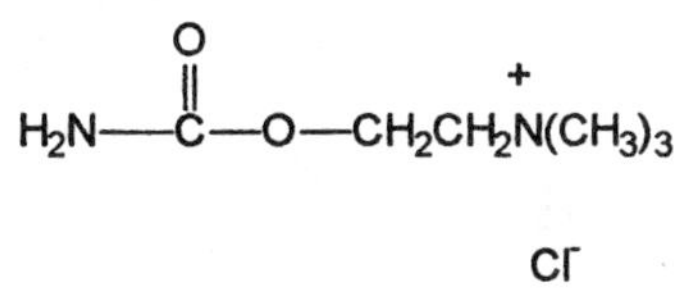

111.1

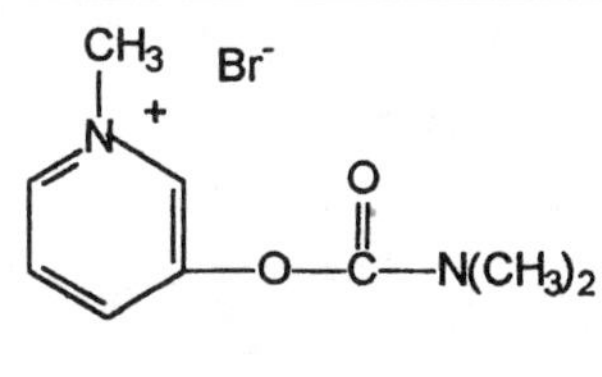

111.2

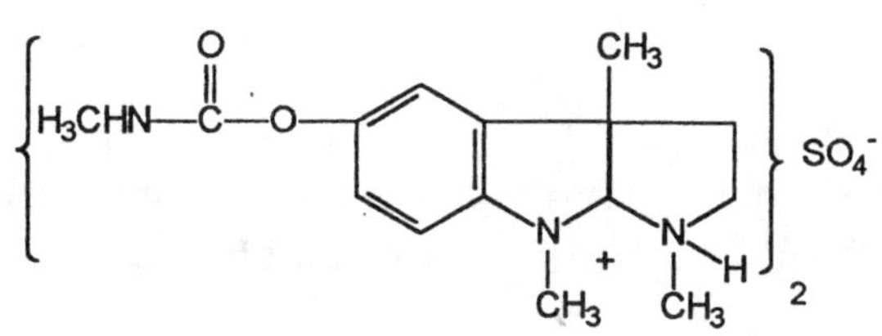

111.3

1. **What are the therapeutic and personal needs of KK and her family?**

2. How well do the structural features of the drug candidates provided (111.1–111.3) meet these needs?

3. Which of the three cholinesterase inhibitors (111.1–111.3) would you recommend?

4. At what pH range would your drug of choice be most stable? Would your drug product benefit from the addition of sodium bisulfite?

Case 112

Submitted by Victoria F. Roche

A physician telephones you regarding an asthmatic patient with underlying cardiovascular disease who has just immigrated from Europe. She is currently on a bronchodilating medication not available in the U.S., and the physician wants your input on suitable replacement therapy. The physician wants the medication administered by inhalation, and you consider the structural candidates provided (112.1–112.4).

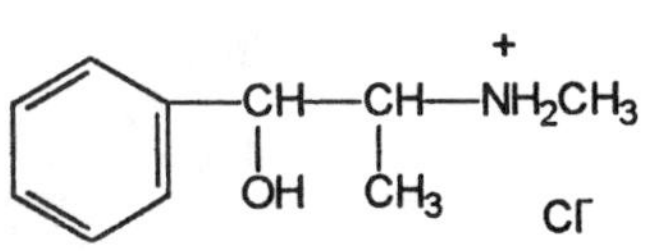

112.1

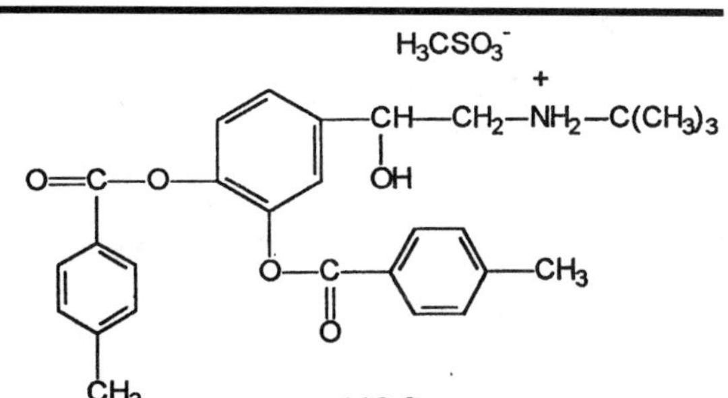

112.2

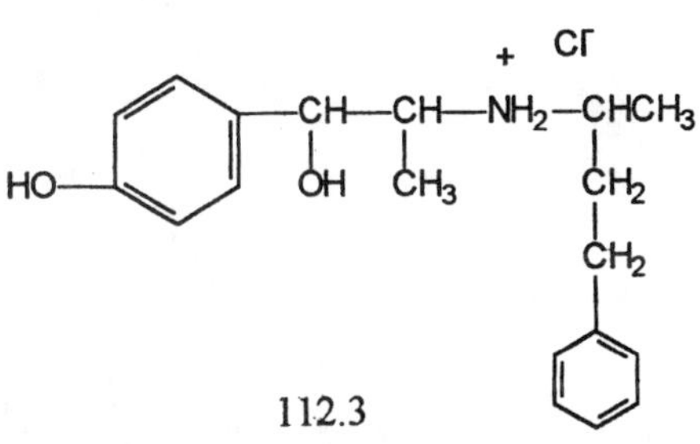

112.3

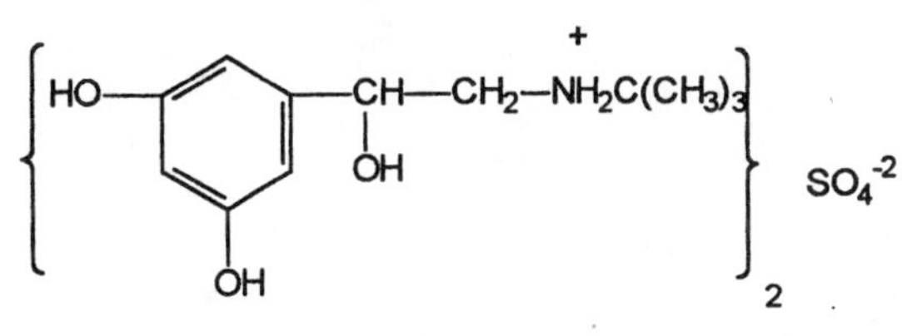

112.4

1. **Which of the bronchodilating structures provided (112.1–112.4) would be viable candidates for inhalation therapy? What structural features of these agents support your answer?**

2. **Of the molecules you select, which would be expected to have the longest duration of bronchodilating action? Why?**

3. **Of the compounds you originally selected for inhalation therapy which, if any, would be suitable for po therapy? What chemical features are responsible for your answer?**

Case **113**

Submitted by Victoria F. Roche

A physician telephones your pharmacy to inquire about appropriate therapy for an asthmatic patient, MF, who is predisposed to ventricular tachycardia. She wants to prescribe po medication, as this patient responds to inhalation therapy with violent coughing. You consider the structures of three bronchodilating molecules (113.1–113.3).

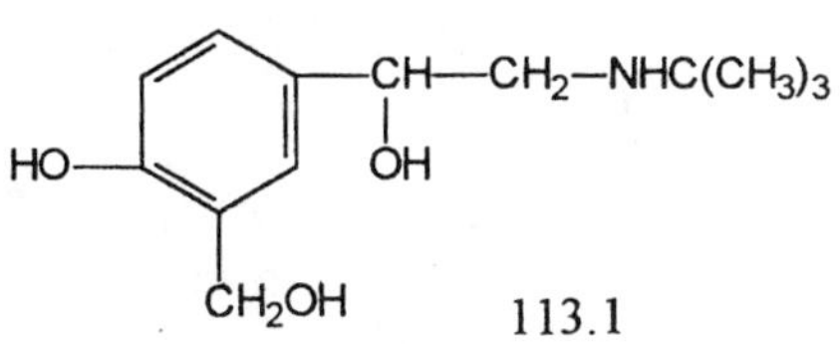

113.1

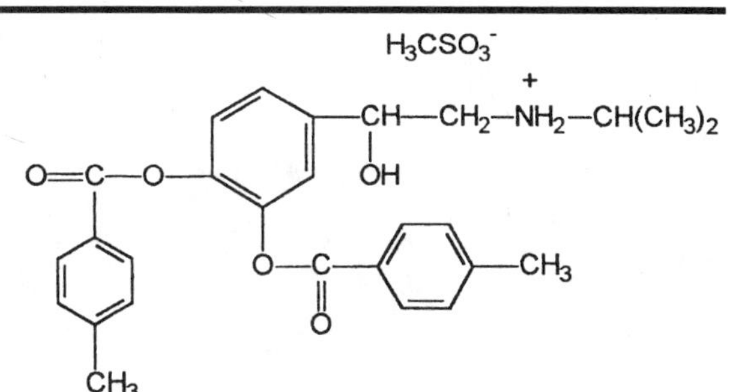

113.2

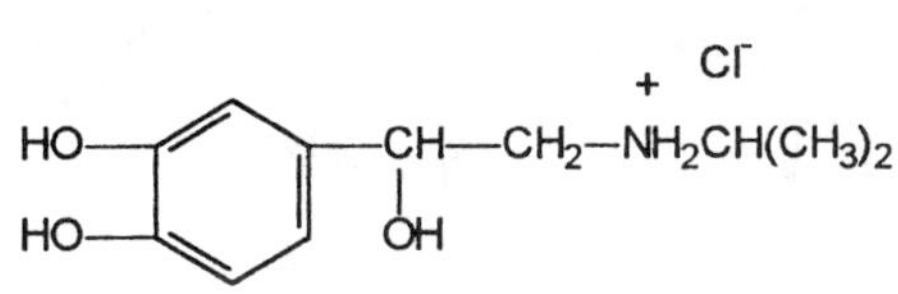

113.3

1. **Which of the bronchodilator molecules (113.1–113.3) being contemplated has good po bioavailability? How do their chemical structures promote po stability?**

2. **What other therapeutic needs of MF must be kept in mind? Do the structures of the po active agents provide pharmacologic activities that meet these needs?**

3. **Which of the three agents (113.1–113.3), if any, would you recommend in this situation?**

Case 114

Submitted by Victoria F. Roche

Dr. Iris Brown, an ophthalmologist, is seeing a patient with corneal inflammation. She wishes to initiate topical antimuscarinic therapy, and is considering the merits of the compounds shown (114.1–114.5).

Dr. Brown wishes to formulate a potent antimuscarinic in a standard lanolin-based ointment and asks your advice on which product to choose.

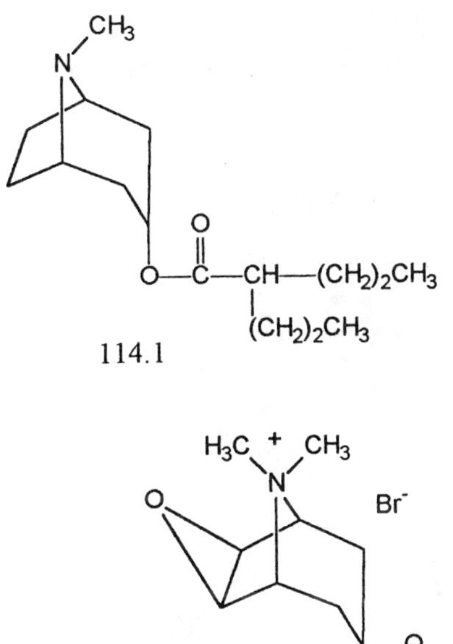

114.1

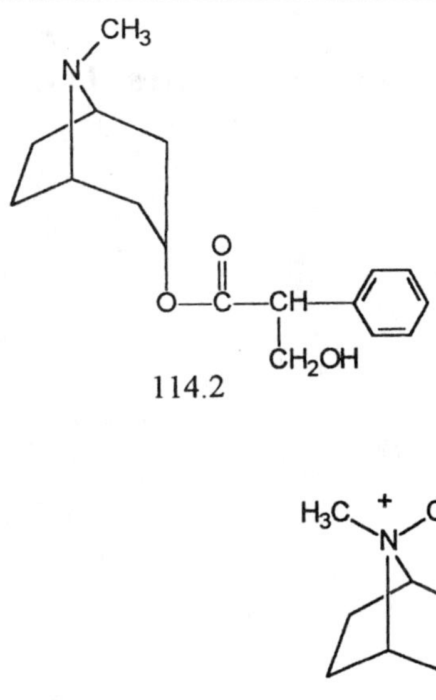

114.2

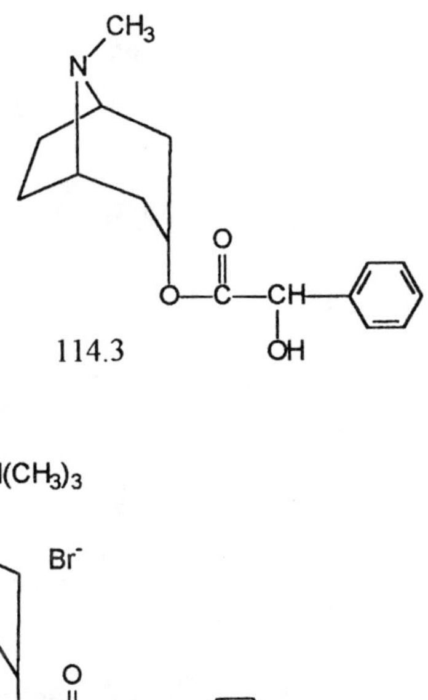

114.3

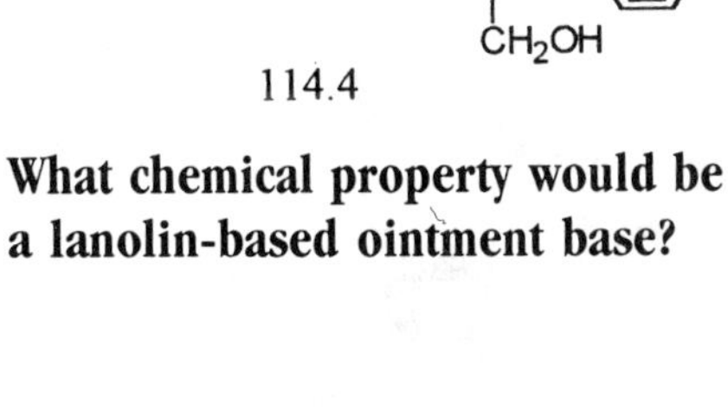

114.4

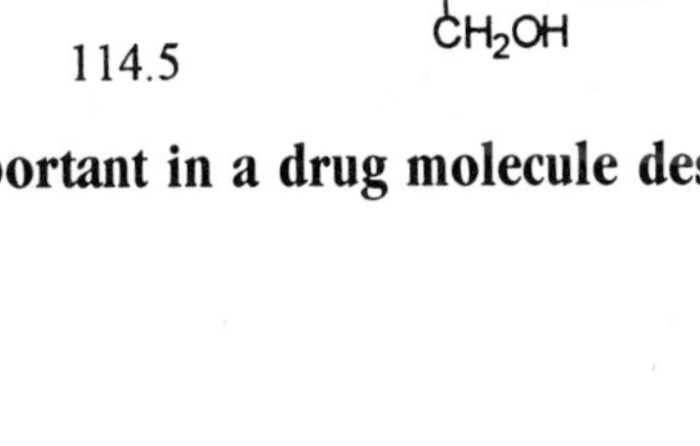

114.5

1. **What chemical property would be most important in a drug molecule destined for a lanolin-based ointment base?**

2. **What structural features contribute to high antimuscarinic potency in Solanaceous alkaloid molecules?**

3. **Which of the five contemplated structures (114.1–114.5) would you recommend to Dr. Brown?**

4. **If Dr. Brown's patient returned to her clinic after a number of days on your drug of choice complaining of dry mouth and blurred vision, how might this be explained?**

5. **Which of the original five molecules (114.1–114.5) would you have selected if Dr. Brown had wanted to prepare an ophthalmic solution for her patient? Could you have easily modified the structure of the drug you selected for an ointment formulation to make it appropriate for delivery in aqueous solution?**

Case 115

Submitted by Arnold R. Martin and Richard N. Herrier

BF is a 49-year-old man admitted to the coronary care unit for cardiac catheterization and workup for a marked increased incidence of anginal pain. While in the hospital, BF complains of difficulty in sleeping. He states that he has always found it hard to fall asleep and remain asleep when he is in a strange place. BF insists that at home he usually sleeps 7 hours/night and wakes up refreshed and alert. His wife confirms his difficulty sleeping in unfamiliar environments and reports that at home BF sleeps normally, with no observable limb twitches, snoring, or respiratory pauses during the night. BF has taken flurazepam (115.5) in the past, but complains that it causes him to sleep late and wake up groggy in the morning.

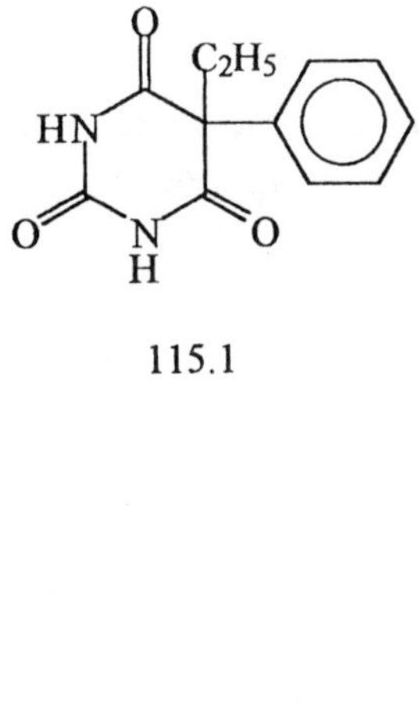

115.1

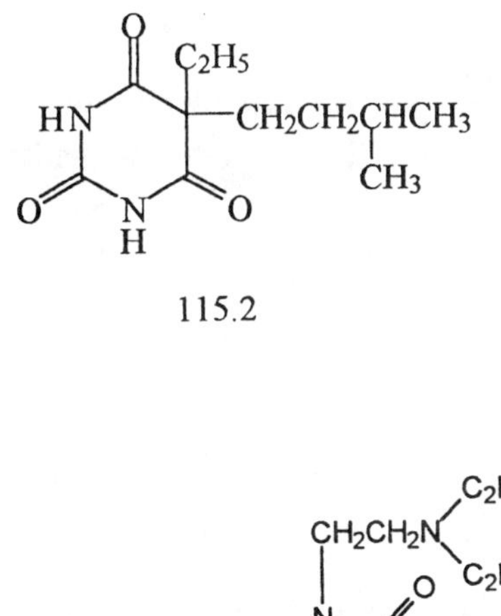

115.2

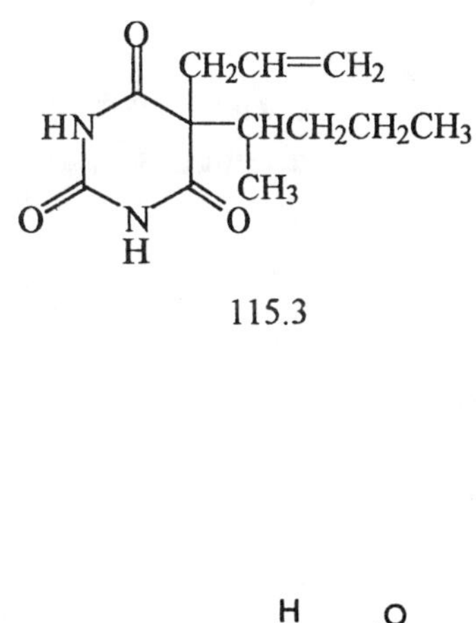

115.3

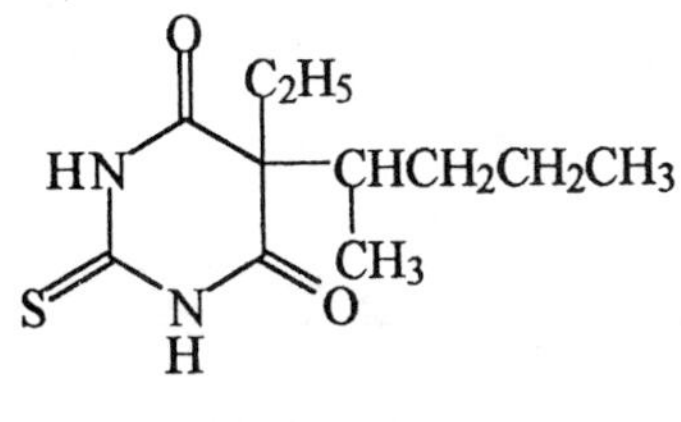

115.4

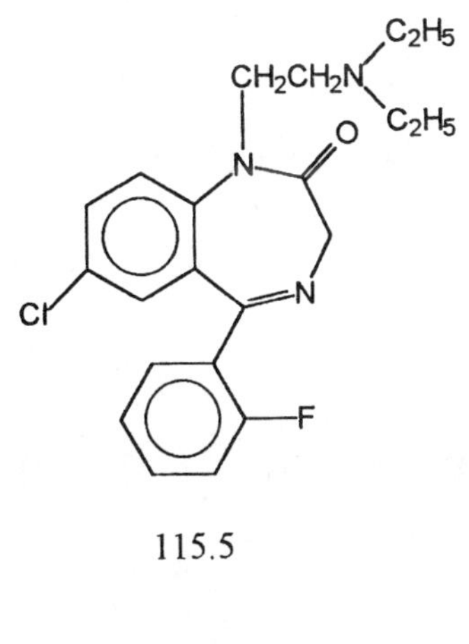

115.5

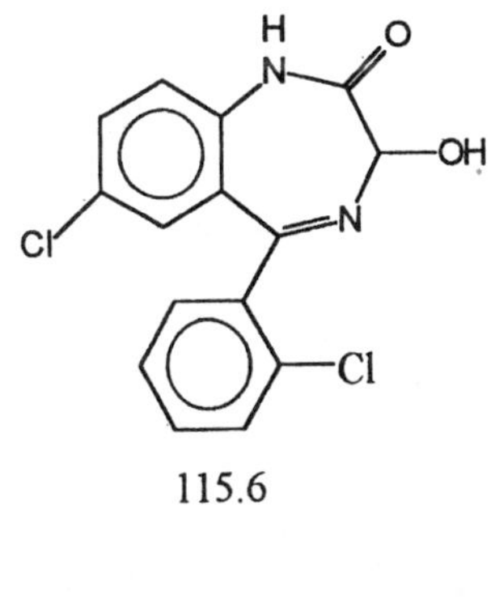

115.6

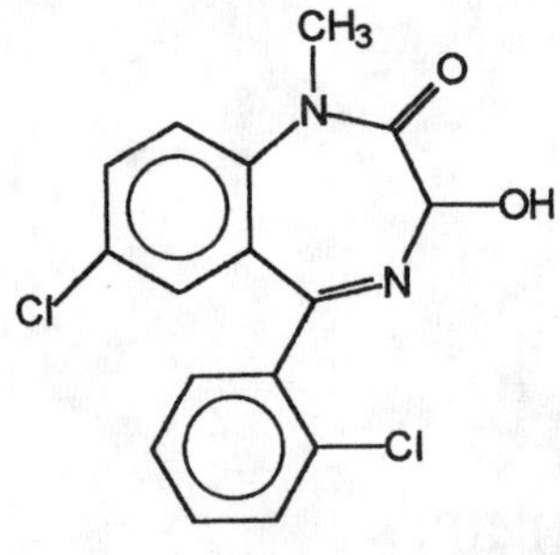

115.7

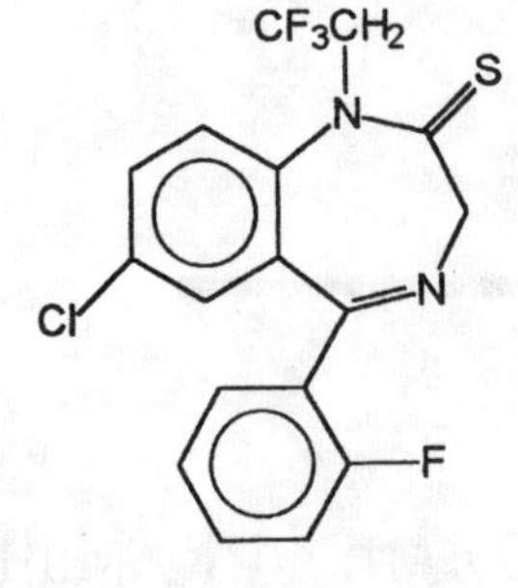

115.8

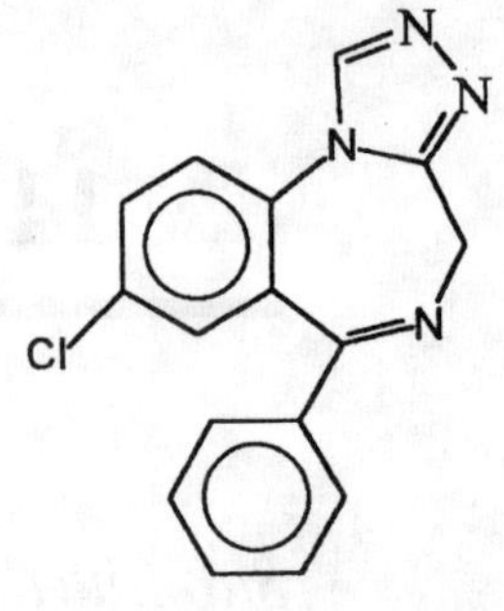

115.9

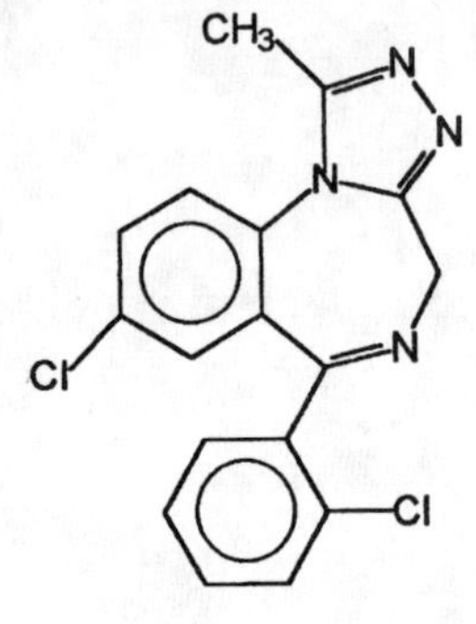

115.10

1. **Select an appropriate sedative/hypnotic from the barbiturate (115.1–115.4) structures provided to treat BF's insomnia. Provide a rationale for your choice on the basis of structure/potency and duration of action relationships. Eliminate any inappropriate choices on the same basis.**

2. **A benzodiazepine sedative/hypnotic could be employed instead of a barbiturate. What advantages might a benzodiazepine have over a barbiturate in this case? Explain.**

3. **Select an appropriate benzodiazepine sedative/hypnotic from the structures (115.6–115.10) provided to treat BF's insomnia. Provide a rationale for your choice on the basis of structure/potency and duration of action relationships. Eliminate any inappropriate choices on the same basis.**

Case 116

Submitted by Bruce L. Currie

BT, a 45-year-old man, approaches you at the prescription counter at ClinPharm Association with a prescription for an angiotensin converting enzyme (ACE) inhibitor, captopril (116.2), written by Dr. Jones. In reviewing your patient profile records, you find that BT was previously treated for hyperthyroidism 2 years ago with propylthiouracil (PTU) (116.1) by Dr. Zik. At that time, BT complained of a bad taste in his mouth (like really bad burnt garlic) within 2 hours of taking his PTU, and he also developed a severe rash on his forearms. His thyroid condition has since been corrected with surgery.

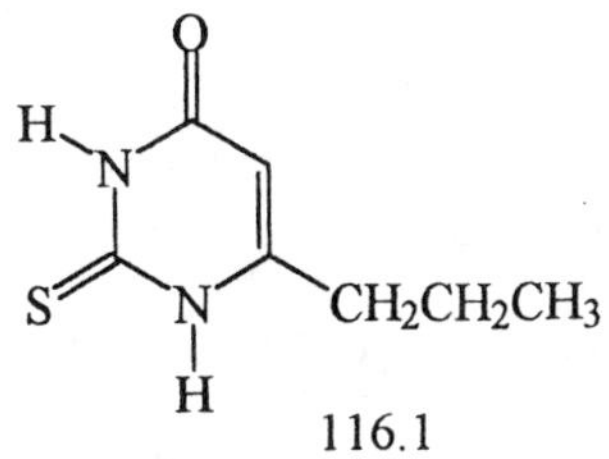

116.1

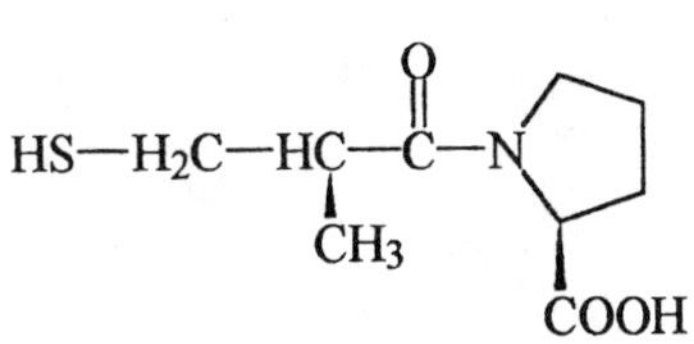

116.2

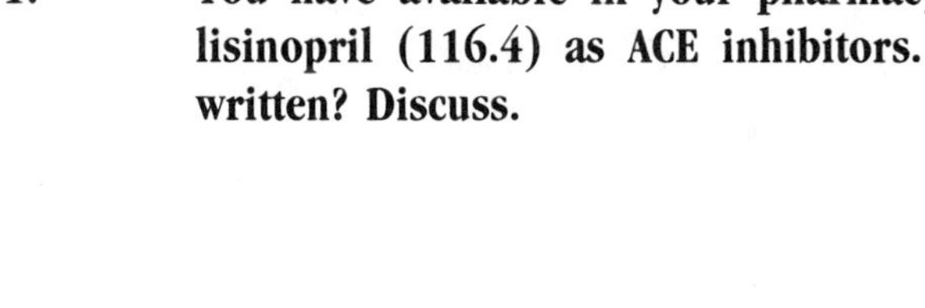

116.3

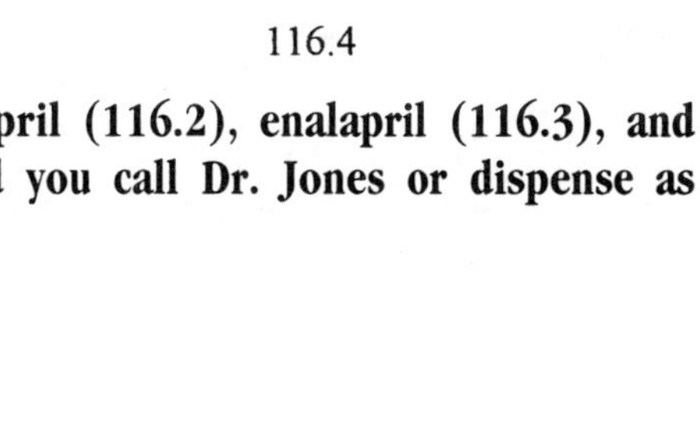

116.4

1. **You have available in your pharmacy captopril (116.2), enalapril (116.3), and lisinopril (116.4) as ACE inhibitors. Should you call Dr. Jones or dispense as written? Discuss.**

2. If you decide to call Dr. Jones, what will be your recommendation?

3. Discuss briefly the role of an ACE inhibitor with regard to the renin-angiotensin system. In the structures of the three ACE inhibitors you have available, circle the functional group that is most critical for the action as an ACE inhibitor.

Case 117

Submitted by Robin M. Zavod

LB, a 63-year-old man, was diagnosed with hypertension nearly 3 years ago. His physician prescribed enalapril (117.1), and LB's hypertension was largely brought under control. Both Mr. and Mrs. B were bothered by a couple of side effects caused by the enalapril (117.1), even after several years of continuous therapy. One of the side effects was easily managed by an OTC anti-inflammatory agent, but the other side effect did not respond to any OTC or prescription medication. LB was sufficiently annoyed that he mentioned it to his physician at his next office visit. The physician was not surprised at all by LB's complaint, and he suggested switching from enalapril (117.1) to imidapril (117.2), a drug that has a similar mechanism of action, but does not have the same side effects. Wanting to better control LB's hypertension, his physician also suggested that LB add to his daily regimen a recently approved drug, losartan (117.3). It was explained to LB that addition of this drug would help to more completely control his hypertension.

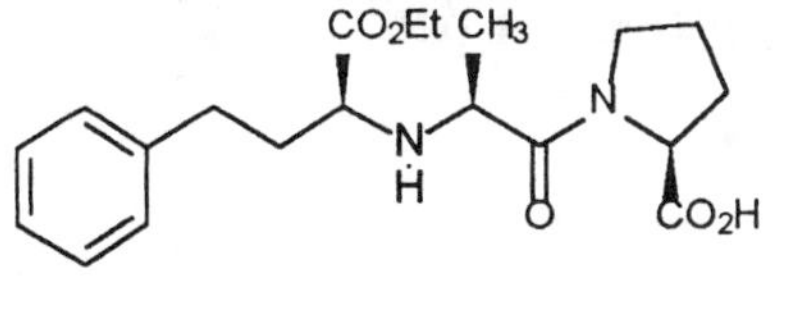

117.1

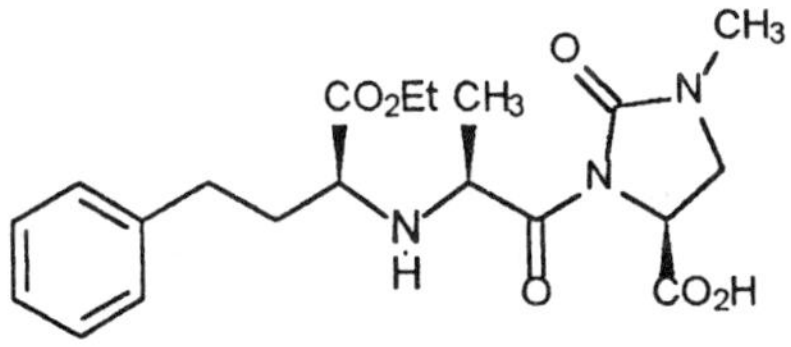

117.2

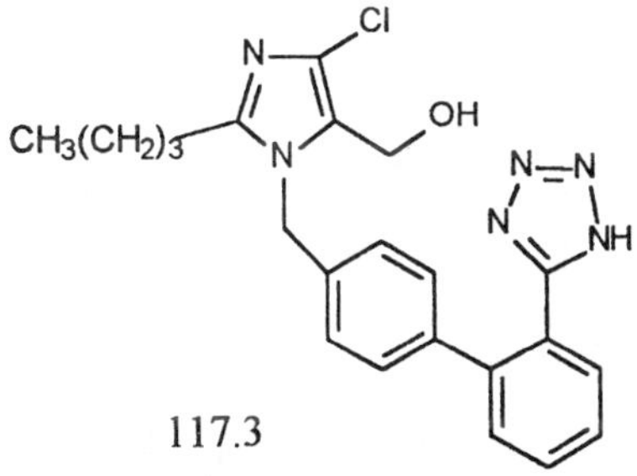

117.3

1. **Identify the mechanism of action of enalapril (117.1) and explain why this type of agent is useful in the treatment of hypertension.**

2. **Which side effect characteristic of ACE inhibitors was not treatable with either OTC or prescription medication? What is the proposed biochemical pathway causing these side effects?**

3. **Explain why imidapril (117.2) is a good substitute for this patient. Include in your answer any relevant biochemical pathways.**

4. **Identify the mechanism of action of losartan (117.3) and explain why the addition of this agent will improve the control of LB's hypertension.**

Case **118**

Submitted by Jeffrey J. Christoff

IB, a 40-year-old man, presents a prescription to your pharmacy for Inderal LA 160 mg (Propranolol) (118.4), i cap. qd x 30 d, refills x 6 from Dr. Dayton. He informs you that this prescription is an additional medication for treatment of his high blood pressure. A quick review of his patient profile provides the following:

Drug	**Dose**	**Treatment**	**Physician**
theophylline (118.1)	300 mg tid	asthma	Dr. Van Buren
hydrochlorothiazide (118.2)	50 mg qd	hypertension	Dr. Dayton
ibuprofen	prn	muscle aches	R.Ph.
terfenidine	60 mg q12h prn	allergies	Dr. De Stefano
terbutaline (118.5) inhaler	ii puffs q6h prn	asthma	Dr. Van Buren
methyldopa (118.3)	500 mg q6h	hypertension	Dr. Dayton

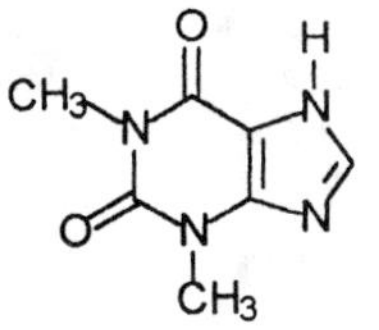

118.1

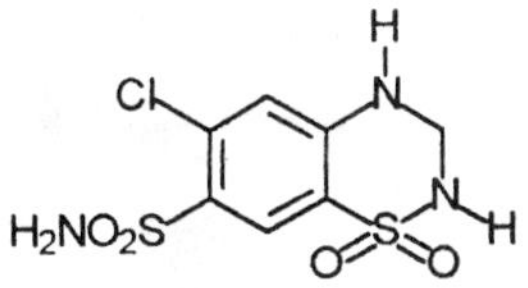

118.2

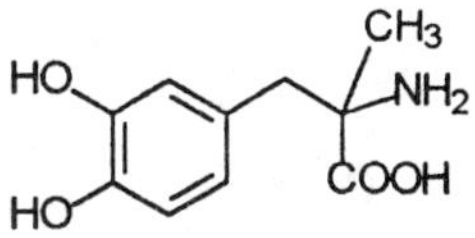

118.3

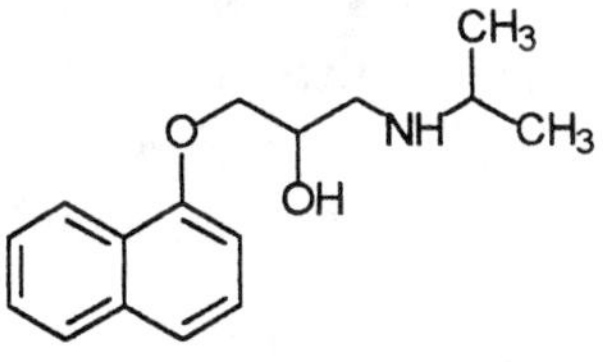

118.4

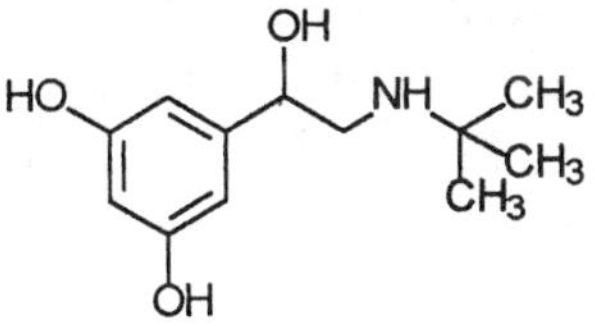

118.5

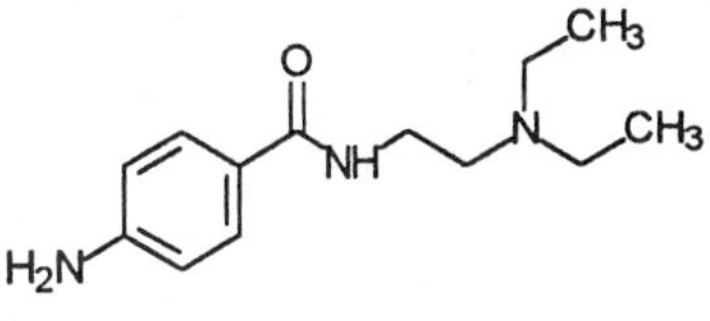

118.6

1. What is the potential problem with IB's prescription?

2. **What is the mechanism of action of each drug affecting the cardiovascular system?**

3. **Why is methyldopa (118.3) used in tablets, but its ester salt, methyldopate hydrochloride, used for intravenous mixtures?**

4. **IB informs you that he has been suffering from leg cramps. He has followed Dr. Dayton's advice and eats a banana and drinks a glass of orange juice every day, but this doesn't seem to help. In order to alleviate this problem, what options are available for recommendation? Describe how each of these recommendations could correct this problem?**

5. **What classes of drugs can be substituted for propranolol (118.4) for use with HCTZ (118.2) and methyldopa (118.3) in treating this patient for hypertension?**

6. **Suggest an alternative calcium channel blocker (CCB) for use with HCTZ (118.2) and methyldopa (118.3) in treating this patient for hypertension. Provide a rationale for your drug selection.**

7. **Which CCB is most likely to cause edema, flushing, headaches, and reflex tachycardia? Which CCB has the potential to produce bradycardia?**

8. **Several years ago, IB was admitted to the hospital for the treatment of a heart attack he suffered while playing basketball. Which class I antiarrhythmic drug would the paramedics probably initiate at the scene if therapy were needed? How does it produce its effects? If the patient were converted to another drug with the same kinetic profile for action at the Na^+ channel, what are the choices? Why would they be more desirable? What drug with a slightly different kinetic profile would be a good choice for prophylaxis of ventricular tachycardia?**

Case 119

Submitted by Jeffrey J. Christoff

ME is a 62-year-old man with a 3-year history of congestive heart failure. Your inspection of ME's profile reveals that he is taking the following medications:
Nitroglycerin ointment (119.1), 1'' q6h
Hydralazine (119.2), 50 mg q6h
Digoxin (119.4), 0.25 mg qd
Furosemide (119.3), 40 mg qAM
Micro-K (KCl supplement) 10 mEq bid

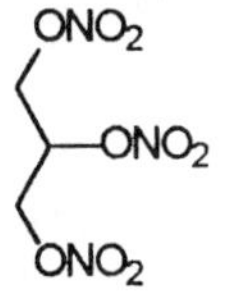

119.1

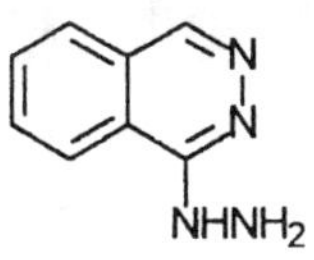

119.2

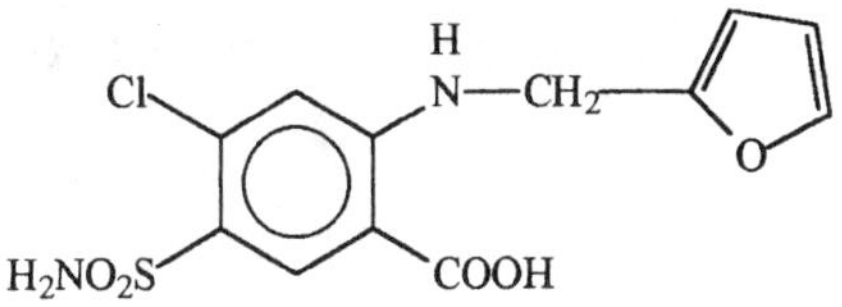

119.3

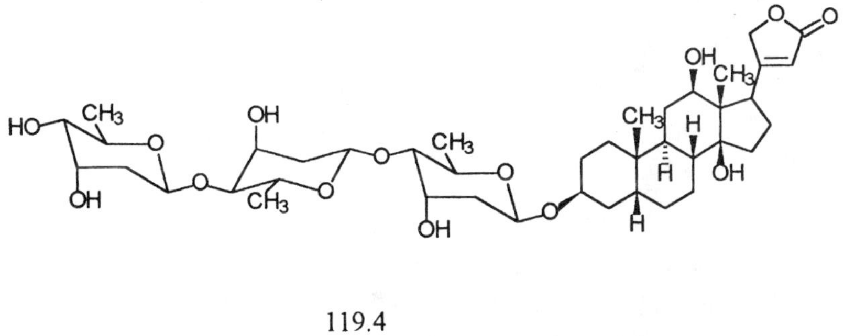

119.4

1. **Compare and contrast the production of nitric oxide ($\dot{N}O$) from an exogenous source (eg, nitroglycerin [119.1]) and its endogenous source.**

2. **What is the metabolic step for hydralazine (119.2) that can be influenced by a patient's genetic phenotype? In what organ is this metabolic enzyme primarily located? What is the significance of the location of this enzyme?**

3. **Describe the formation of the predominant metabolite for hydralazine (119.2) and describe the importance of metabolism as related to its activity?**

4. **In general, rationalize what type of dosage recommendations could be necessary for hydralazine (119.2) based on the patient's genetic phenotype? Cover the dosage options necessary for the predominant genetic phenotypes.**

Case 120

Submitted by Jeffrey J. Christoff and Mary Chavez

SB is a 72-year-old woman who has a 31-year history of hypertension, but no other significant cardiovascular disease, and a 5-year history of schizophrenia. Last month, the patient was moved from an extended care facility to her niece's household because her family could not afford her health care in the nursing home. SB is currently being treated with captopril (120.1), risperidone (120.2), Dyazide (hydrochlorothiazide [HCTZ] [120.3], and triamterene hydrochloride [120.4]). Ever since SB moved to her niece's home, she has been losing weight. She has lost 6 lb in the last 4 weeks, dropping to 105 lb on her 1/15/95 doctor's office visit. SB claims, "My niece can't cook and the food tastes like garbage!" Her niece shows you the following patient care report from the nursing home.

Shady Acres Senior's Assistance Community
Patient Medication Profile
SB
Admission Date 9/23/94 Discharge Date 12/18/94

Drug	Start Date	Stop Date	Dose	Notes
10% dextrose	12/15/94	12/15/94	75 mL/h IV	supplement inadequate food intake, patient removed IV line
captopril	12/12/94		12.5 mg po bid	MD change from nifedipine
risperidone	9/23/94		2 mg bid	patient on med over 6 months PTA
HCTZ/triamterene 25/50	9/23/94		i po qAM	patient on med over 2 years PTA
nifedipine XL	9/23/94	12/12/94	30 mg po qd	patient on med over 2 years PTA; D/C patient chews pill against instructions

Family Notes on Discharge: Encourage adequate food intake. Consult dietitian about supplemental nutrition if patient continues to limit food intake.

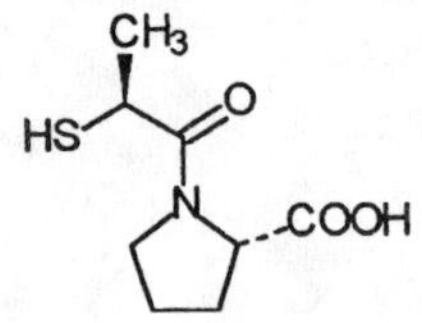

120.1

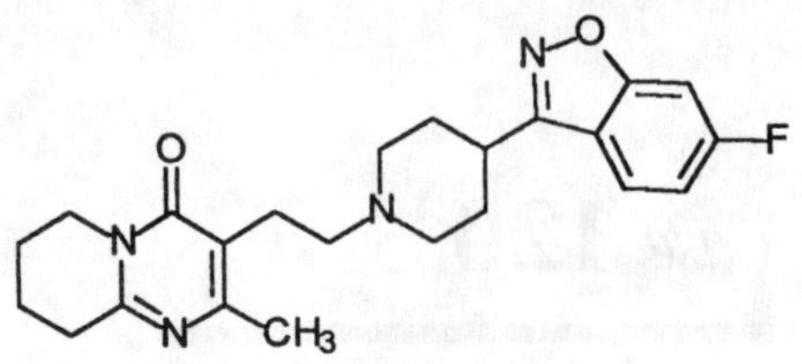

120.2

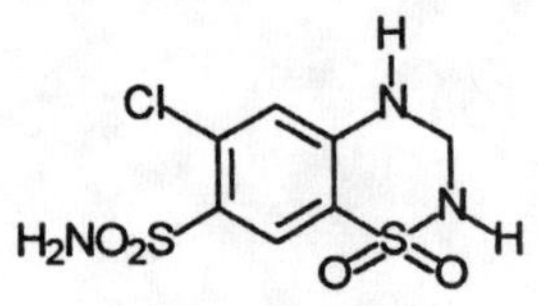

120.3

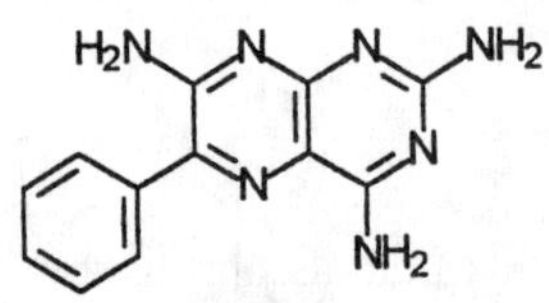

120.4

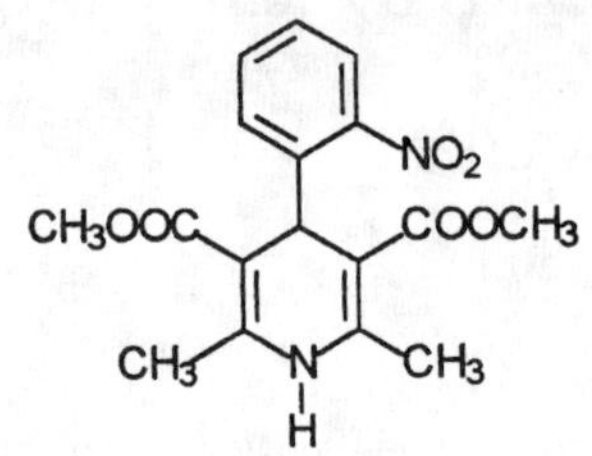

120.5

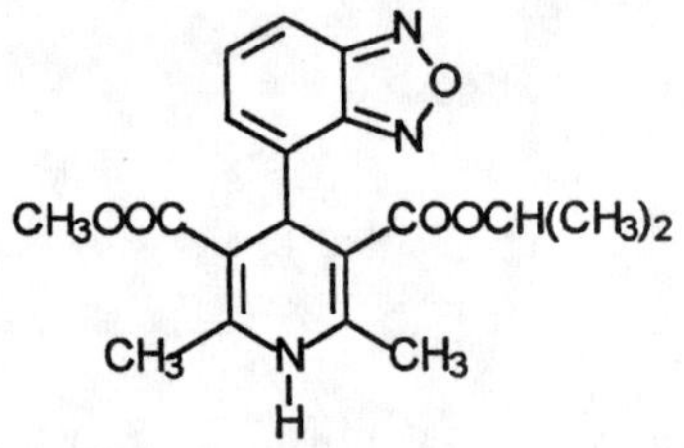

120.6

1. **Describe the mechanism of action for each component of the combination drug (120.3 and 120.4). Why are these two drugs used in this combination?**

2. **What are the mechanisms of action for the other drugs on SB's profile? Are there any potential drug interactions? Explain.**

3. **Are any of SB's medications potentially causing her to lose weight? Should medication therapy be changed? Why, or why not?**

4. **SB's physician has called in a new prescription for her. The new prescription is for DynaCirc CR 5 mg qd (isradipine [120.6] controlled-release). Are there any potential problems with the new prescription? Explain.**

Case 121

Submitted by Arnold R. Martin and Richard N. Herrier

As the staff pharmacist at the local hospital, you have been called in to consult on a case involving a 63-year-old itinerant farm worker who was brought in with symptoms of "insulin shock," that is, confusion, disorientation, and convulsions. His blood glucose was 30 mg/100 mL. The symptoms were gradually reversed with IV glucose (5%). From a review of the record, you learn that the patient had been hospitalized 1 month earlier with acute pulmonary aspergillosis. This was treated with amphotericin B (slow IV) over a 2-week period. The patient was then released with a prescription for oral itraconazole (121.1) (two 100-mg tablets twice daily) for 2 months. Inspection of the chart also revealed that the patient is an adult onset diabetic, who has been taking Tolbutamide (121.2) (500 mg four times daily) for the past 5 years.

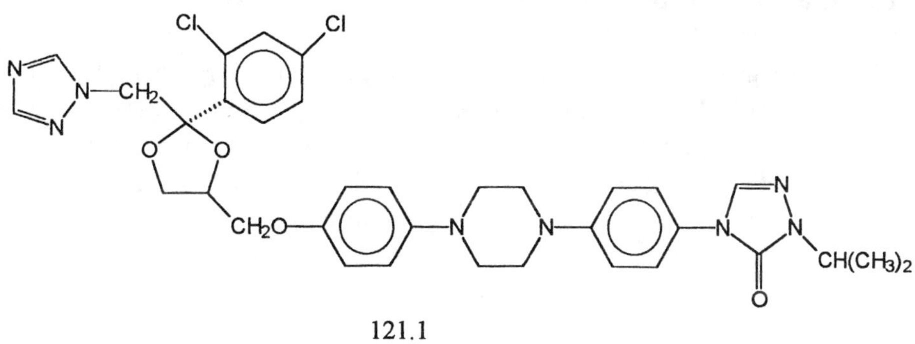

121.1

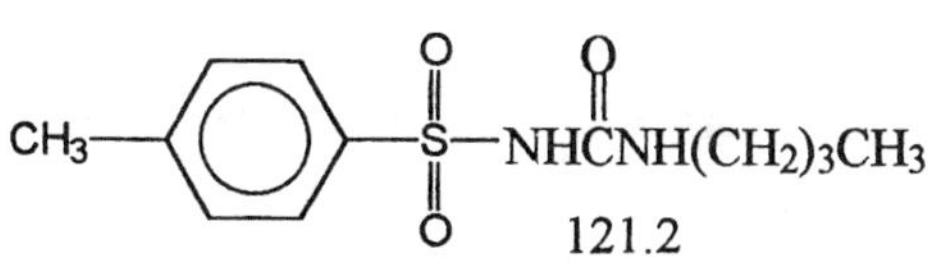

121.2

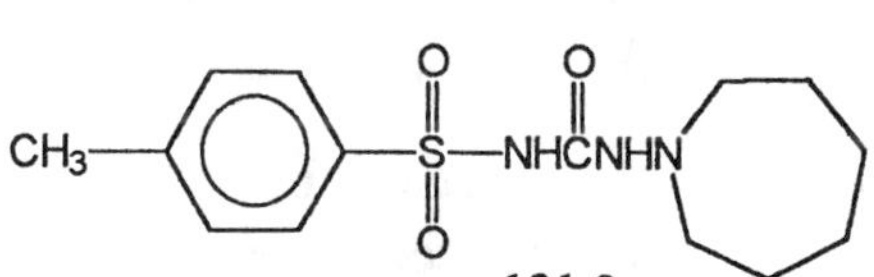

121.3

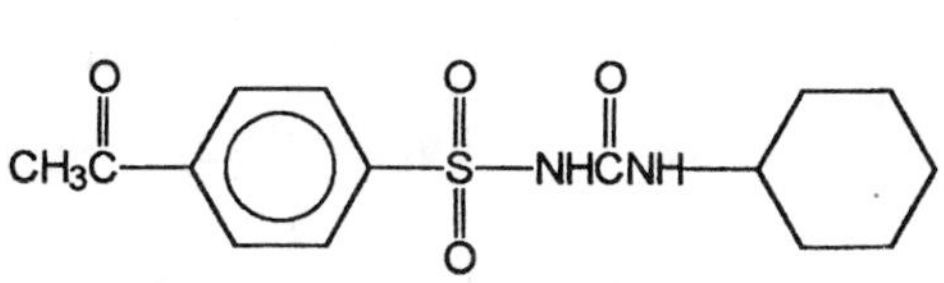

121.4

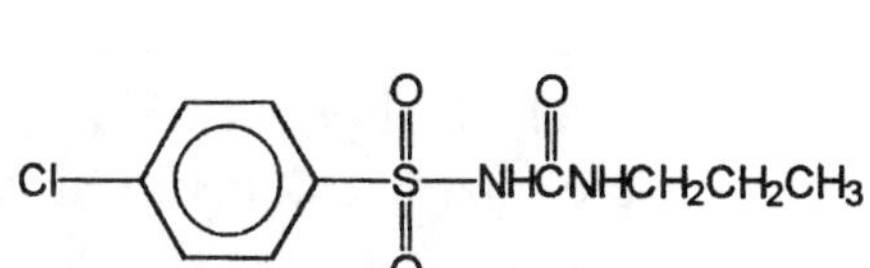

121.5

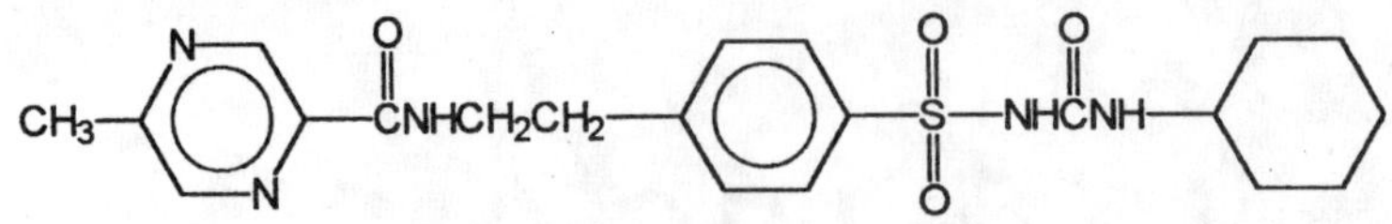

121.6

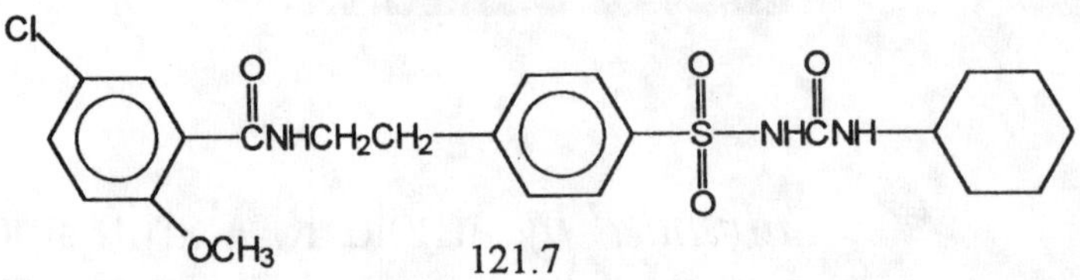

121.7

1. What was the likely cause of the hypoglycemic crisis? Explain fully.

Answer Tolbutamide (121.2) is a short-acting agent due to rapid inactivation by cytochrome P-450 oxygenases (the aryl methyl group is oxidized to the corresponding alcohol and acid; these are conjugated and excreted). Itraconazole (121.1) inhibits cytochrome P-450 oxygenases, thereby causing the hypoglycemic action of tolbutamide (121.2) to be greatly enhanced and prolonged.

2. Suggest approaches by which the type II diabetes can be controlled and the prophylaxis of the aspergillosis continued on an outpatient basis.

Answer Two, or three possibilities exist.

1. Control the diabetes using the appropriate dosage regimen of insulin during the course of the itraconazole (121.1) prophylaxis. (This will require that the patient self-administer his insulin).
2. Select a sulfonylurea derivative that is not inactivated by cytochrome P-450 oxygenases.
3. Maintain the patient on a rigorous, low-carbohydrate, low-fat diet. (This may not be practical, depending on the severity of the diabetes and the compliance behavior of the patient.)

3. Assume that the attending physician prefers to maintain the diabetes using an oral hypoglycemic drug (perhaps with dosage adjustment), while maintaining itraconazole (121.1) prophylaxis. Select the sulfonylurea oral hypoglycemic drug from the structures provided (121.3–121.7) to substitute for tolbutamide (121.2) that would require the least dosage adjustment: acetohexamide (121.4), chlorpropamide (121.5), glipizide (121.6), and glyburide (121.7). Justify your choice by discussing the pharmacokinetics of each drug.

Answer Acetohexamide (121.4) is the best choice since it is a relatively short-acting agent that is not inactivated by cytochrome P-450 oxygenases. It is, instead, bioactivated by reduction to the alcohol, which is then inactivated by O-glucuronidation. Chlorpropamide (121.5) is a long-acting drug that is extensively protein bound to plasma proteins and slowly converted to both active and inactive metabolites. Since itraconazole (121.1) is also extensively plasma protein bound, there is a danger that the two may compete for binding sites, leading to altered hypoglycemic effects. Tolazamide (121.3) and glipizide (121.6) each have aromatic methyl groups that are attacked by cytochrome P-450 oxygenases to undergo inactivation. Glyburide (121.7) is oxidatively O-demethylated by cytochrome P-450 oxygenase to less active or inactive conjugated metabolites.

Case 122

Submitted by Bruce L. Currie

A recent literature report has described the discovery of a new form of the peptide hormone, luteinizing hormone-releasing hormone (LH-RH), that was obtained from chickens. This new form is also a decapeptide (structure shown). "Chicken LH-RH" was shown to have 100% of the potency of human LH-RH when tested in vitro on chicken anterior pituitary tissue (location of its receptors) when compared with human LH-RH. However, when tested in vitro on rat anterior pituitary tissue, "chicken LH-RH" had only 1% of the potency of human LH-RH on rat receptors.

Human LH-RH: pGlu-His-Trp-Ser-Tyr-Gly-Leu-Arg-Pro-Gly-NH_2

Chicken LH-RH: pGlu-His-Trp-Ser-Tyr-Gly-Leu-Gln-Pro-Gly-NH_2

1. Identify any structural differences between "chicken LH-RH" and "human LH-RH."

Answer Chicken LH-RH has a Gln (glutamine) at position 8, whereas human LH-RH has Arg (arginine). Arginine is protonated and therefore positively charged at physiologic pH, but glutamine is neutral at all pHs and not ionizable.

2. Based on the information given, does the structural change influence binding <u>or</u> intrinsic activity for "chicken LH-RH" on the rat receptor?

Answer There is only a change in potency; thus, the structural change influences binding, not intrinsic activity.

3. **What do these observations imply about the structure of the "chicken LH-RH receptor" in the chicken anterior pituitary gland? Explain briefly using the structural information along with the results of the biologic evaluation.**

Answer The information previously given implies that there is no complementary ionic binding for the residue at the 8 position, that is, a negative charge on the receptor. The human or rat receptor, however, has a negative charge that is complementary to the protonated arginine that contributes substantially to the binding affinity.

4. **Discuss any changes that are expected for the "chicken LH-RH" in terms of stability toward proteolytic degradation as a result of the structural change that has occurred relative to "human LH-RH."**

Answer A change from arginine to glutamine would ordinarily confer stability toward trypsin and trypsin-like enzymes that are specific for basic side chains. However, the arginine in human LH-RH is before a proline, and thus, the Arg-Pro linkage is already quite resistant to proteolytic cleavage. Thus, little change in susceptibility to proteolytic cleavage would be expected.

Case 123

Submitted by S. William Zito

A new mother with a strong desire to breast-feed her pride and joy has stopped lactating. As a caring and enterprising formulation pharmacist, you decide to develop a nasal delivery preparation for the prescribed oxytocin. You believe penetration enhancers to be the best approach to this problem.

---

1. What general types of penetration enhancers are available for this project?

2. Give the general chemical structure of each type and propose a mechanism of penetration enhancement for each type.

Case 124

Submitted by S. William Zito

As a practicing pharmacist you will often be asked to explain highly technical theories to your patients. For example, a diabetic patient who has been receiving pork insulin for years has just been changed to human insulin. This patient asks you about the differences between the two sources of insulin and is particularly worried about taking a product that has been produced in a genetically altered organism.

---

1. Clearly explain how you would present recombinant DNA theory so that this patient can understand it and feel safe taking this new product.

Case **125**

Submitted by Victoria F. Roche

TK, an elderly man from a small rural Nebraska community, has just been diagnosed with Addison's disease secondary to destructive adrenal lesions. Concomitant malabsorption syndrome has made oral therapy impossible, and since the town clinic and pharmacy recently closed, TK must depend on his daughter to drive him 50 miles to North Platte for therapy.

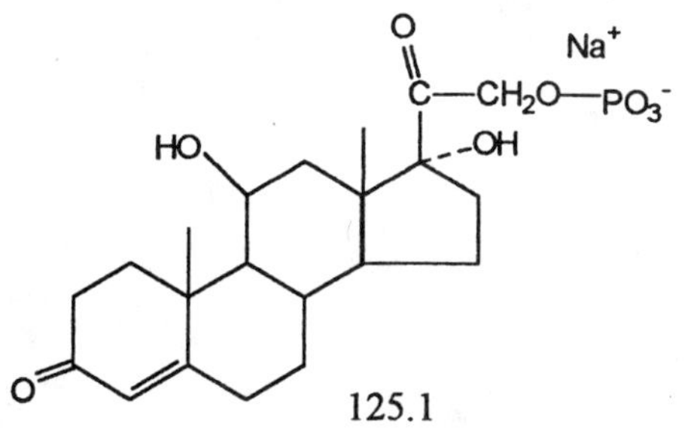

125.1

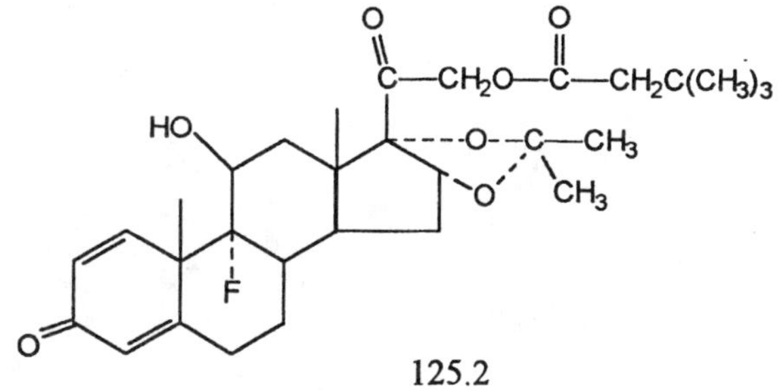

125.2

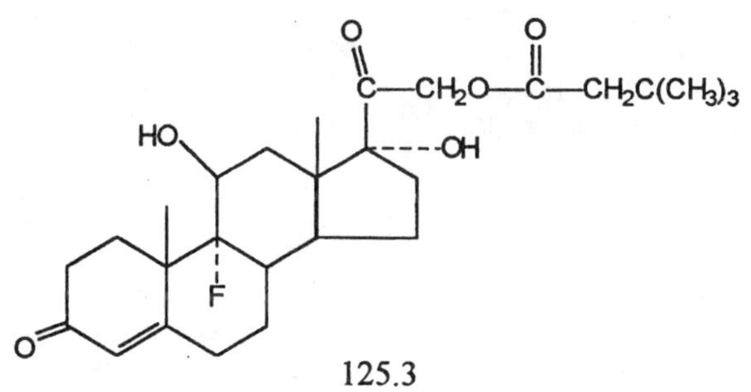

125.3

1. What are the therapeutic needs of TK and his family?

2. **How well do the structural features of the steroid structures provided (125.1–125.3) meet these needs?**

3. **Which steroid agent of the three shown, if any, would you recommend?**

Case 126

Submitted by Arnold R. Martin and Richard N. Herrier

HB is a 59-year-old moderately obese man who is referred by his primary care physician to a diabetes clinic for evaluation of adverse reactions related to his oral hypoglycemic agent therapy. HB was diagnosed 18 months ago with non–insulin-dependent diabetes. Despite several attempts, HB was unable to control his blood glucose levels with diet and exercise alone. His primary care physician then prescribed chlorpropamide (126.4) (250 mg qd) 6 weeks ago. On five occasions during the past 2 weeks, HB experienced symptoms of hypoglycemia (tremulousness, sweating, hunger, confusion), which he recognized and self-treated with orange juice. HB also complains of incidences of dizziness and flushing of the face after consuming a glass or two of his favorite red wine on several occasions during the past month. (HB is quick to point out that he exercises great moderation in his consumption of alcoholic beverages because of his diabetic condition.) A random blood glucose drawn 2 hours after breakfast is 70 mg %, and a glycosylated hemoglobin A_{1C} is 7.6%, which was done at the same time. A thorough review of the patient's medical history reveals that HB had tuberculosis as a teenager. He was cured with a combination of isoniazid, ethambutol (2 years), and streptomycin (1 year) therapy. As a result, HB suffered hearing impairment. Suspecting that the extended therapy with streptomycin could also have damaged the patient's kidneys, the attending physician has renal function tests performed. The results reveal proteinuria and a creatinine clearance of 40 mL/min.

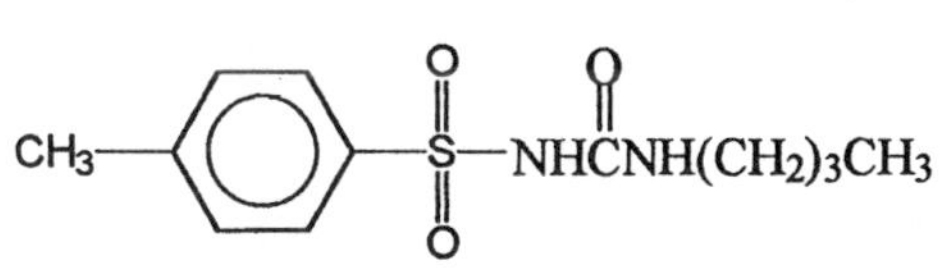

126.1

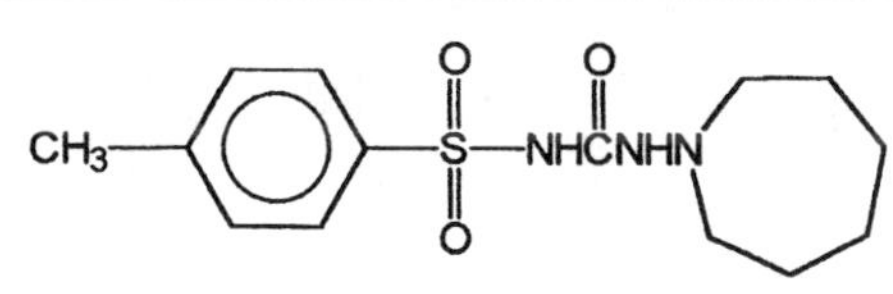

126.2

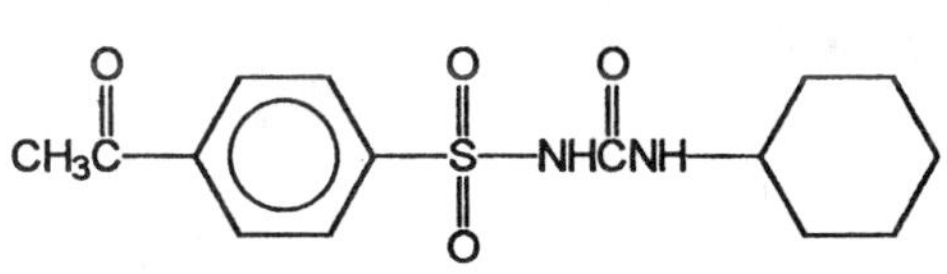

126.3

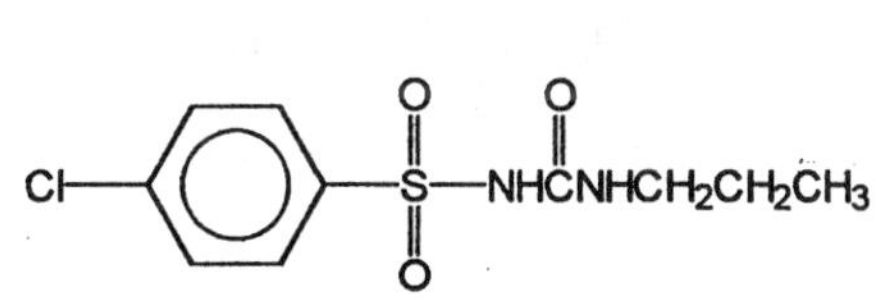

126.4

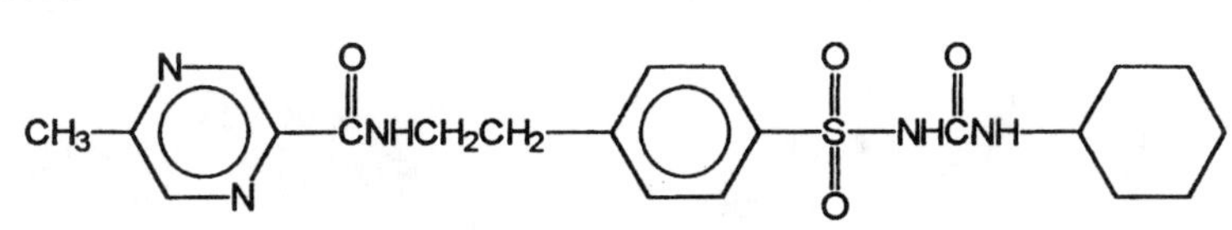

126.5

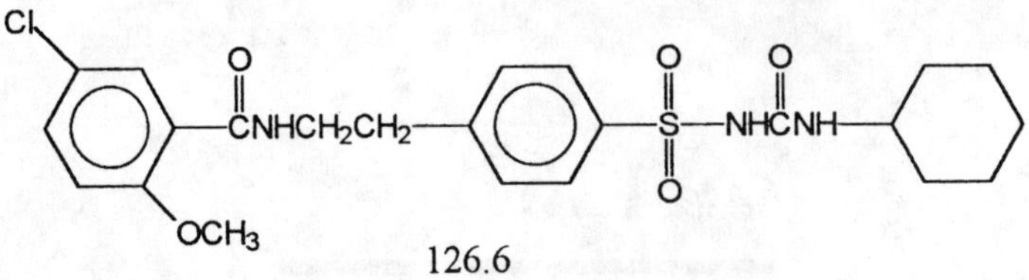

126.6

1. **The attending physician decides to discontinue chlorpropamide (126.4) and replace it with another oral hypoglycemic agent. Explain why.**

2. **As the staff pharmacist, you are consulted regarding a suitable replacement for chlorpropamide (126.4). Make a selection from the available remaining oral hypoglycemic drugs (see structures [126.1–126.6] provided). Justify your choice and explain in detail why each of the remaining choices were rejected.**

3. **What is the likely cause of HB's intolerance to alcohol?**

4. **What advice, if any, would you give to the patient with regard to his drinking habits? Explain.**

Case 127

Submitted by Robin M. Zavod

VS, an 80-year-old woman, has a 20-year history of Type II diabetes. She has been successful in managing the disease with one or another sulfonylurea-type hypoglycemic agent since she was first diagnosed. Although she has been healthy and relatively stress free recently, she has noticed that her blood glucose values have been gradually increasing. Her husband reports that she feels miserable and lethargic almost all of the time. Finally, she decides that she needs to visit her endocrinologist to find out how to "get better." (Drug History: tolbutamide [127.1] or chlorpropamide [127.3] for the last 15 years; now taking glyburide [127.4].)

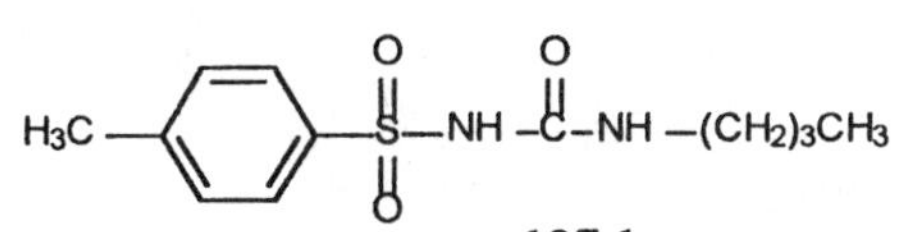

127.1

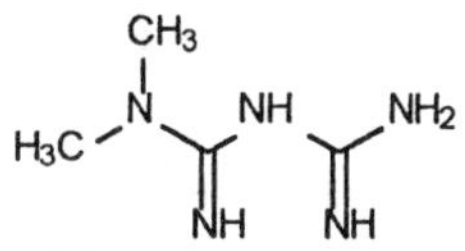

127.2

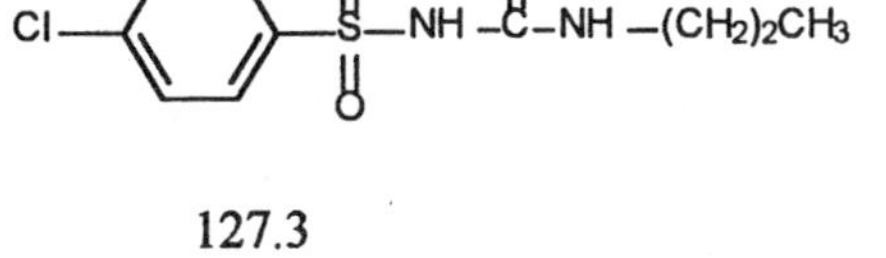

127.3

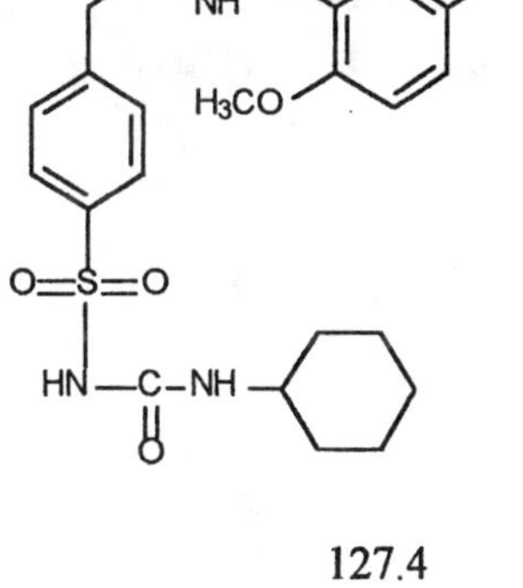

127.4

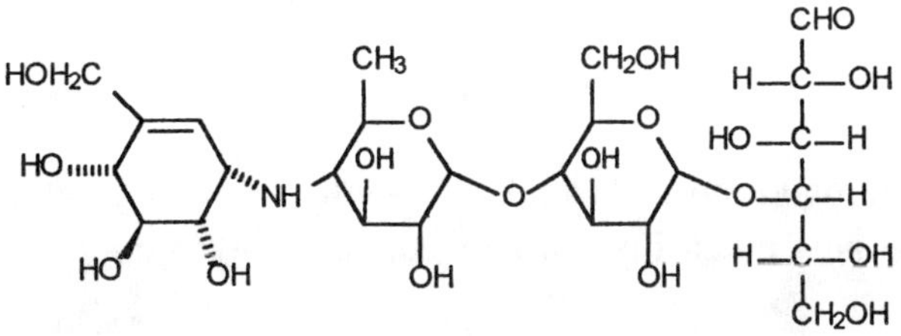

127.5

1. **What are several physiologic reasons that might explain why this therapy seems to be failing?**

2. **What nonpharmacologic adjunct therapies might be causing the drug failure?**

3. **What are some of the proposed mechanisms of action of the sulfonylurea class of hypoglycemic agents?**

4. **The endocrinologist suspects that VS is not producing adequate insulin. How can he determine how much insulin she is producing?**

5. **What type of drug therapy might the physician suggest if it is determined that the patient is not producing sufficient insulin?**

Case 128

Submitted by Robin M. Zavod

SH, a 22-year-old active male marathoner, is physically fit and trains regularly. After a particularly nasty winter with several serious bouts of the flu, SH finds himself craving pounds of Oreo cookies, drinking a couple of gallons of fluid daily, and getting up hourly at night to urinate. He is diagnosed with Type I diabetes and is immediately prescibed an insulin regimen. After education classes, SH is able to handle injections twice daily with a combination of two types of insulin: Regular and Lente. He is instructed to inject his Insulin before his morning and evening meals.

---

1. **Based on the duration of action of these two types of insulin, explain whether SH can eat lunch or not? Explain the chemical reasons why these two types of insulin differ in their duration of action.**

2. **SH is unable to eat a noon meal at a regular time during the work week and often becomes hypoglycemic in the middle of the afternoon. How could his insulin therapy be modified to allow flexibility in the timing of the noon meal? Should a different type of insulin be administered? If so, what type?**

3. **Based on the observation that SH becomes hypoglycemic in the middle of the afternoon if he doesn't eat his noon meal early enough, what might you expect to happen if SH does not consume sufficient carbohydrates during his evening meal?**

4. **What type of combination of insulin therapies might provide an adequate basal level of insulin, yet allow flexibility in the timing of all of SH's insulin injections?**

Case 129

Submitted by Arnold R. Martin and Richard N. Herrier

DJ is a 23-year-old woman who was started on LoOvral (norgestel [129.3] 0.3 mg and ethinyl estradiol [129.2] 30 mg) for contraception 6 months ago. She returns to the clinic to see her nurse practitioner (NP). The NP calls you, the staff pharmacist, and explains that DJ has developed progressively worsening acne since starting "the pill." She has also gained 10 pounds and has noted a decreased desire for sex. The NP assumes that these problems are due to the LoOvral and requests your recommendation for change.

Norethindrone acetate (129.1) 1.5 mg Ethinyl estradiol (129.2) 30 mcg	LoEstrin
Norgestrel (129.3) 0.3 mg Ethinyl estradiol (129.2) 30 mg	LoOvral
Medroxyprogesterone (129.4)	Depo-Provera
Norethindrone (129.5) 1 mg Ethinyl estradiol (129.2) 35 mcg	Ortho Novum 1/35
Norgestimate (129.6) 0.25 mg Ethinyl estradiol (129.2) 35 mcg	Ortho Cyclen

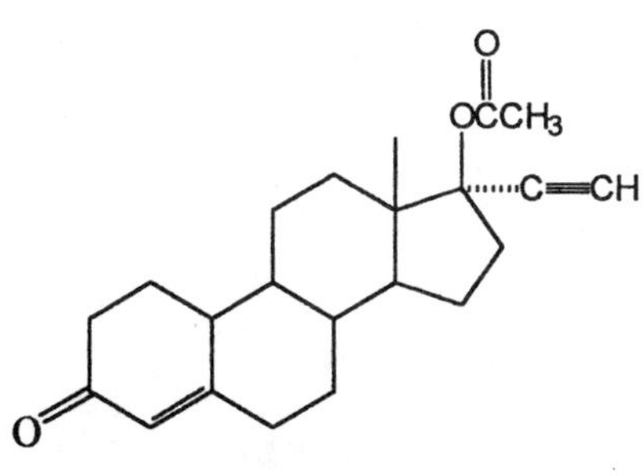

129.1

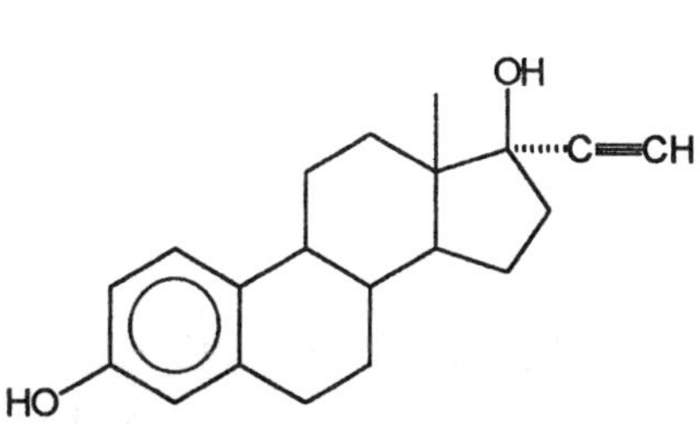

129.2

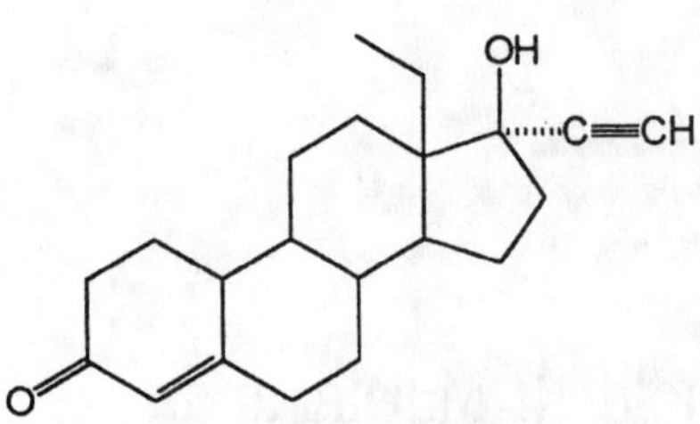

129.3

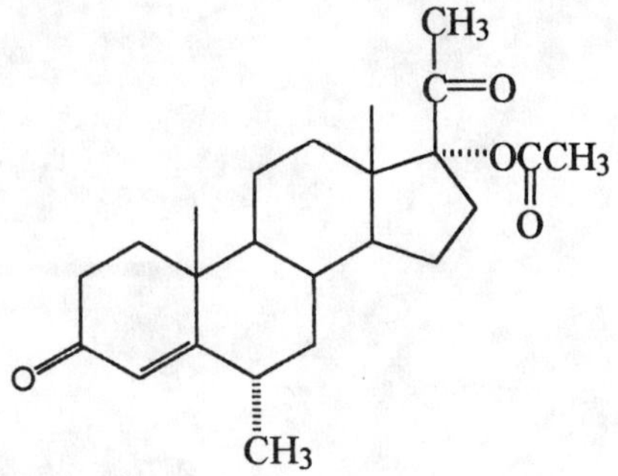

129.4

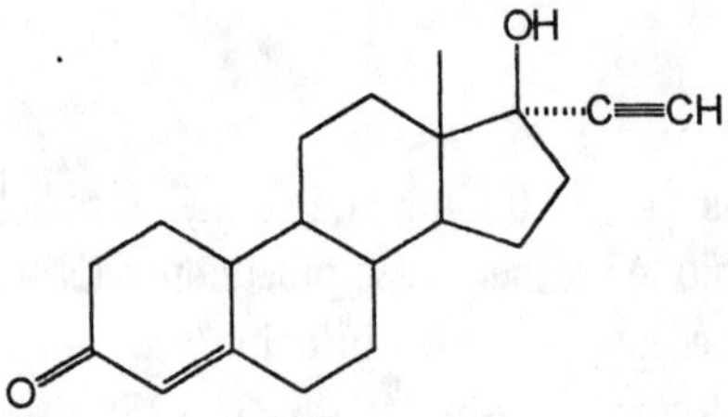

129.5

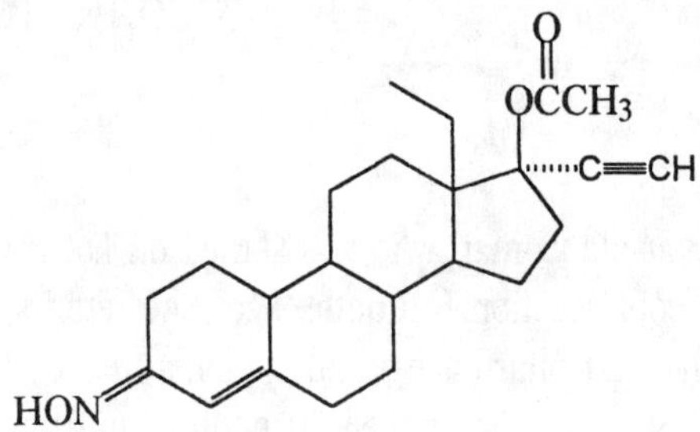

129.6

1. **Which component of LoOvral is the likely cause of DJ's complaints?**

2. **Select the most suitable contraceptive treatment from the list provided. Justify your choice on the basis of the properties of the ingredient(s) present. Discuss each of the remaining possible choices and explain why they were not chosen.**

Case 130

Submitted by Bruce L. Currie

IG, a 62-year-old man, has just been diagnosed with advanced prostate cancer by his urologist. He has been given three choices for treatment of his condition: (1) surgical castration and chemotherapy, (2) extensive radiation, or (3) drug therapy with flutamide (130.1) and leuprolide (130.2). He has chosen the drug therapy and you, the pharmacist at the urology clinic, are to explain to IG why the medication prescribed is appropriate for his prostate cancer.

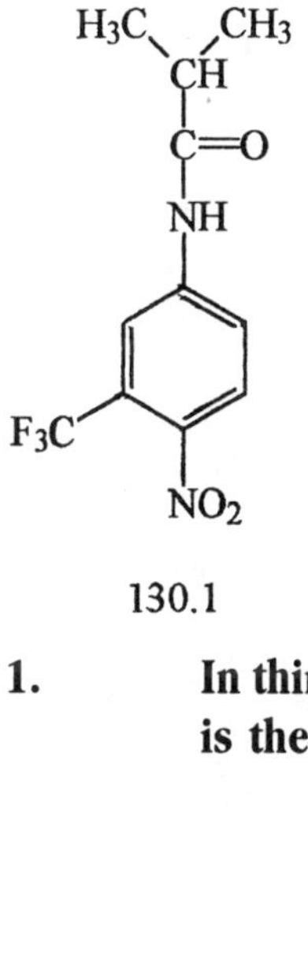

130.1

pGlu - His - Trp - Ser - Tyr - D-Leu - Leu - Arg - Pro - NHEt

130.2

1. **In thinking about these two drugs (flutamide [130.1] and leuprolide [130.2]), what is the mechanism of action of each?**

2. **Is it appropriate that these drugs be used in combination therapy in this situation? Discuss.**

3. **In your opinion, did IG make a good choice among the three options given by his physician? Discuss.**

Case 131

Submitted by Victoria F. Roche

You are consulting with a physican regarding CR, a 37-year-old female AIDS patient. CR has developed systemic candidiasis, and the physician wishes to prescribe ketoconazole (131.1), an orally active antifungal agent. Like many AIDS patients, CR has achlorhydria, and her stomach pH is approximately 4.4.

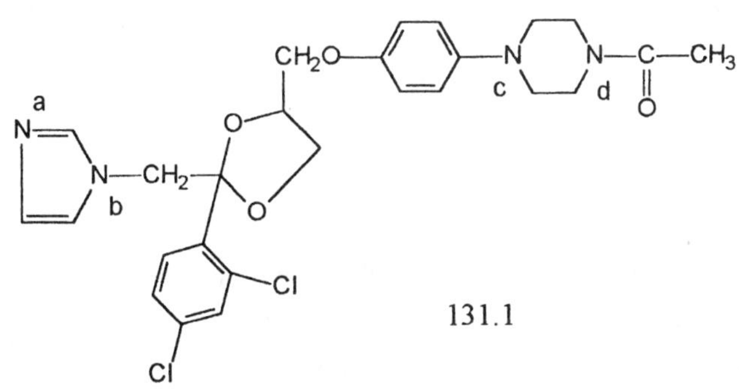

131.1

1. Ketoconazole (131.1) has two pKas (2.9 and 6.4). Match each to the appropriate functional group on the ketoconazole (131.1) structure. The nitrogen atoms have been labeled a-d for you.

Answer Only two of ketoconazole's (131.1) four nitrogens have acid-base character. Nitrogen b is neutral, because its lone pair of electrons contributes to the aromaticity of the imidazole ring. Nitrogen d is also neutral, because its lone pair of electrons is participating in resonance delocalization with the adjacent carbonyl group (the functional group involving an sp^2 hydridized nitrogen bonded to a carbonyl group is an amide). Nitrogen a is sp^2 hybridized, and the lone pair of electrons is found in an sp^2 hydridized orbital. The lower directionality associated with an sp^2 hybridized orbital normally means a decrease in basic strength for an occupying lone pair compared with a lone pair in an sp^3 hydridized orbital (like nitrogen c). But nitrogen c is conjugated with a phenyl ring, and the lone pair of electrons can be delocalized into the ring current, which diminishes its availability for protonation (compare the pKa values for pyridine and aniline). Nitrogen a is actually the most basic nitrogen of the molecule and would have the higher of the two pKa values (6.4). The pKa of nitrogen c would be 2.9. This is lower than the pKa of aniline because electron deficient nitrogen d is withdrawing electrons from nitrogen c through negative induction.

2. **What is the anticipated solubility of this agent in normal gastric fluid (pH = 1.5)? How would you expect the achlorhydria experienced by CR to influence po bioavailability of this antifungal? How would you explain an answer different from the one you have offered?**

Answer In normal gastric fluid, you would predict a high degree of protonation of ketoconazole. As an unionized base, this molecule will take on a proton in acidic media to become the cationic conjugate acid. The ratio of ionized to unionized forms that exist is dependent upon media pH and drug pKa as shown in the Henderson-Hasselbalch equation: pH - pKa = log [basic conjugate]/[acidic conjugate]. For an unionized base like ketoconazole (131.1), this equation becomes:

pH - pKa = log [unionized form]/[ionized form].

At a pH of 1.5, the i/u ratio of ketoconazole (131.1) would be approximately 10^5:1. Since ionized drug forms do not readily penetrate membranes, absorption would be low.

In a patient with achlorhydria, no hydrochloric acid is being produced by the parietal cells of the stomach, and gastric pH increases. In the more basic environment, the basic unionized conjugate of ketoconazole (131.1) would predominate to a greater extent than is observed at normal gastric fluid pHs (i/u ratio would be 10^2:1). Since the relatively more unionized, absorbable form is predominating at this higher pH, po bioavailability could increase. If it does not, there is an explanation. Ketoconazole (131.1), in unionized form, is highly water insoluble. Tablet dissolution cannot occur if the pH of the gastric fluid is too high. The absorption of ketoconazole (131.1) is actually decreased in the presence of H_2 antagonists, which inhibit gastric acid secretion and can raise the pH of gastric fluid up to 7.4.

3. **The *Candida* infection has now spread to CR's vaginal tract (pH = 4.5). Would you expect ketoconazole (131.1) to distribute from the systemic circulation into the vaginal canal? If you say yes, would you expect the drug to distribute back again?**

Answer At a normal plasma pH of 7.4, the i/u ratio of ketoconazole (131.1) would be 1:10. At a vaginal pH of 4.5, the i/u ratio would be approximately 10^2:1. Since unionized ketoconazole, which can cross membranes, is predominating in the plasma, the drug could distribute into the vagina where it would become predominately ionized and, effectively, trapped.

Case **132**

Submitted by Arnold R. Martin and Richard N. Herrier

MJ is a 67-year-old male heart transplantation patient receiving immunosuppressive therapy (cyclosporin and corticosteroids) who was readmitted to the hospital 2 months after his surgery, complaining of fever, malaise, joint pain, and painful oral lesions (white plaques). Cardiac output and renal function tests were normal, indicating that there were no signs of tissue rejection. A chest radiograph revealed diffuse lung infiltrates. Blood and throat cultures yielded *Candida albicans*. Therapy was initiated for MJ with amphotericin B by slow IV infusion. After 20 hours, severe thrombophlebitis developed near the injection site. Although miconazole (132.4) is available for IV administration, the cardiologist was reluctant to use it.

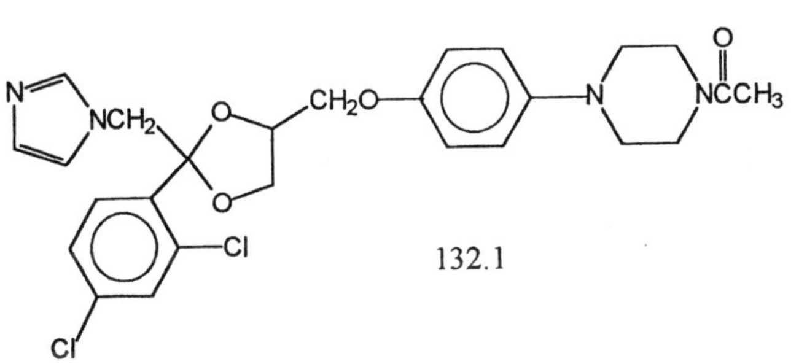

132.1

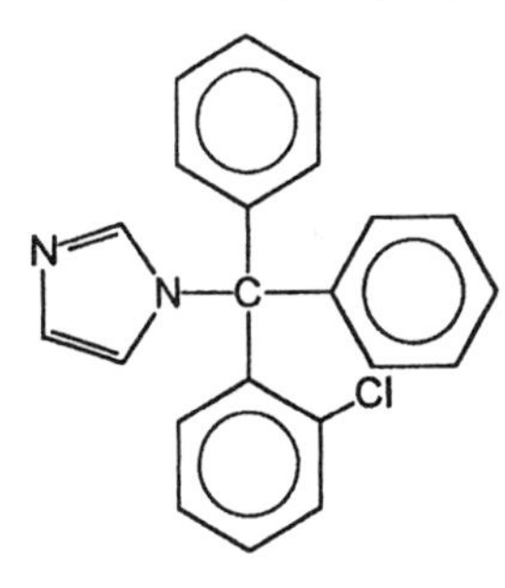

132.2

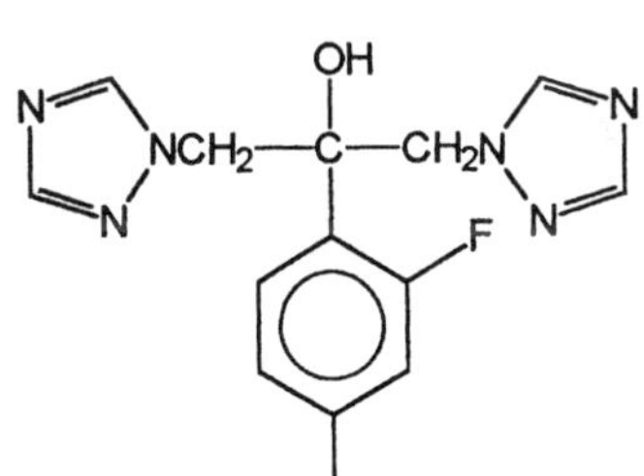

132.3

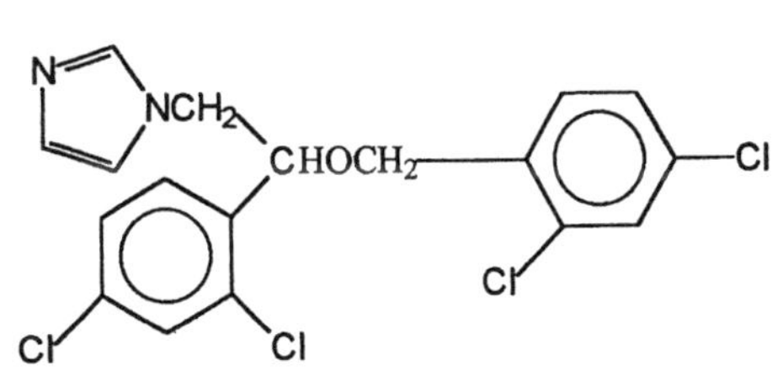

132.4

1. **Describe the injectable form of miconazole (132.4). What is the basis for the physician's reluctance to administer it to this patient?**

Answer Miconazole (132.4) IV is solubilized with polyethylene glycol-40 and castor oil. It is irritating when administered IV and frequently causes thrombophlebitis.

2. **Select, from the imidazole antifungal structures (132.1–132.3) given, the most appropriate one for IV administration to this patient. Justify your choice on the basis of structure/solubility properties.**

Answer Fluconazole (132.3) is available in both injectable and oral dosage forms. Its solubility in water is much greater than the solubilities of the other antifungal imidazoles and triazoles, due to the presence of a tertiary hydroxyl group and two triazole rings.

Case 133

Submitted by Arnold R. Martin and Richard N. Herrier

A 21-year-old college coed has been admitted to the university hospital complaining of chills, fever with pelvic pain, and vaginal discharge. A vaginal examination revealed 4+ (severe) generalized tenderness, with localization and a foul-smelling bloody purulent vaginal discharge. She was diagnosed with septic abortion and was administered IV cefoxitin (133.1) 2 gm every 6 hours and doxycycline (133.2) 100 mg every 12 hours. Culture results 48 hours later revealed *Bacteroides fragilis* as the only bacterial pathogen. Based on sensitivity tests, cefoxitin (133.1) was replaced with IV metronidazole (133.3) 500 mg every 6 hours. As of today, she has been afebrile for 48 hours and is scheduled to be discharged with oral metronidazole (133.3) 500 mg three times daily for 5 days and doxycycline (133.2) 100 mg twice daily for 5 days. While you are bringing her medications to her room, you stop by the nurses' station to review the chart. During the review, you note that the patient has had repeated visits to the emergency room with alcohol-related problems over the past 2 years.

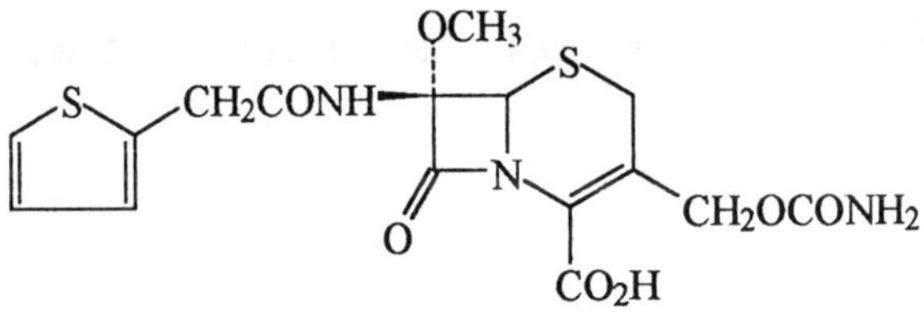

133.1

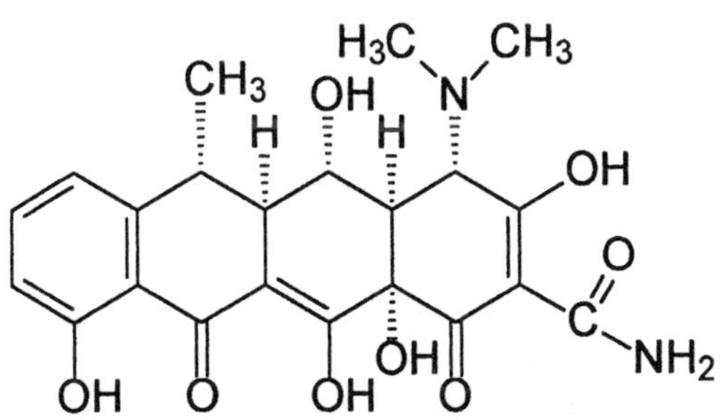

133.2

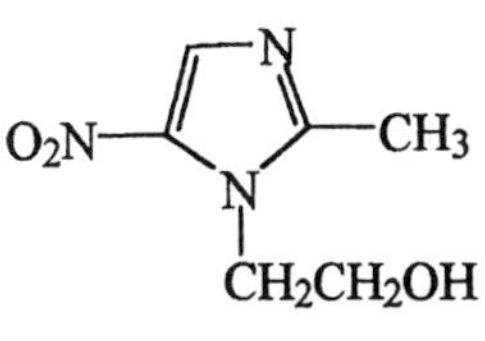

133.3

1. What advice would you give to the patient regarding the medications prescribed? Explain fully.

2. Different chemical forms of metronidazole (133.3) are employed for IV versus oral administration. Explain why.

3. Suggest alternative chemotherapy for the treatment of septic abortion caused by *B. fragilis*.

Case 134

Submitted by Marc W. Harrold

JS visits his physician complaining of a fever, stiffness of the neck and other joints, a low-grade headache, and general fatigue. He further states that the symptoms started about 2 weeks ago and have been constant during that time. He states that he initially thought he might have caught a cold or the flu during his annual camping trip with his old army buddies; however, since the symptoms have not dissipated, he is now concerned that these symptoms could be associated with recently diagnosed problems with his kidneys. A routine examination 4 months before this visit revealed abnormal creatinine levels and a low glomerular filtration rate, along with other laboratory results indicative of the early stages of renal failure. At present, his renal status is being monitored monthly, and he has been placed on Amphogel (aluminum hydroxide suspension), 15 mL po with meals. JS's physician suspects that JS's current symptoms are unrelated to his renal problems but could be due to Lyme disease, which could have been contracted during his camping trip. The use of tetracyclines to treat early stages of Lyme disease has been shown to decrease the duration of symptoms and to prevent any major complications.

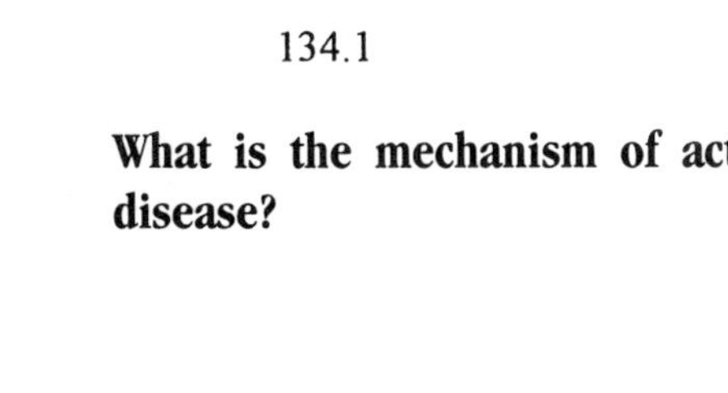

134.1

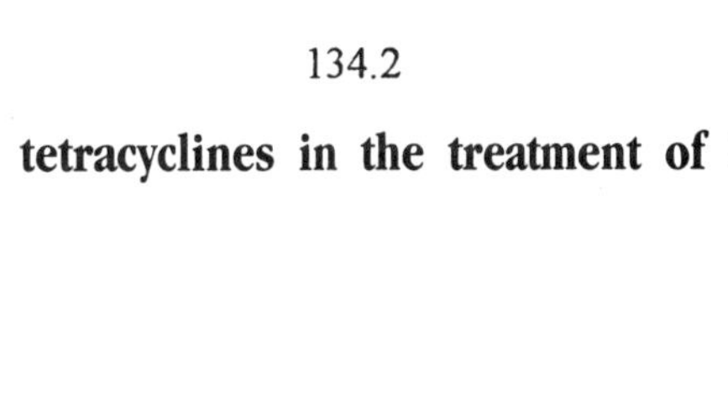

134.2

1. **What is the mechanism of action of the tetracyclines in the treatment of Lyme disease?**

2. **Which of the two tetracyclines shown (134.1–134.2) would be the best choice for this patient? Explain your answer based on the chemical differences between these two compounds.**

3. **What is the purpose of using Amphogel in this patient?**

4. **Why does the addition of either tetracycline (134.1) or doxycycline (134.2) to this patient's drug regimen require the pharmacist to help the patient establish a proper dosing schedule?**

Case **135**

Submitted by Victoria F. Roche

RK, a 15-year-old adolescent girl with a severe case of acne, comes to your community pharmacy with a prescription for oxytetracycline (135.1). Upon consulting her patient profile card, you find she also has a gastric hyperacidity disorder that causes intermittent distress severe enough to warrant medication with an antacid (Mylanta, which contains both aluminum hydroxide and magnesium hydroxide, is her OTC drug of choice). Compliance has been a problem with this young woman when taking medications in the past, and she claims she "just doesn't remember to take her pills when she's supposed to." You check your inventory and find tetracycline products 135.1 to 135.4 on hand.

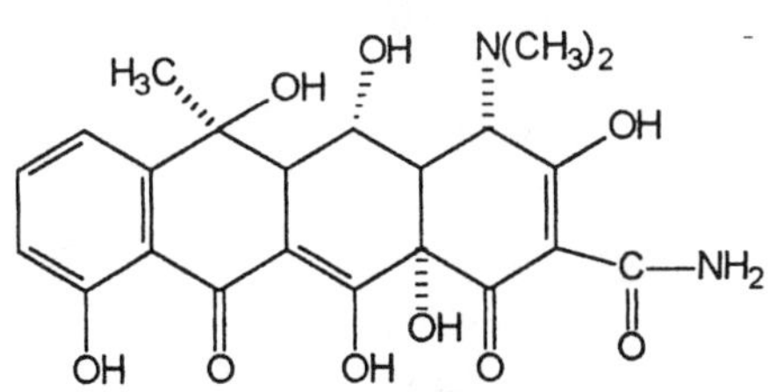

135.1

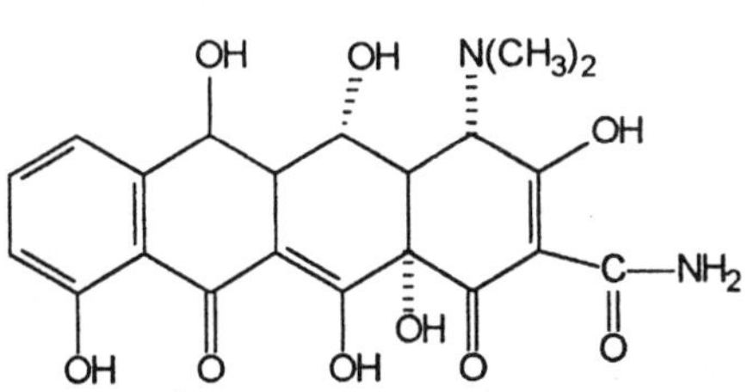

135.2

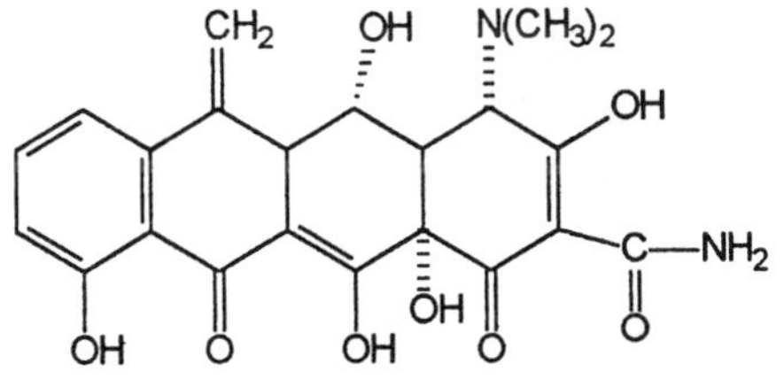

135.3

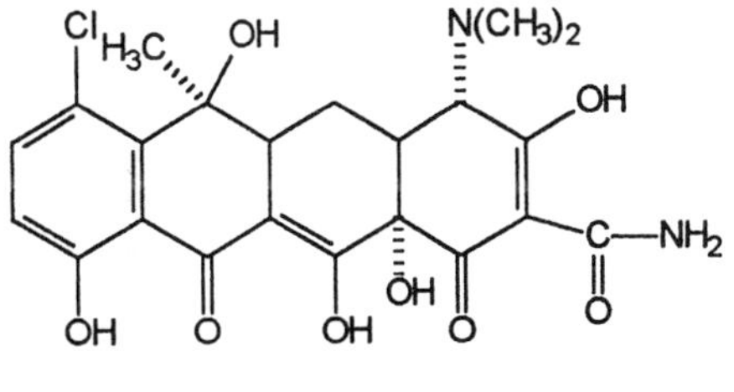

135.4

1. **Discuss the antibiotic potency and duration profiles of each of these molecules based upon their chemical structures.**

2. **Discuss the potential drug-drug interaction between any of these tetracycline structures and the Mylanta taken intermittently by this patient. What professional advice would you provide if both drugs needed to be taken?**

3. **Which tetracycline structure (135.1–135.4) would be most stable in this patient's stomach during a hyperacidity 'flare-up'? For one molecule not selected, diagram the mechanism of acid-catalyzed drug decomposition and show the final inactive product that could form.**

4. **Based on all of the preceding, would you dispense the original agent or call the physician to discuss substituting one of the alternative agents?**

Case 136

Submitted by Arnold R. Martin and Richard N. Herrier

PN is a 23-year-old woman who comes to the clinic with complaints of "having to run to the bathroom" frequently and pain and burning on urination. Since PN has a history of allergic reactions to a number of medications (penicillin and sulfonamides), her internist decides to treat the urinary tract infection with a tetracycline. A culture of PN's urine grew *Escherichia coli* and *Proteus mirabilis*.

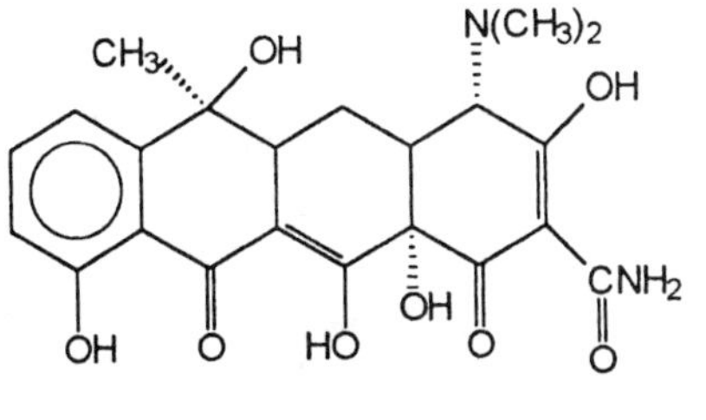

136.1

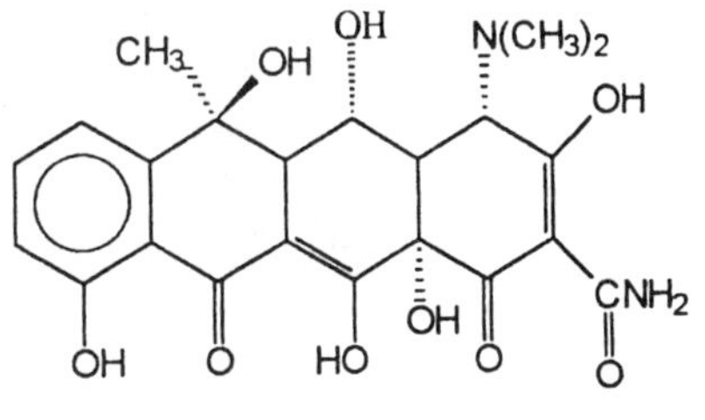

136.2

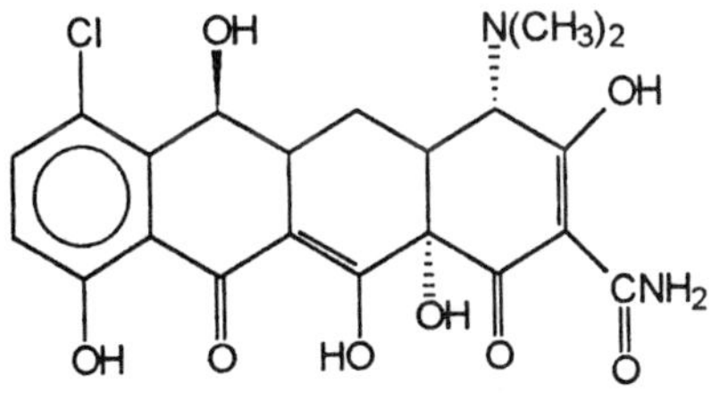

136.3

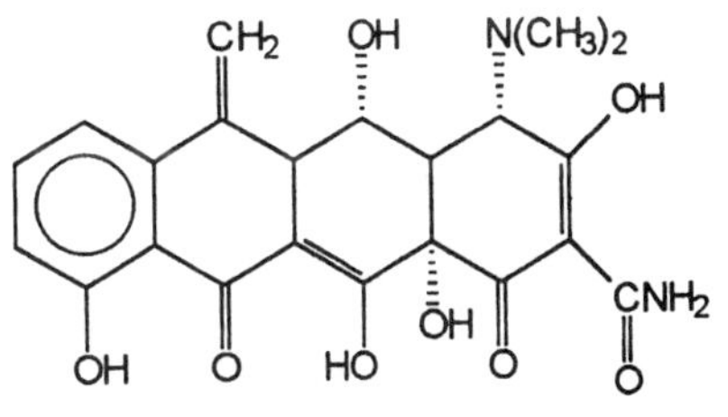

136.4

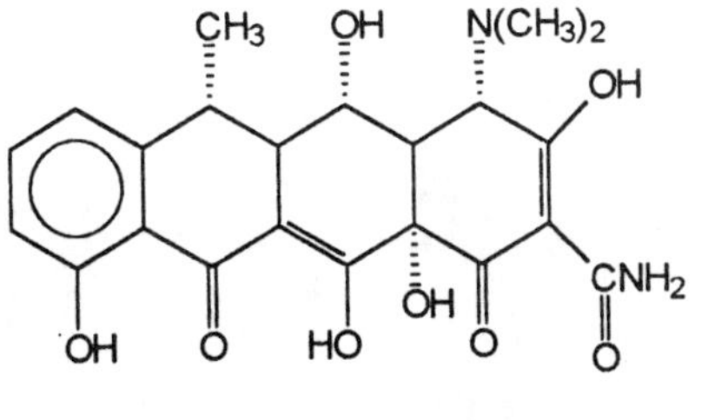

136.5

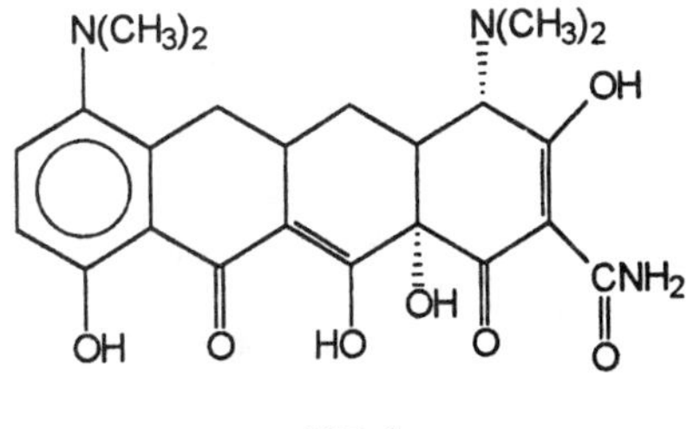

136.6

1. **Select a tetracycline derivative from the structures (136.1–136.6) provided that would be the most appropriate for the treatment of PN's acute urinary tract infection. Justify your choice on a pharmacokinetic basis. Explain the basis for your rejection of the remaining structures.**

Case **137**

Submitted by Arnold R. Martin and Richard N. Herrier

BN is a 17-year-old woman with a 3-year history of severe, treatment-resistant acne. Topical applications of both benzoylperoxide (10% cream) and tetracycline (137.1) have met with limited results. Bacterial culture of the comedone yielded *Propionibacterium acnes*, a common cause of acne. BN's dermatologist decides to attempt systemic treatment using a tetracycline analogue that has good tissue penetration.

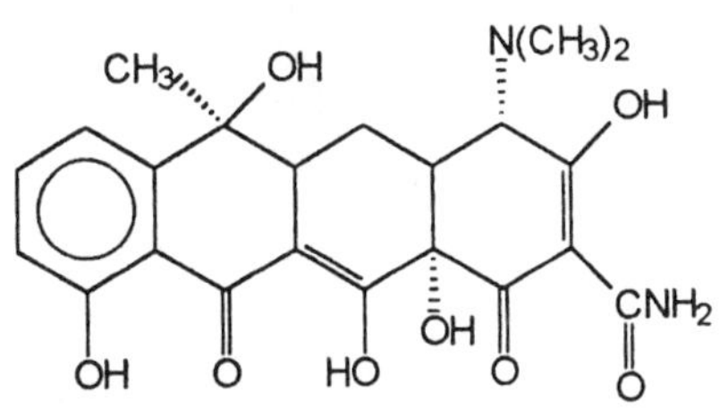

137.1

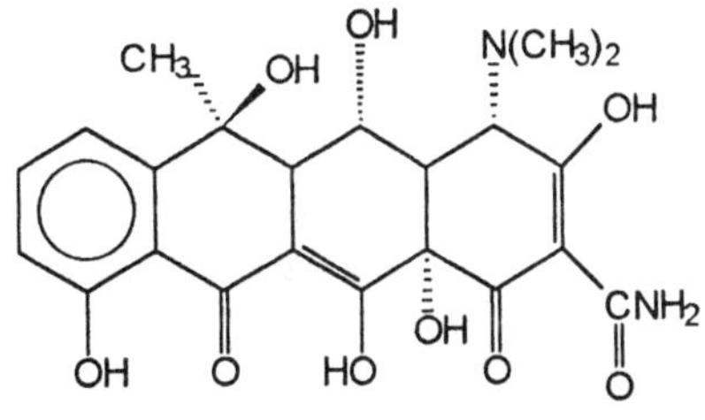

137.2

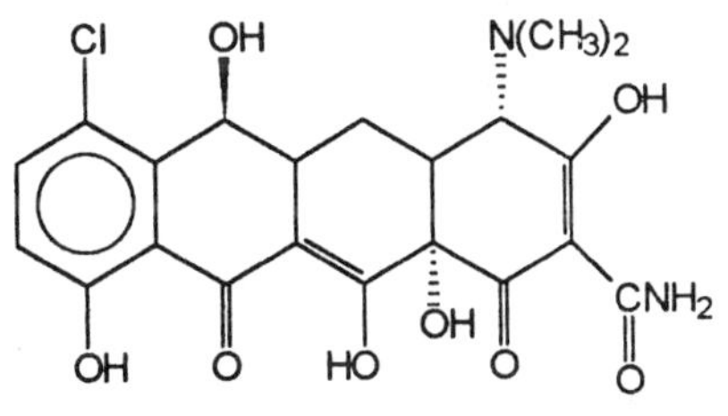

137.3

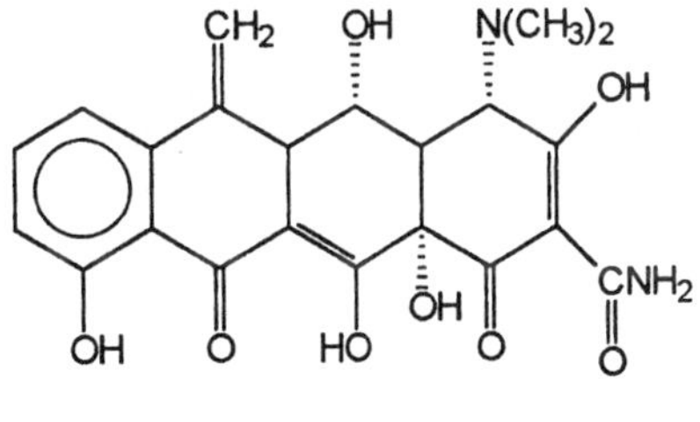

137.4

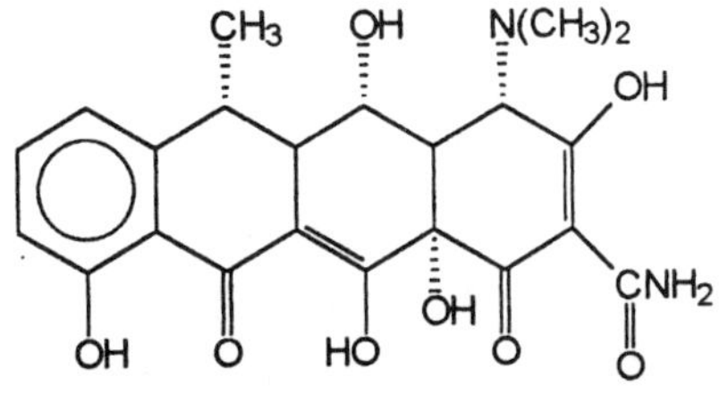

137.5

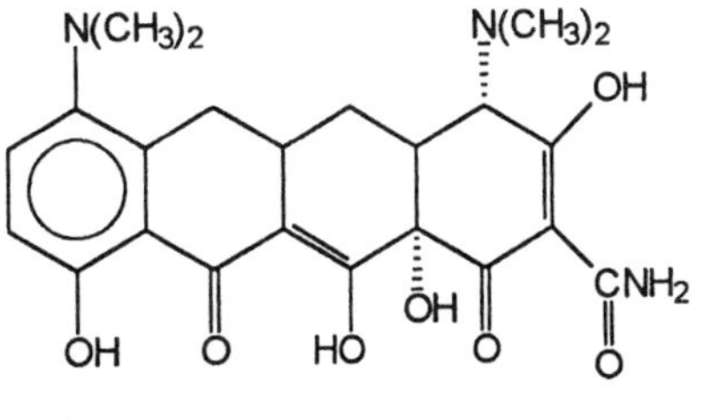

137.6

1. **Select the tetracycline derivative from the structures (137.1–137.6) provided that would be expected to provide the best treatment for BN's acne. Defend your choice based on the distribution properties of the various tetracyclines.**

Case 138

Submitted by Arnold R. Martin and Richard N. Herrier

A woman in her early 20s who is a regular customer comes into your pharmacy with a prescription for doxycycline (138.1) 100 mg with the directions "take one capsule daily for 2 weeks beginning the first day of the tour." The prescription is for her recently widowed grandmother, who plans to go on a 2-week tour to Egypt and East Africa. While waiting for you to fill the prescription, the young lady selects a bottle of Pepto-Bismol from the shelf of gastrointestinal remedies in front of the prescription counter. She comments that her grandmother requested that she pick up a bottle of the "pink stuff" when she got the prescription filled. She further states that her grandmother does not want to suffer like she did during her vacation in Mexico last year.

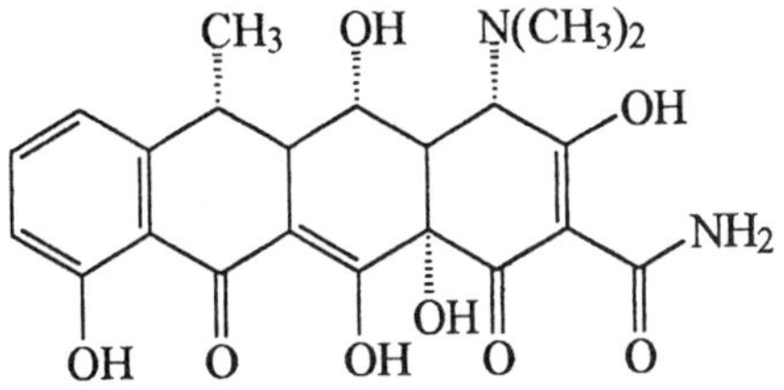

138.1

1. **What advice, if any, should be given to pass on to the patient? Explain fully.**

2. **What is the active ingredient in Pepto-Bismol? Does it have any value in the treatment of infectious diarrhea? Explain.**

Case 139

Submitted by Arnold R. Martin and Richard N. Herrier

A prominent senior citizen, BG, comes into the chain-store pharmacy in La Jolla, California, where you work, with a prescription for his wife AG. The prescription calls for twelve 500-mg ciprofloxacin (139.1) tablets, with the directions to take two tablets daily for diarrhea. The customer explains that his wife has just returned from a vacation in Acapulco, only to come down with a severe case of traveler's diarrhea on the last day. You recall that AG is a regular customer and make a quick check of her patient record to see if there are any drug interactions involving her other medications and the ciprofloxacin (139.1). You discover that AG is currently undergoing estrogen therapy (conjugated estrogens) and takes OTC calcium supplements to prevent osteoporosis.

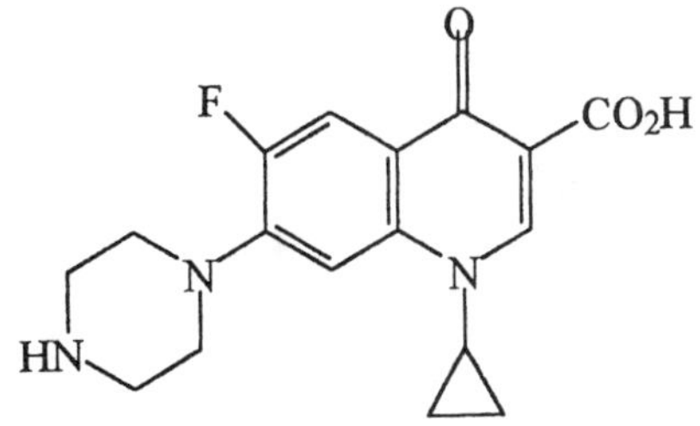

139.1

1. **What advice should you give to BG to convey to the patient regarding the ciprofloxacin (139.1) and the other medications she is taking? Explain fully.**

Case 140

Submitted by Arnold R. Martin and Richard N. Herrier

DG is a 56-year-old female paraplegic with a history of anemia, nephrolithiasis and recurrent urinary tract infections (UTIs). Her last three UTIs have been due to a combination of *Escherichia coli* or *Enterobacter cloacae* with indole-positive *Proteus* species. Today she returned for follow-up of her last UTI. A midstream clean catch urinalysis with culture and sensitivity (MSCC UA with C&S) done 2 days ago via a suprapelvic tap urine specimen is negative, but the urinary pH was 8.2. An unspun specimen today reveals no bacteria upon microscopic examination, but has a strong ammonia odor and a pH of 8.0. To prevent recurrence, the doctor decides to begin Methenamine hippurate (140.1) prophylaxis 1 gram twice daily with food and acetohydroxamic acid (140.2) 250 mg 4 times daily. DG takes ferrous sulfate 325 mg three times daily for her anemia.

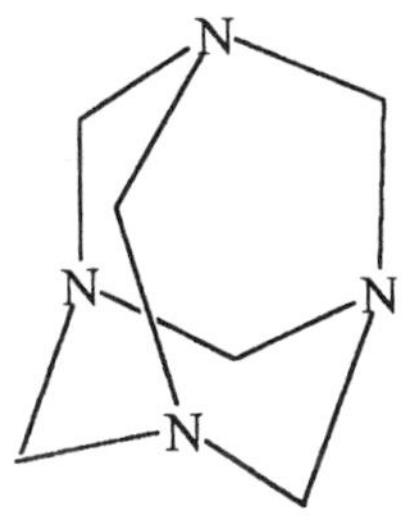

140.1

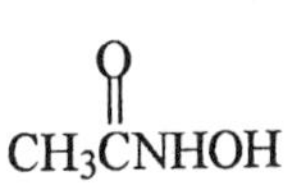

140.2

1. **What is the most likely causative organism of the chronic UTI? Explain.**

2. **What is the mechanism of antibacterial action of methenamine hippurate (140.1)? What advice should be given to the patient regarding its use?**

3. **What is the purpose of acetohydroxamic acid (140.2)? Explain fully.**

4. **Are all of the drugs compatible?**

5. **Provide a means by which any identified incompatibility can be avoided without compromising the health of the patient.**

Case 141

Submitted by Marc W. Harrold

TL is a 62-year-old woman who has periodic asthmatic attacks. She has responded very well to theophylline treatment; however, she is now complaining of a fever, a chronic cough, and a periodic shortness of breath. Her physician suspects a respiratory infection, and in the absence of any culture results wants to cover a wide spectrum, including *Hemophilus influenzae, Streptococcus pneumoniae,* and *Pseudomonas aeruginosa*. He wants to prescribe one of the following agents and has asked for your opinion.

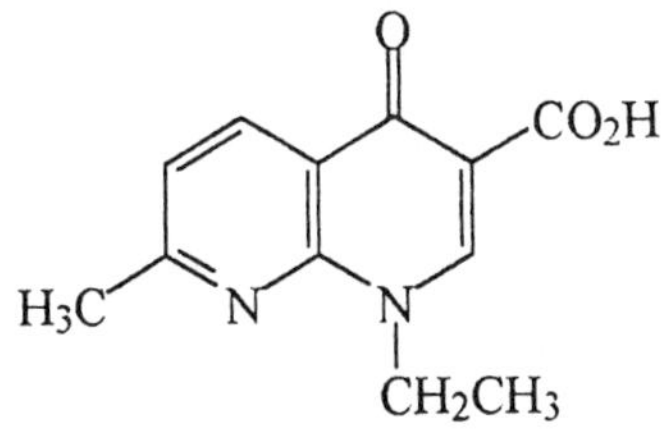

141.1

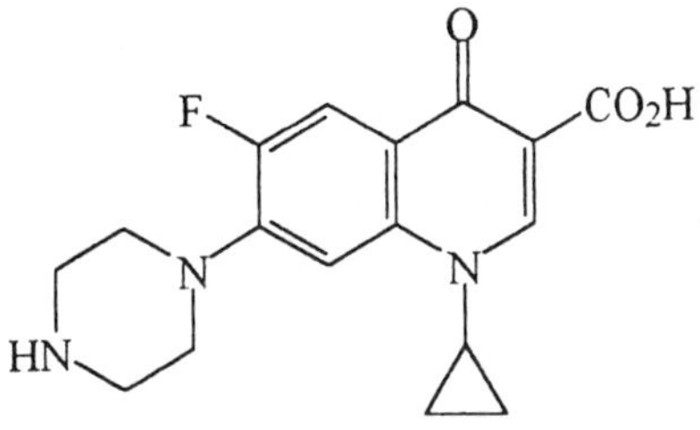

141.2

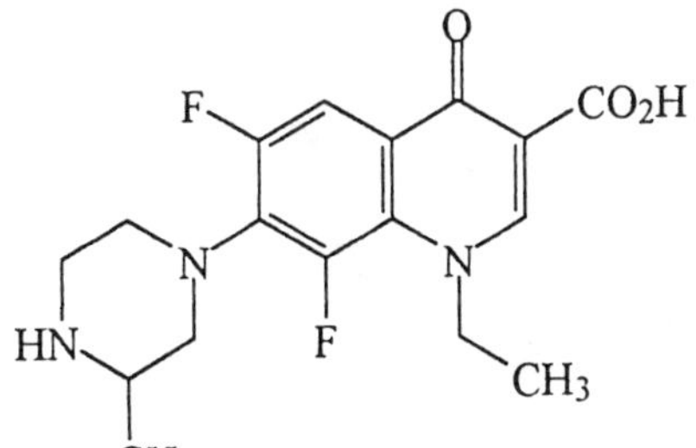

141.3

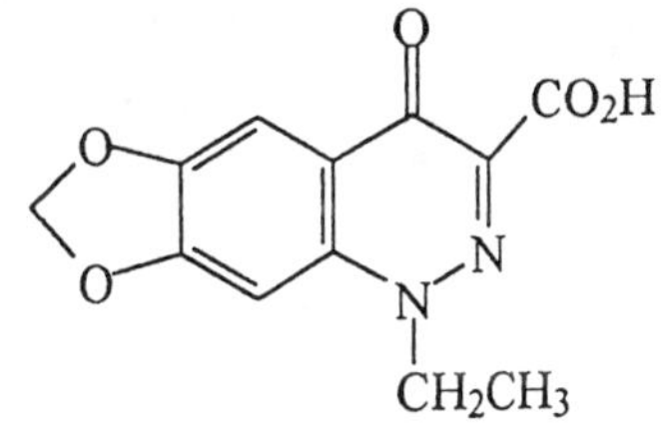

141.4

1. Provide the chemical classification and mechanism of action for these compounds (141.1–141.4).

Case **142**

Submitted by Bruce L. Currie

SR, a 42-year-old man, has recently returned from filming a new movie in the jungles of Cambodia. After several days of alternating high fever and chills, he presented to the clinic at the studio. The physician noted the recent travel in taking the patient's history and suspected malaria or other tropical disease. Blood culture confirmed the diagnosis of chloroquine-resistant malaria (*Plasmodium falciparum*). The following treatment regimen was begun immediately:

quinine sulfate (142.1)	650 mg q 8h X 3 d
pyrimethamine (142.2)	75 mg single dose
sulfadoxine (142.3)	1500 mg single dose

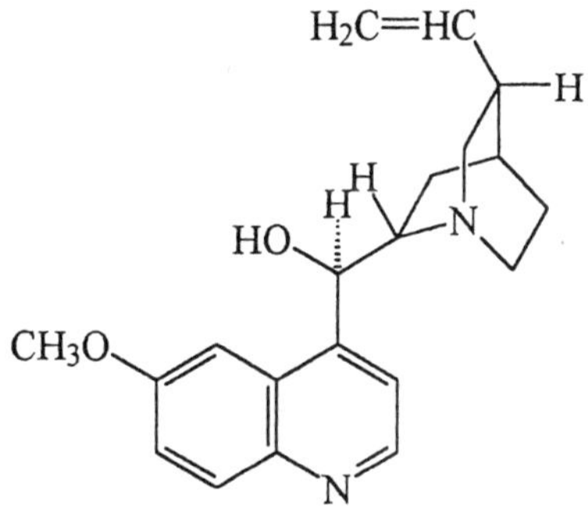

142.1

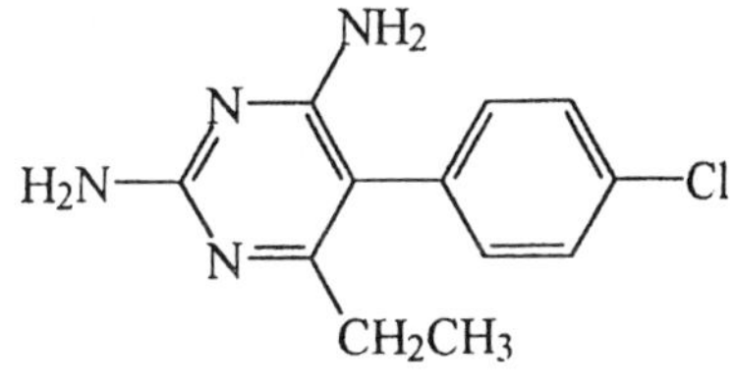

142.2

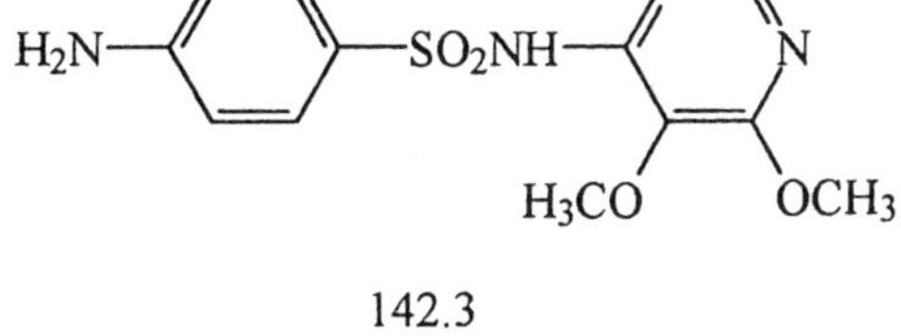

142.3

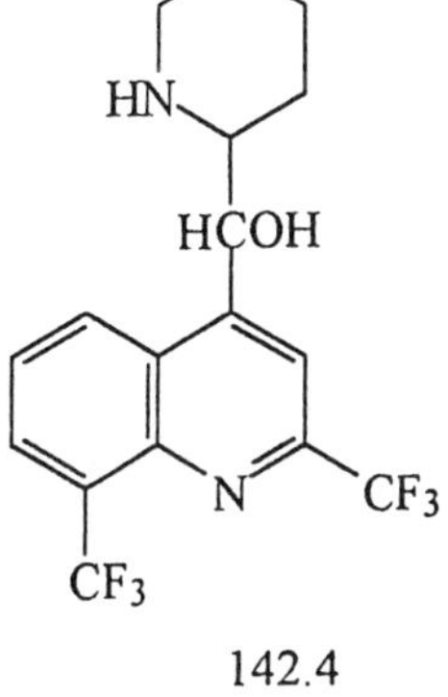

142.4

1. **What are the mechanisms of action of pyrimethamine (142.2) and sulfadoxine (142.3) with regard to the treatment of malaria? Discuss whether these two drugs are a rational combination together and with quinine (142.1).**

2. **What could SR have done (using prescription drugs) to avoid contracting malaria while making this movie? Suggest at least two approaches and discuss each briefly.**

Case 143

Submitted by Arnold R. Martin and Richard N. Herrier

GT is a 12-year-old African-American male who presents to the emergency room with severe pain and tenderness in the left lower leg. He also complains of pain in the left knee, but states that the leg hurts more. His leg pain has become progressively worse during the past week. GT has a history of sickle cell anemia and has been admitted to the hospital in the past for supportive treatment and management of sickle cell crisis. GT remembers that he fell on the playground while playing basketball about 2 weeks ago. He thought the pain he had been experiencing was due to his disease. The acetaminophen that he had been taking helped to relieve his pain initially, but it does not appear to be effective now. Blood cultures grew *Salmonella typhimurium*. Antibiotic sensitivity results will be available in 24 hours. An x-ray of the left tibia was suggestive of osteonecrosis, and a 4-phase bone scan confirmed osteomyelitis at that site.

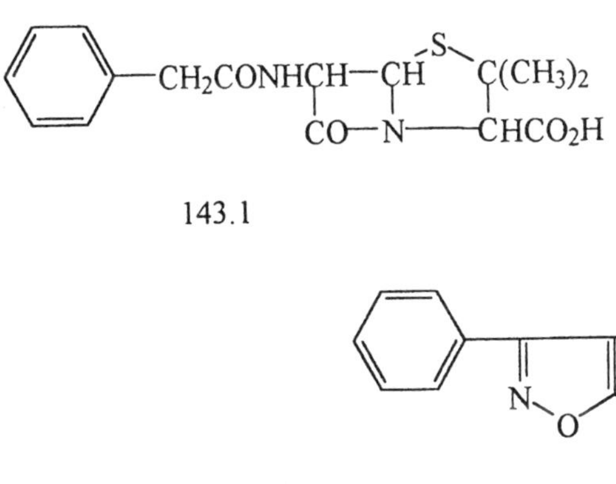

143.1

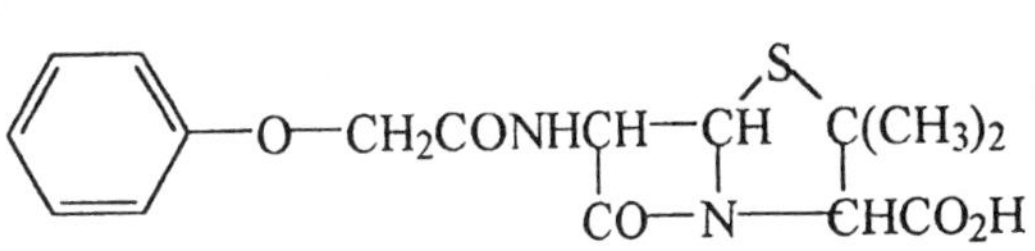

143.2

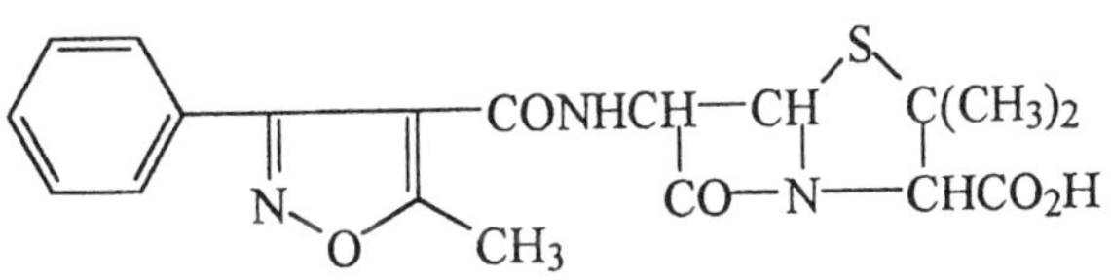

143.3

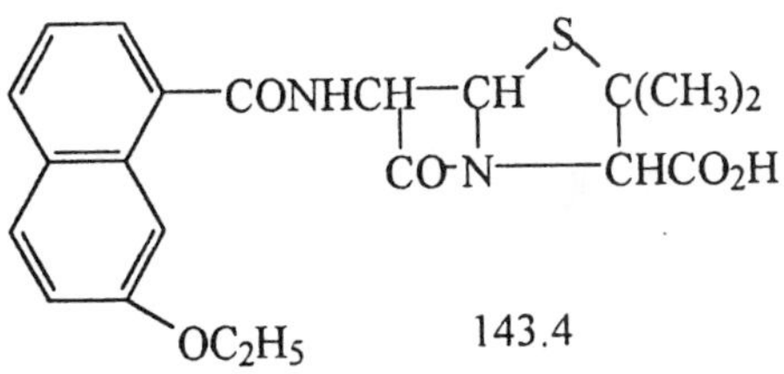

143.4

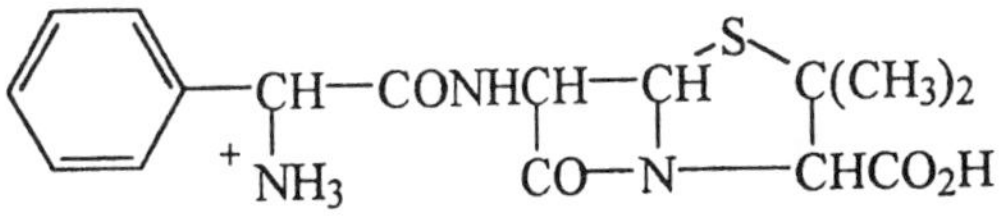

143.5

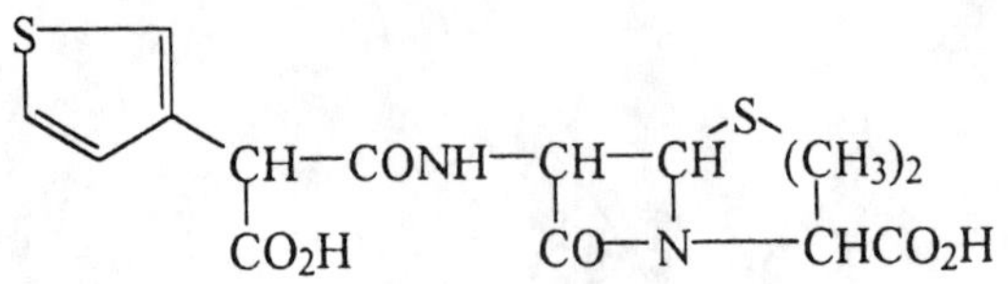

143.6

1. **Select an appropriate penicillin derivative (143.1–143.6) to treat GT's osteomyelitis along with the preferred route of administration and chemical form. Justify your choice on the comparative microbiologic properties of the chosen agent *vs* the remaining drugs.**

2. **Assume that sensitivity tests later reveal that your initial choice is ineffective. Explain how this could happen. Would this mean that no penicillin derivative, under any circumstances, would be effective in this case? Explain.**

3. **Suggest other antibiotics or chemotherapeutic agents that could be employed if the patient were allergic to penicillin.**

Case 144

Submitted by Arnold R. Martin and Richard N. Herrier

A 56-year-old male chemistry professor, who has recently had a squamous cell carcinoma removed from his upper torso at the university dermatology clinic, returns to the clinic complaining that his surgical wound has become infected. The wound is tender, red and swollen, and, on manipulation, releases large amounts of creamy, yellow pus. A Gram's stain reveals large numbers of gram-positive cocci in clumps. Culture and sensitivity results will be available in 48 hours. Since the professor is otherwise healthy and insists that he has classes to teach, the attending dermatologist decides to treat him on an outpatient basis using a penicillin derivative.

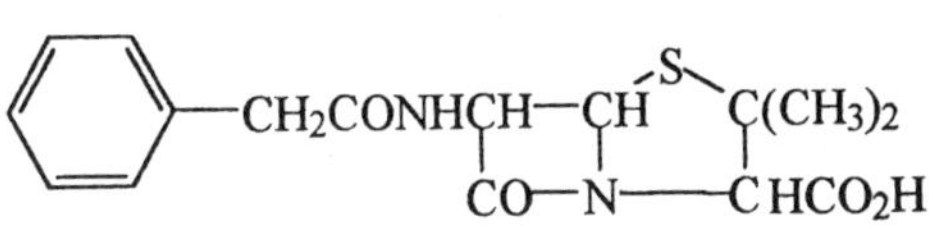

144.1

144.2

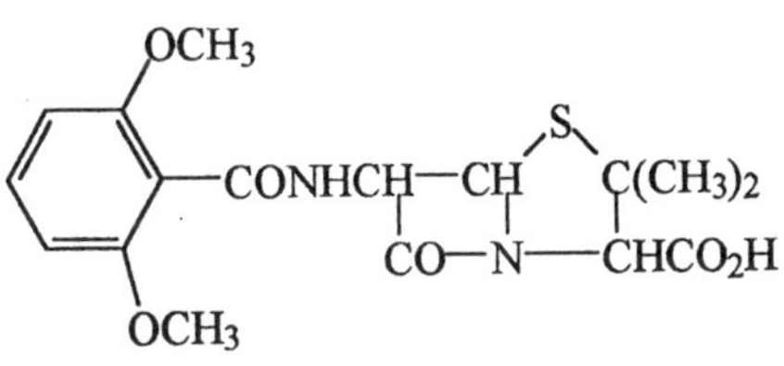

144.3

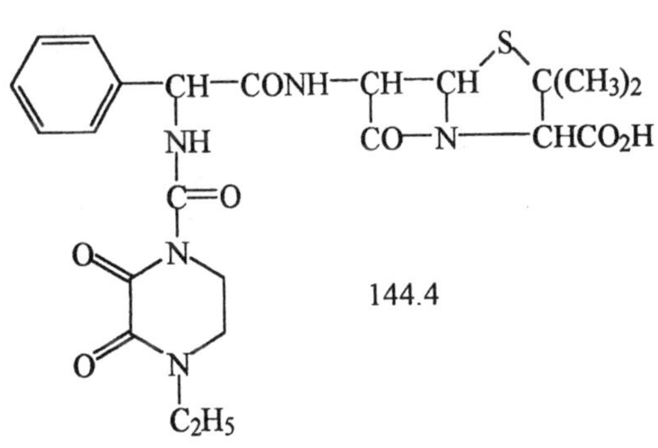

144.4

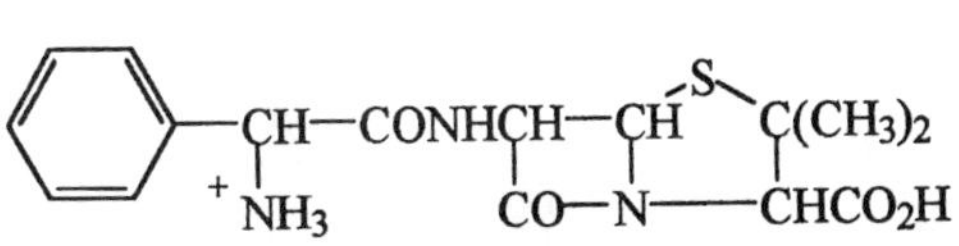

144.5

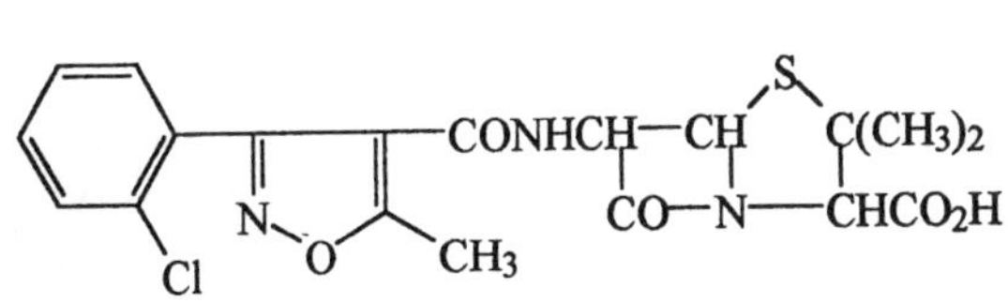

144.6

1. **What is the likely causative organism in this case?**

2. **What initial treatment should be instituted? Select the most appropriate penicillin from the structures (144.1–144.6) provided. Justify your choice on the basis of the properties of each.**

Case 145

Submitted by Victoria F. Roche

SH, a 50-year-old man, visits your clinic with a severely infected wound that, upon culture, reveals the presence of penicillinase-producing *Staphylococcus aureus*. The clinic physician wishes to initiate po antibiotic therapy, and seeks your advice. Upon taking a history, you find that SH had previously developed fever and a severe rash when given IM potassium penicillin G.

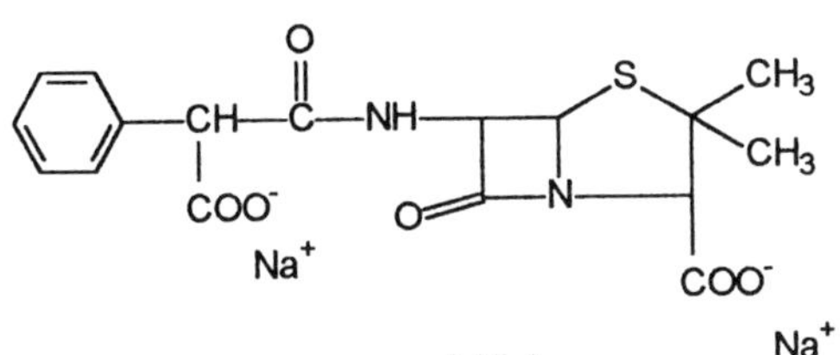

145.1

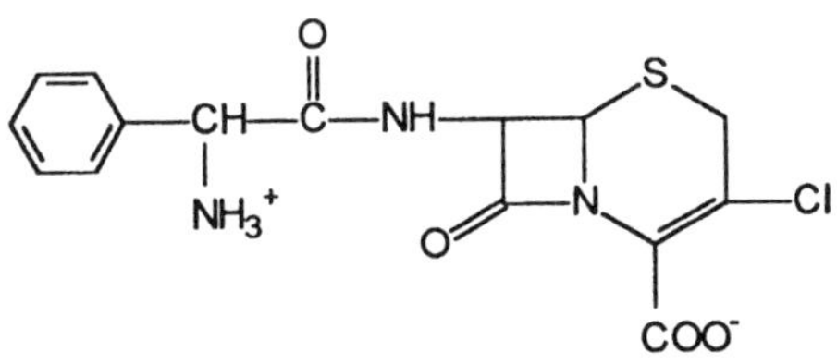

145.2

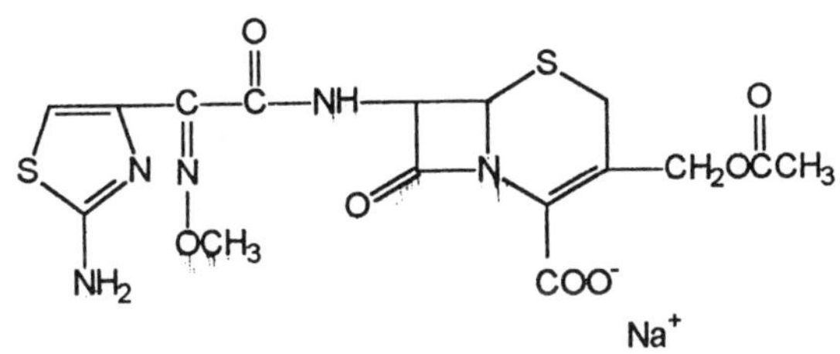

145.3

1. What is the significance of the fever and rash that SH had previously experienced?

2. **Which antibiotic structure (145.1–145.3) of the three provided would you recommend? What structural considerations led to your answer? What precautions would you observe when initiating therapy?**

3. **Why not recommend a tetracycline antibiotic for SH?**

Case **146**

Submitted by Arnold R. Martin and Richard N. Herrier

BJ is a 56-year-old man with chronic obstructive pulmonary disease (COPD) and a history of heavy alcohol use who presents to the emergency room with fever, severe right-sided chest pain, shortness of breath (SOB), and an increased heart rate.

BJ has a long history of COPD, related to his 40 years of smoking. He uses two inhalers, albuterol and ipratroprium, when he is short of breath. BJ was last hospitalized for an exacerbation of his COPD 2 months ago and has been taking oral prednisone 30 mg daily ever since. BJ joined Alcoholics Anonymous 4 years ago after suffering several attacks of delirium tremens. It was at an AA meeting where he met his second wife Alice, who was his inspiration to quit alcohol and who now is helping him to kick the smoking habit. Everything was going well until 2 weeks ago, when BJ's leg began to swell. A deep vein thrombosis (DVT) was diagnosed, and BJ was treated and released with a prescription for oral Warfarin 5 mg daily.

Today BJ presents to the emergency room after 2 days of spiking fevers, chest pain, SOB, malaise, and tachycardia. Physical assessment reveals BP, 140/90; RR, 28; HR, 125; T, 39C. A Gram's stain of the sputum culture showed short, plump gram-negative rods, later identified as a non-β-lactamase producing strain of *Klebsiella pneumoniae*. The attending physician elects to treat BJ's pneumonia with an injectable cephalosporin, but is uncertain as to choice because BJ is currently taking Warfarin to prevent DVT.

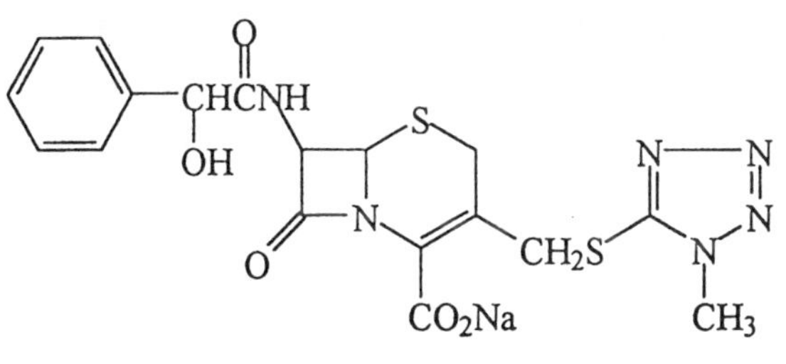

146.1

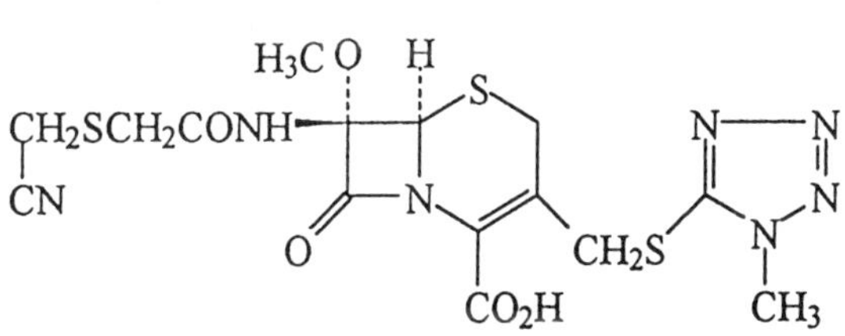

146.2

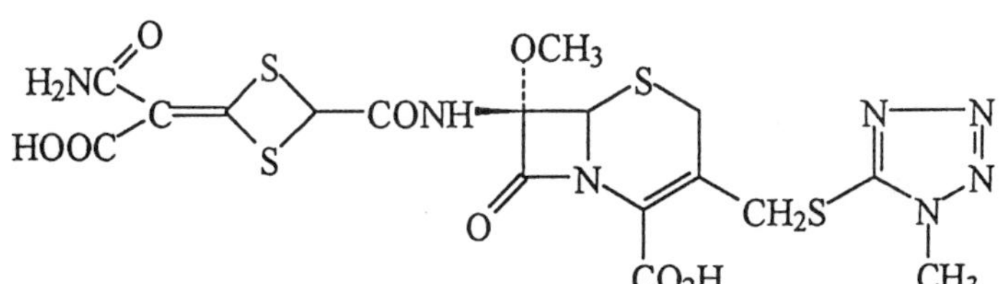

146.3

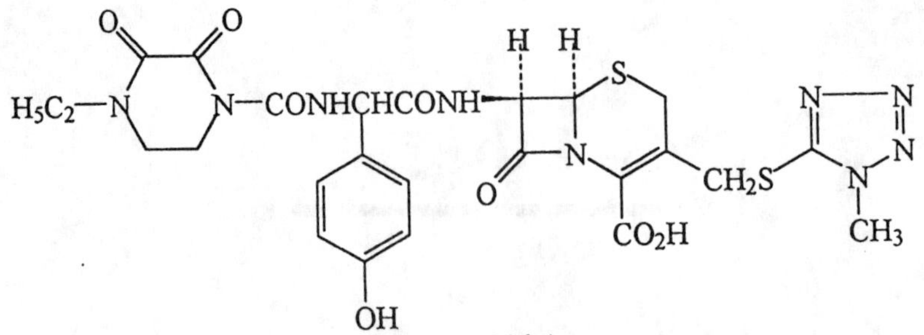

146.4

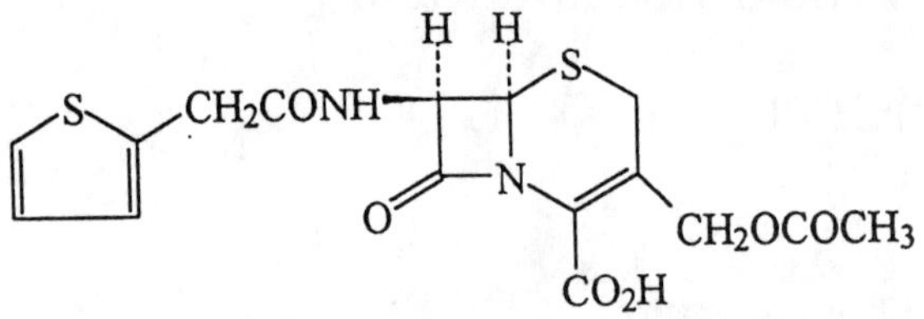

146.5

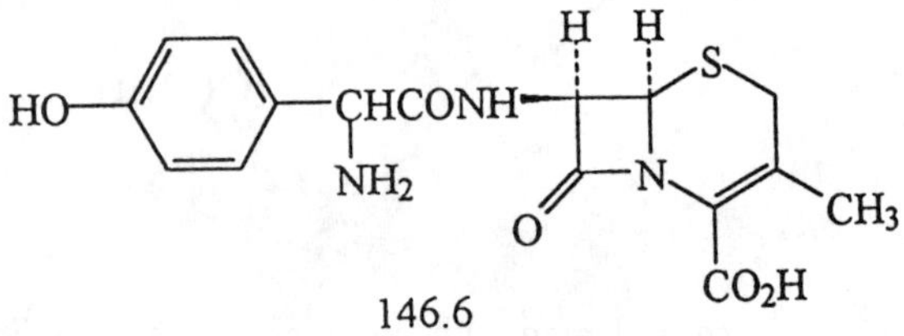

146.6

1. **What is the basis for the physician's concern with regard to the concomitant use of certain cephalosporins with Warfarin? Explain fully.**

2. **Select a cephalosporin from the structures (146.1–146.6) provided that can be safely and effectively employed in this case. Justify your choice, eliminate inappropriate choices, and explain why they were eliminated.**

Case 147

Submitted by Denise L. Howrie and Paul L. Schiff, Jr.

SA is a 4-year-old hospitalized with fever, petechiae, stiff neck, and confusion and treated with ceftriaxone sodium (Rocephin) (147.1) for presumed *Neisseria meningitidis* meningitis. Confirmatory cultures are obtained within 24 hours, and the child's mother and father are prescribed Rifampin (147.2) 300 mg po bid x 2 days as a prophylactic measure. As their pharmacist, you determine that both parents are in generally good health. Mr. A has a history of seizures since a car accident last year; the seizures are well-controlled with phenytoin 300 mg po qd. Mrs. A has no medical problems but wears contact lenses and is taking Ortho-Novum 1/50 q d.

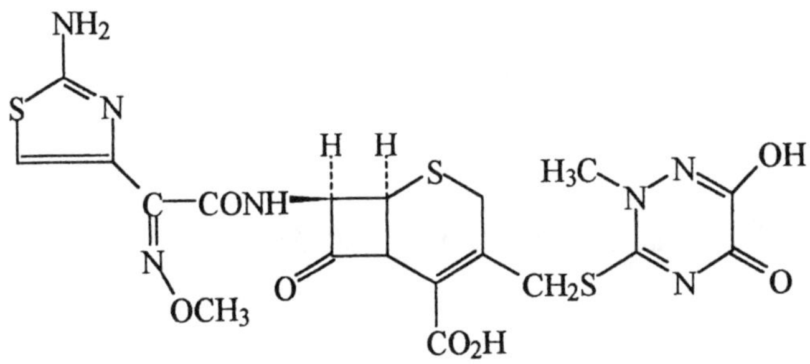

147.1

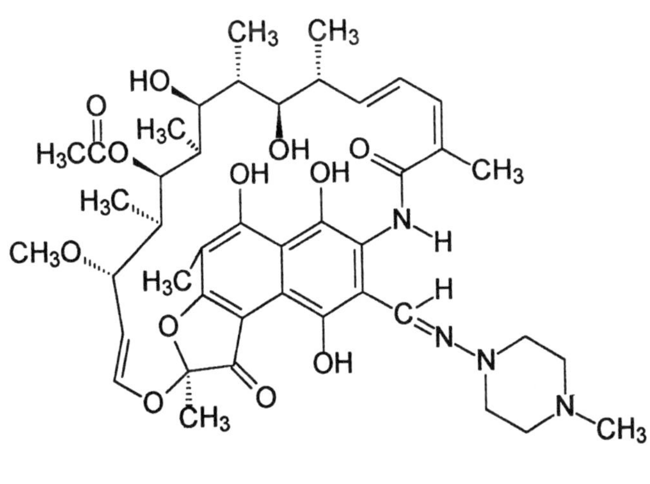

147.2

1. What specific patient counseling would you provide as the family's pharmacist?

2. What potential problems can you foresee from this limited medication history?

3. Is there any significance to the concurrent drugs?

Case 148

Submitted by Arnold R. Martin and Richard N. Herrier

JD is a 27-year-old prostitute who is admitted to the hospital with fever and severe abdominal pain. She has experienced generalized pain for the past 3 days and has lost her appetite. In addition to the pain in the lower abdomen, JD reports a brown, foul-smelling vaginal discharge. JD reveals that she has been a prostitute for the past 8 years and has been treated on three different occasions in the past 18 months for sexually transmitted diseases: twice for gonorrhea and once for syphilis. Laboratory results show gram-negative cocci, and tests for *Chlamydia* are pending. A diagnosis of pelvic inflammatory disease (PID) is made, and the doctor decides to administer ceftriaxone (148.7) (250 mg IM) to JD and release her with a prescription for an oral tetracycline derivative.

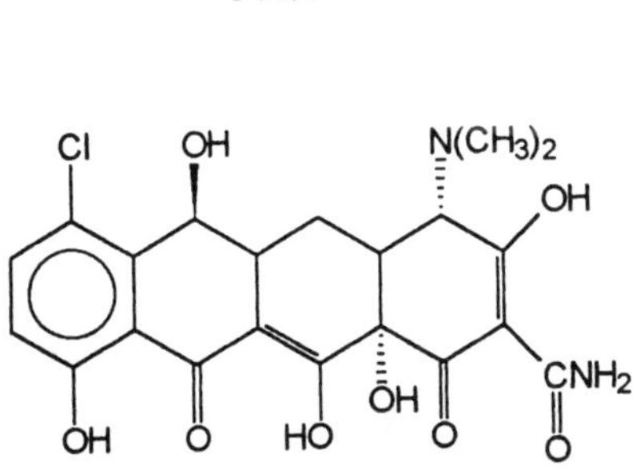

148.1

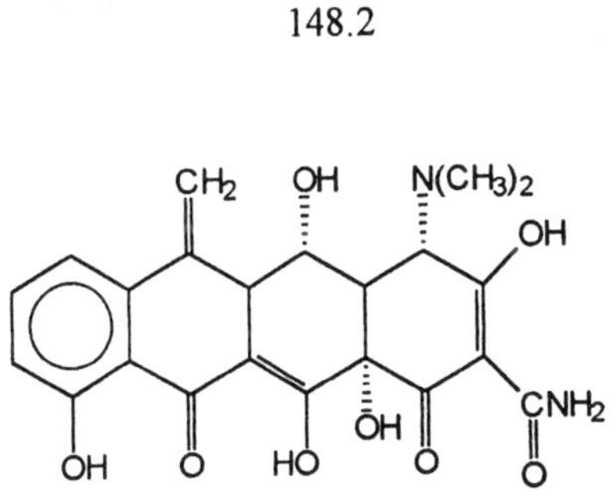

148.2

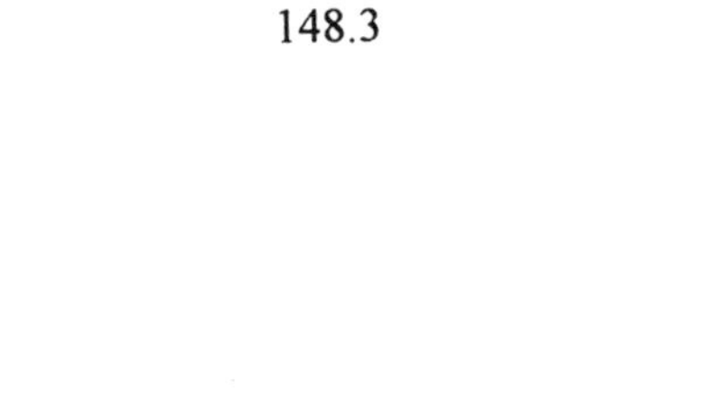

148.3

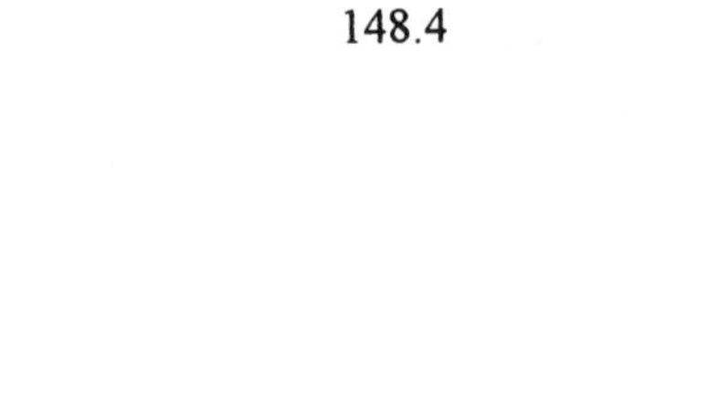

148.4

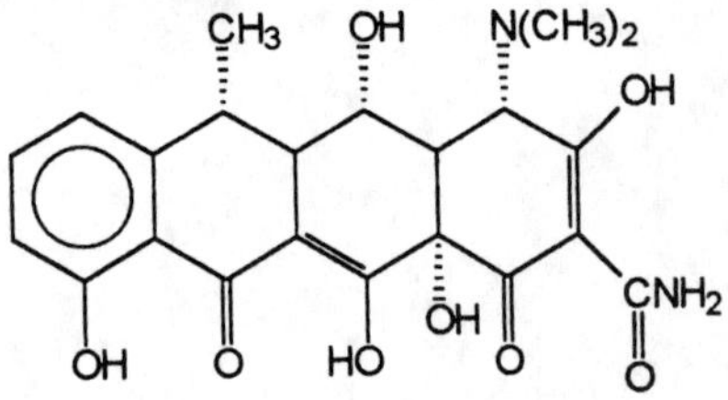

148.5

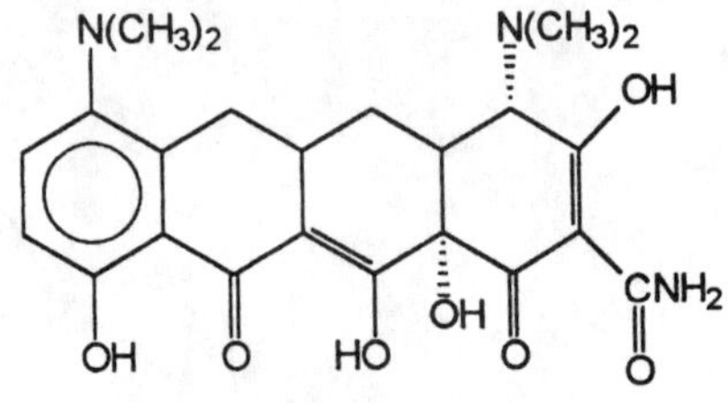

148.6

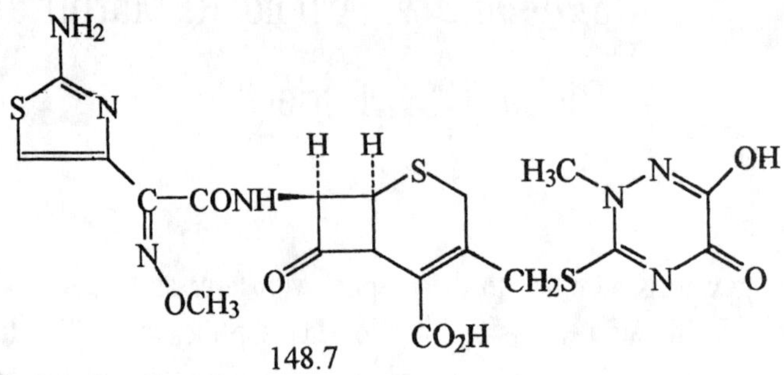

148.7

1. **What are the most likely causative microorganisms of JD's PID based on the available information?**

2. **Why is ceftriaxone specifically employed in this case?**

3. What is the purpose of the oral tetracycline derivative?

4. Select the most appropriate tetracycline from the structures (148.1–148.6) provided. Justify your choice on the basis of the pharmacokinetic properties of the various tetracyclines available.

Case 149

Submitted by Arnold R. Martin and Richard N. Herrier

You, as a staff pharmacist at the VA hospital, are called in to consult on a case involving a 55-year-old municipal bus driver, who has been admitted with an acute urinary tract infection. Past medical history includes rheumatic fever with eventual mitral valve replacement and a history of UTIs due to a congenital inflammation of the urinary tract that is not surgically correctable. Blood and urine cultures confirm group D *Streptococcus* (enterococcus). Because of the microorganism's inherent resistance to single drug treatment and the potential for subacute endocarditis, a combination of penicillin with an aminoglycoside is indicated. Five years ago, the patient experienced a gentamicin-induced elevation in serum creatinine while being treated for a urinary tract infection. Current serum creatinine level is 1.3, with a creatinine clearance of 75 mL/min.

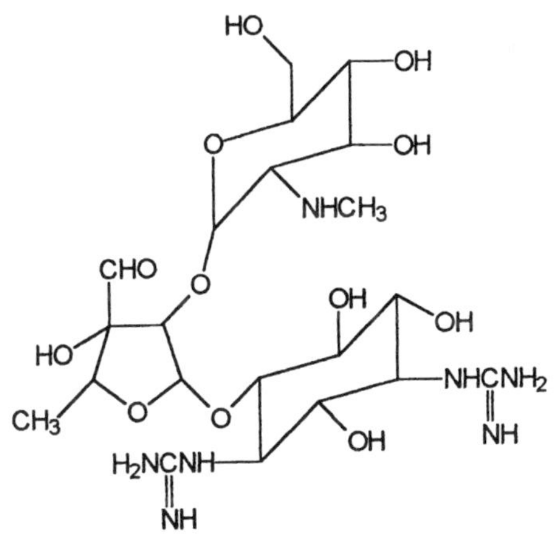

149.1

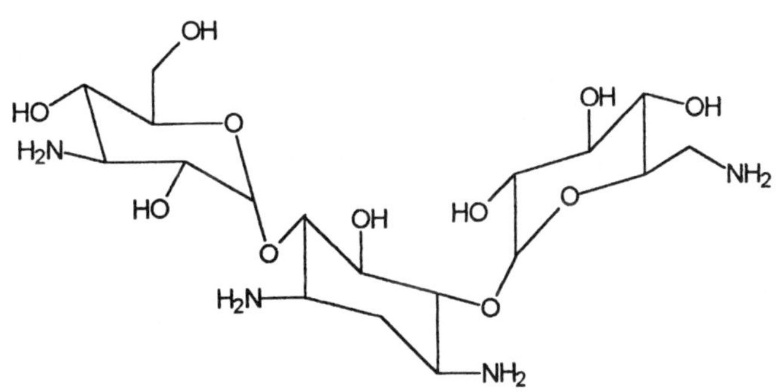

149.2

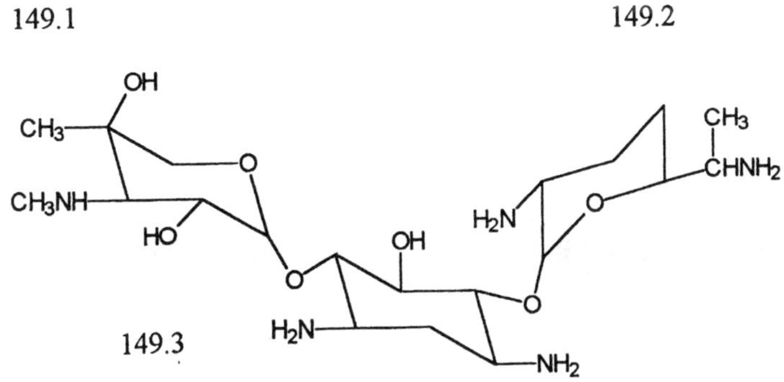

149.3

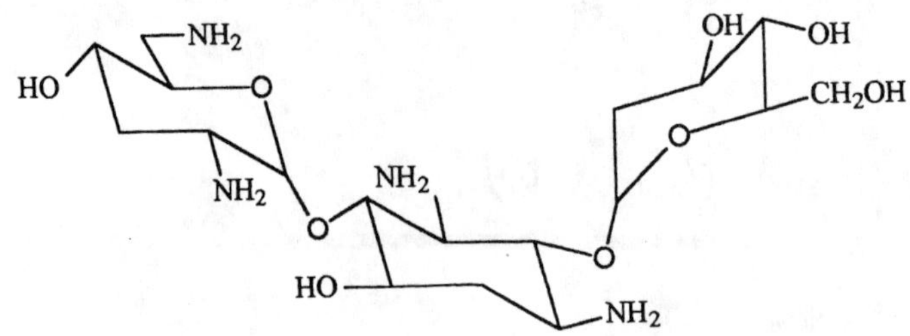

149.4

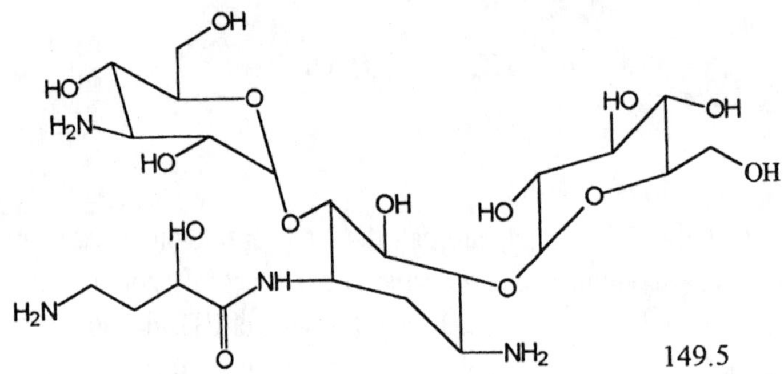

149.5

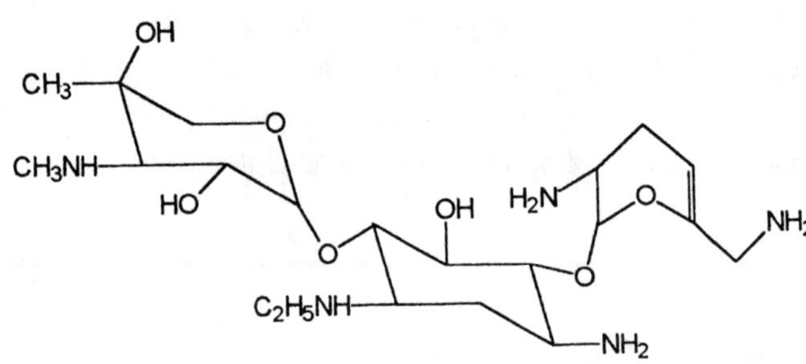

149.6

1. **Why is a combination of antibiotics indicated for this patient? Explain on the basis of the mechanism(s) of action of the antibiotics.**

2. **What are the causes for the physician's concern? Are these concerns justified?**

3. **Select an aminoglycoside from the structures provided; streptomycin (149.1), kanamycin (149.2), gentamicin (149.3), amikacin (149.5), netilmicin (149.6), tobramycin (149.4), that when combined with penicillin G, would maximize the benefit to the patient and minimize the risks. Provide a rationale for your choice based on the properties of the aminoglycoside chosen and explain why each of the remaining aminoglycosides were rejected.**

4. **Will dose adjustment(s) of the two antibiotics be required? Explain.**

Case **150**

Submitted by Arnold R. Martin and Richard N. Herrier

DN is a 24-year-old chemistry graduate student who was admitted to the burn unit 1 week ago with severe burns over 60% of his body, resulting from an explosion and fire in the laboratory. Because of lung involvement, he was immediately intubated after admission. His burns were treated topically with silver sulfadiazine cream and parenteral antibiotics, Vancomycin IV (because his medical history revealed allergies to penicillin and cephalexin), and tobramycin (150.4) IM prophylactically. DN remained afebrile for 5 days, but became increasingly disoriented, and his fever spiked to 104°F. Sputum cultures grew *Staphylococcus aureus* that was methacillin-resistant but vancomycin-sensitive, and *Pseudomonas aeruginosa* that was resistant to Cotrimoxazole (trimethoprim/sulfamethoxazole), tobramycin, and amikacin. Given the resistance pattern to the aminoglycosides, you, as the consulting pharmacist, suspect enzyme inactivation, and a quick check of the literature on your Micromedex computer program reveals that bacterial strains of certain *Enterobacteriaceae* and *Pseudomonas* species elaborate a 4,4′-nucleotidyl transferase that inactivates both tobramycin (150.4) and amikacin (150.5).

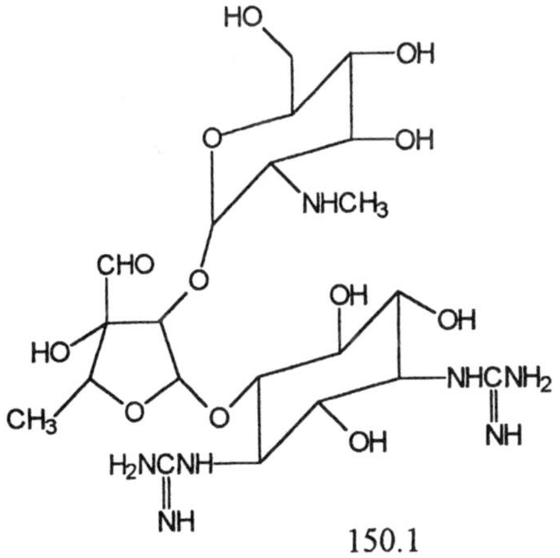

150.1

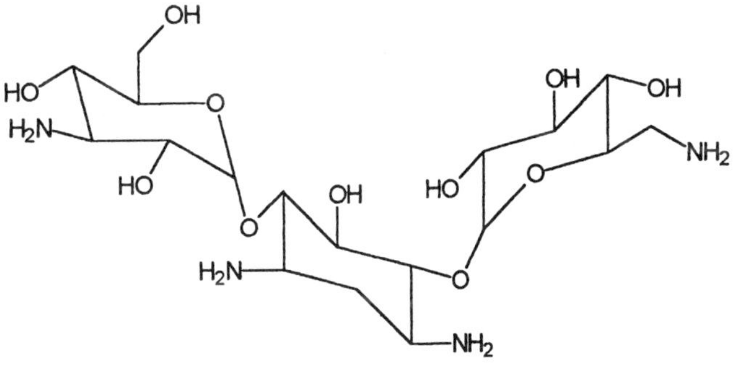

150.2

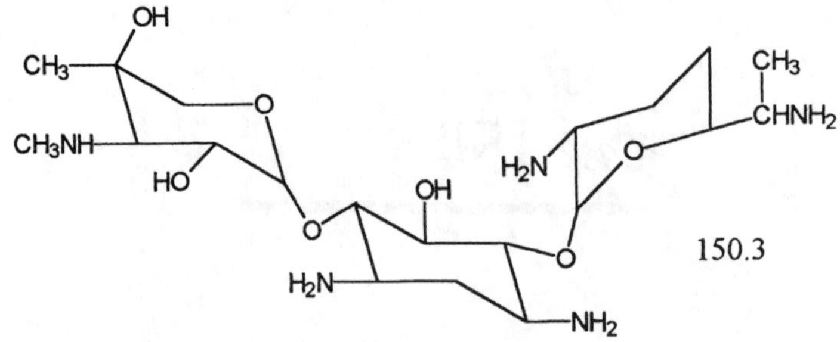

150.3

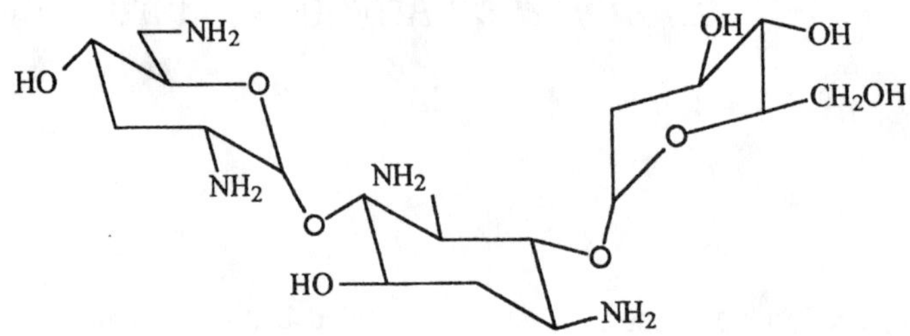

150.4

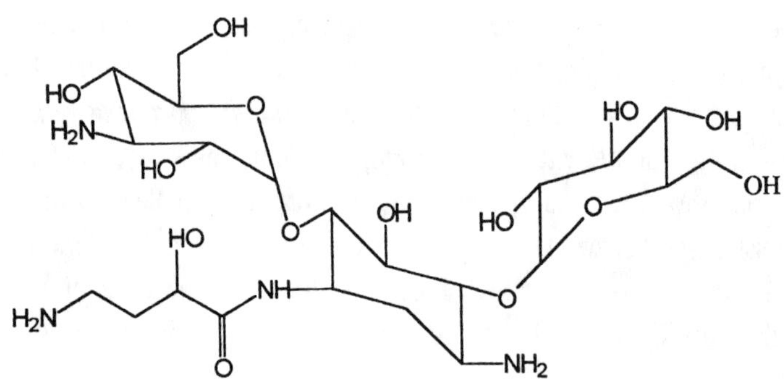

150.5

1. **Select an aminoglycoside (AG) antibiotic from the structures provided; streptomycin (150.1), kanamycin (150.2), gentamicin (150.3), tobramycin (150.4), amikacin (150.5), that is likely to be effective against the particular strain of *Pseudomonas aeruginosa* infecting DG. Justify your choice on a structural basis. Explain why none of the remaining structures are likely to be effective.**

2. **List additional aminoglycoside inactivating enzymes that can be elaborated by various bacterial strains. Is your choice of antibiotic likely to be effective against these strains? Explain fully.**

Case **151**

Submitted by Denise L. Howrie and Paul L. Schiff, Jr.

TW is a 14-year-old boy diagnosed in a local emergency room with "walking pneumonia" and prescribed erythromycin (151.1) 250 mg po qid. TW's father, Mr. W, is concerned about cost, as the family has no prescription plan and must pay for all drugs. He is additionally troubled when you explain that nausea, vomiting, abdominal cramps, and diarrhea are relatively common side effects of therapy with erythromycin (151.1), because his son has "a sensitive stomach." You mention two more expensive alternatives, clarithromycin (Biaxin)(151.2) and azithromycin (Zithromax) (151.3), that produce minimal to negligible gastrointestinal upset. As you prepare to dispense erythromycin (151.1), you ask whether TW has any other medical problems. As Mrs. W supervises TW's other medications, Mr. W only knows that his son has asthma and "hayfever" treated with prescriptions prescribed by his pediatrician. A computer check shows TW receives cromolyn sodium (Nasalcrom) 2 puffs bid, albuterol (Ventolin) 2 puffs qid prn, terfenadine (Seldane) 30 mg qhs, and theophylline (Theo-Dur) 200 mg bid.

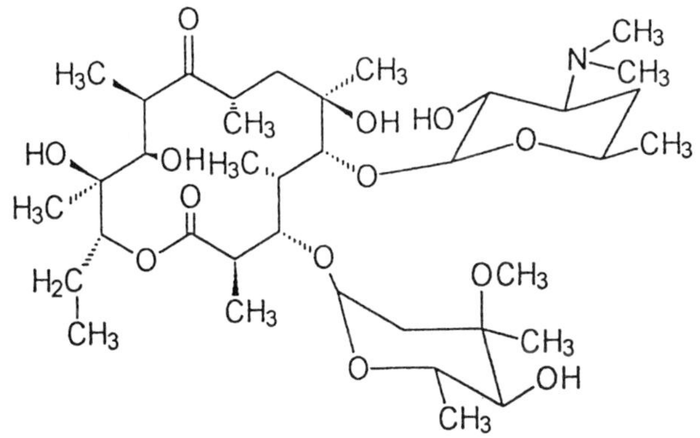

151.1

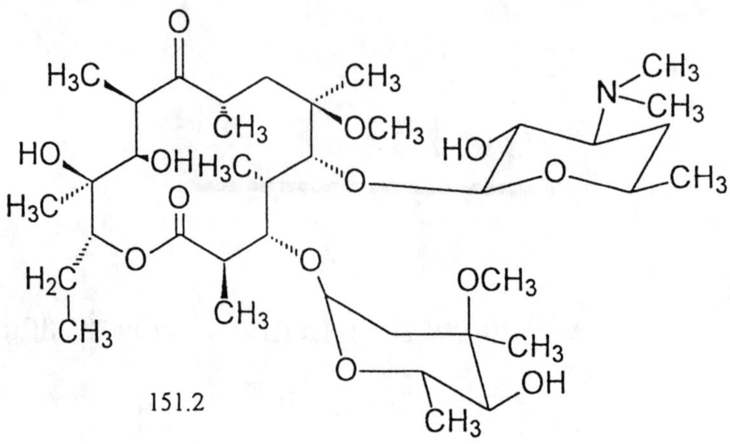

151.2

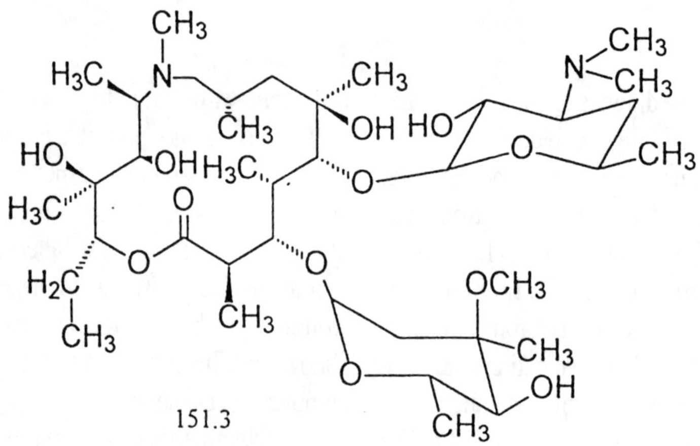

151.3

1. **Discuss structural changes and mechanisms of action of these three macrolides (151.1–151.3) relative to antimicrobial potency under acidic conditions and gastro-intestinal side effects.**

2. **Is the cost differential significant enough for the three agents to warrant a trial of erythromycin (151.1) as prescribed?**

3. **What potential problems can you foresee from this limited medication history? Of what significance, if any, are the concurrent drugs?**

4. **How would you proceed as the family's pharmacist?**

Case **152**

Submitted by Bruce L. Currie

JM, a 68-year-old woman, was recently diagnosed with non-Hodgkin's lymphoma. Immediately after diagnosis, she was treated with the following therapeutic regimen known as M-BACOD:

methotrexate (152.1)	3000 mg/m^2 IV with leucovorin (152.7) Day 14
bleomycin (152.3)	4 mg/m^2 IV Day 1
doxorubicin (152.6)	45 mg/m^2 IV Day 1
cyclophosphamide (152.4)	600 mg/m^2 IV Day 1
vincristine (152.5)	1 mg/m^2 IV Day 1
dexamethasone (152.2)	6 mg/m^2 po Days 1 to 5

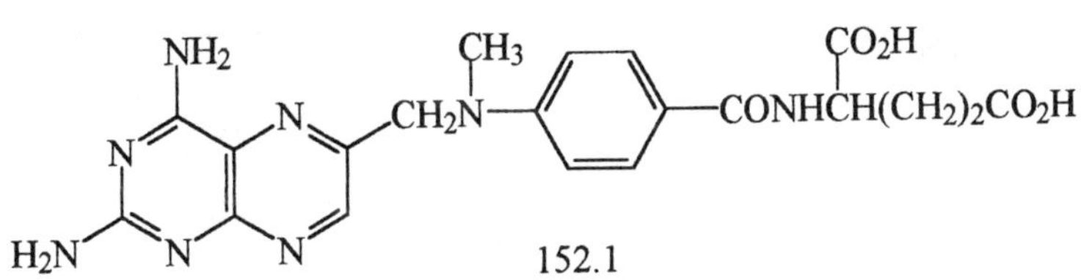

152.1

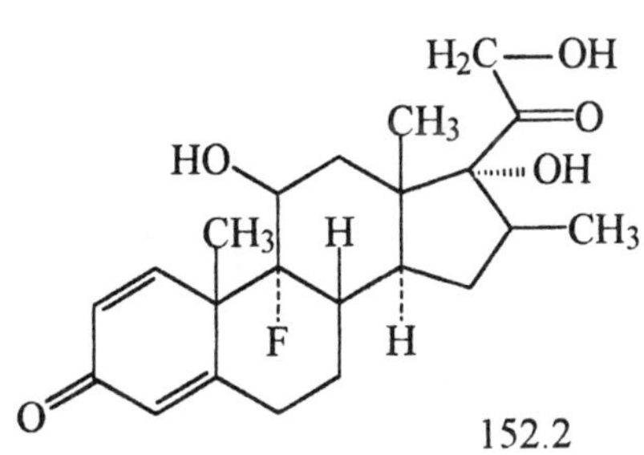

152.2

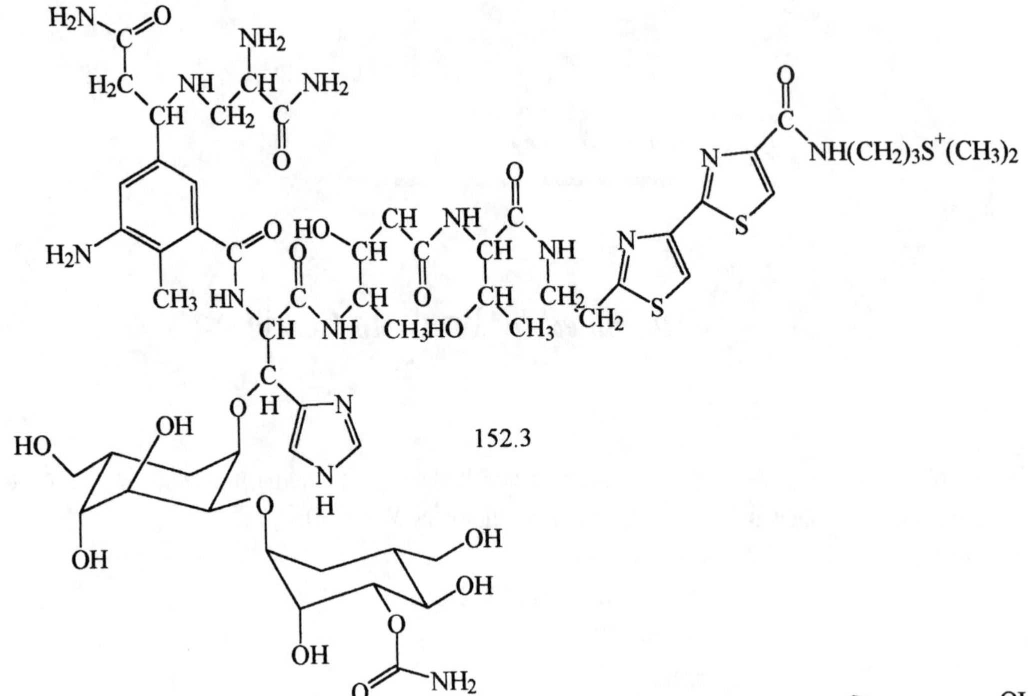

152.3

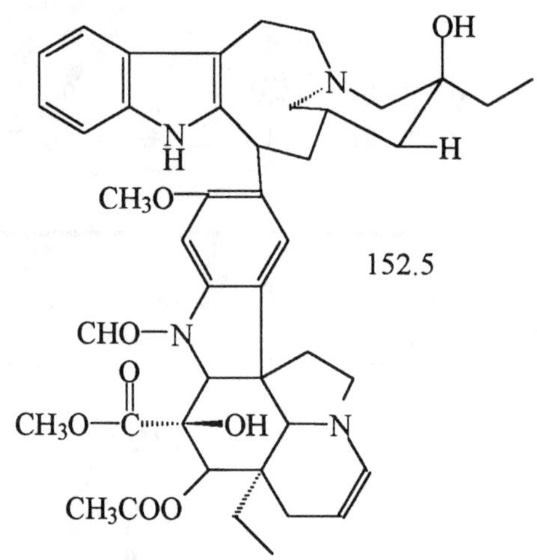

152.5

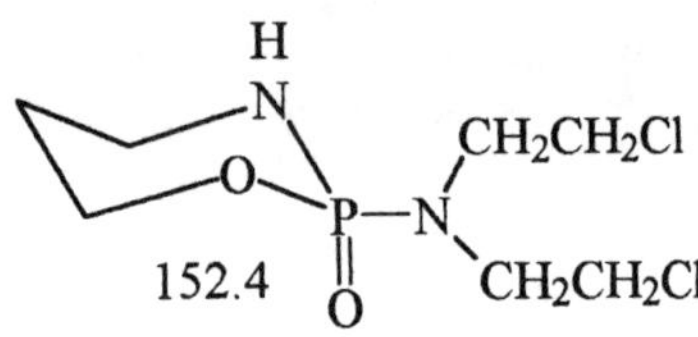

152.4

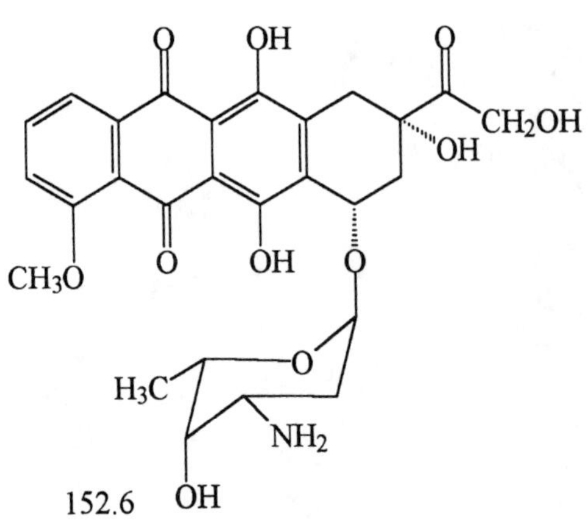

152.6

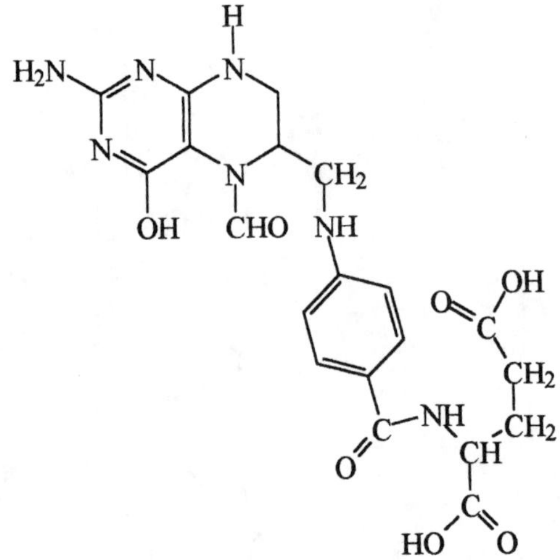

152.7

1. **What is the target enzyme for the action of methotrexate (152.1)?**

Answer Methotrexate (152.1) is a competitive, pseudo-irreversible inhibitor of dihydrofolate reductase.

2. **What effects are produced in cancer cells as a result of the action discussed in 1?**

Answer Inhibition of dihydrofolate reductase severely limits the biosynthesis of tetrahydrofolate, which is necessary for the formation of the various one carbon derivatives of tetrahydrofolate that are critical to the biosynthesis of the purine and pyrimidine nucleotides needed for the biosynthesis of DNA and RNA. The most critical is the need for N^5, N^{10}-methylenetetrahydrofolate, which is needed for the synthesis of thymidine monophosphate (TMP) by thymidylate synthetase.

3. **What is leucovorin (152.7) rescue? Why is it appropriate in this particular case?**

Answer Leucovorin (152.7) is a one carbon derivative of tetrahydrofolate, and it can be used directly by cells for nucleic acid biosynthesis. It bypasses the blockade of dihydrofolate reductase by methotrexate (152.1). It provides a rescue for the normal cells and allows for the use of higher doses of methotrexate (152.1) than could otherwise be used in order to effect maximum killing of neoplastic cells. Thus, it counteracts the toxic effects of methotrexate (152.1) on the normal, non-malignant cells.

Case 153

Submitted by Victoria F. Roche

You are a medicinal chemistry professor in your first year on the faculty of a college of pharmacy. Your research program in antimetabolite antineoplastic agents has generated significant interest among the undergraduates, and many wish to work with you in the laboratory. One of your current students approaches you with the idea of generating analogues of the popular pyrimidine antagonist 5-fluorouracil (153.1). He intends that the analogues will act by the same mechanism (inhibition of thymidylate synthase). Four of the analogues he suggested are shown (153.2–153.5).

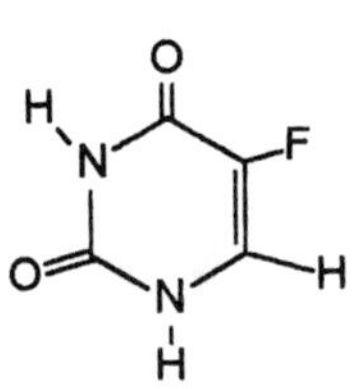

153.1

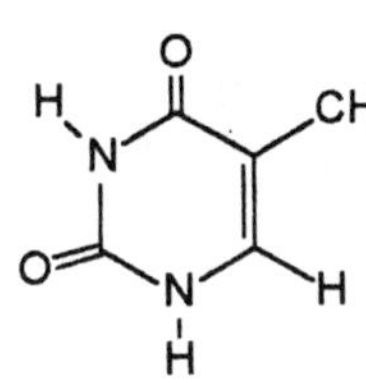

153.2

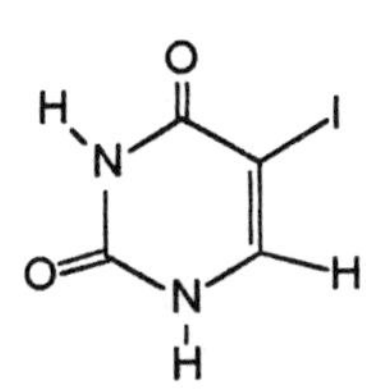

153.3

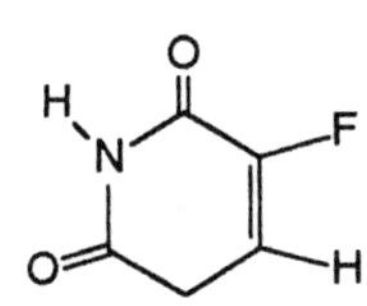

153.4

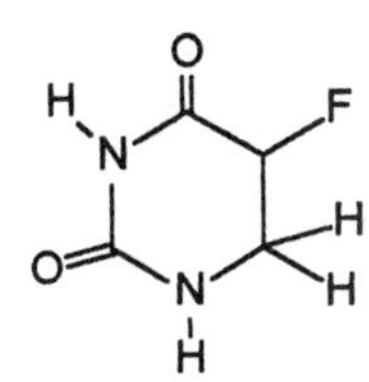

153.5

1. **What is the chemical mechanism by which 5-fluorouracil (153.1) inhibits thymidylate synthase?**

2. **How many of the student's proposed structures (153.2–153.5) would be expected to provide a synthase inhibiting activity at least as potent as 5-fluorouracil's? Defend your answer from a chemical standpoint.**

Case 154

Submitted by Robin M. Zavod,

Jeffrey J. Christoff, and Bruce L. Currie

KO, a 37-year-old premenopausal mother of 3, has recently been diagnosed with breast cancer. A biopsy revealed that the mass detected by routine mammography was malignant. Through further testing, it was also determined that several lymph nodes were involved. Her oncologist suggested surgery to remove the primary tumor and affected lymph nodes, followed by systemic chemotherapy. The chemotherapy prescribed included a combination of 5-fluorouracil (154.1) and cyclophosphamide (154.2). The patient was educated about the side effects of this type of drug therapy and given a series of appointments for the chemotherapy.

5-fluorouracil (154.1) 600 mg/m^2 IV Days 1 and 8 (cycle 28 days)

cyclophosphamide (154.2) 100 mg/m^2 po Days 1 to 14 (cycle 28 days)

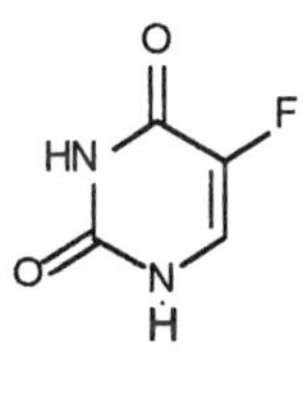

154.1

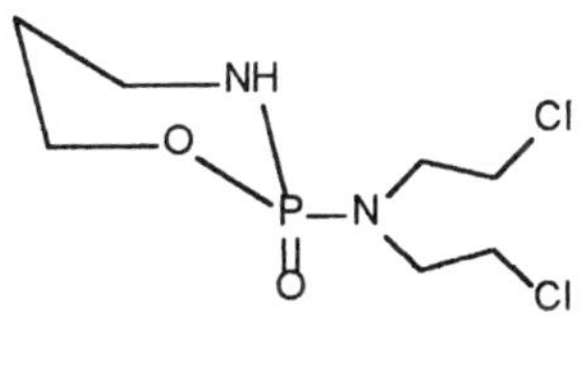

154.2

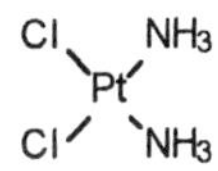

154.3

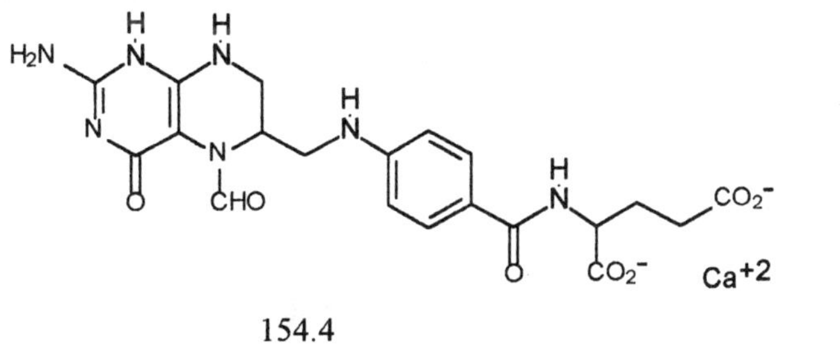

154.4

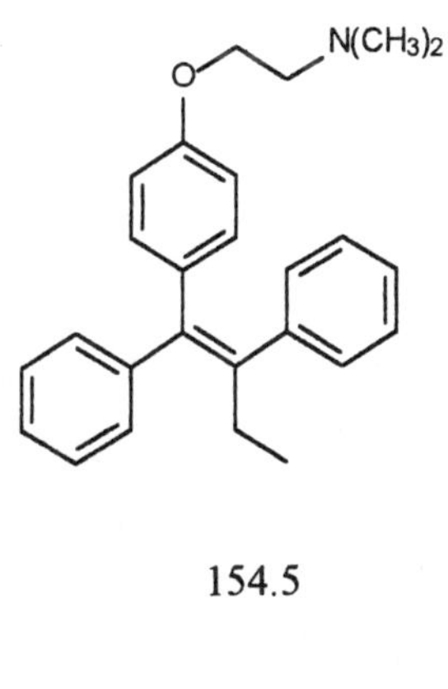

154.5

1. **What is the mechanism of action for each of these chemotherapeutic agents? Include in your answer the appropriate chemical steps relevant to the activation and mechanism of action of each drug.**

2. **In the primary literature it has been shown that cisplatin (154.3) increases intracellular folate concentrations. If cisplatin (154.3) was added to this combination, would this affect either of the drugs prescribed? If so, describe how each drug might be affected.**

3. **The oncologist decided to add leucovorin (154.4) (500 mg/m^2 IV over 2 hours) to the original 5-fluorouracil (154.2)/ cyclophosphamide (154.2) combination. What is the purpose of using 5-fluorouracil (154.1) and leucovorin (154.4) in combination? In your answer be sure to include the mechanism of action of leucovorin (154.4) with respect to this patient.**

4. **The results of the initial biopsy were also used to determine whether the cancer cells were estrogen receptor positive or estrogen receptor negative. Would hormone therapy be appropriate if the result came back ER+? What if the result was ER-? Explain the rationale for each of your answers and include relevant drug therapies.**

Case **155**

Submitted by Marc W. Harrold

TL was diagnosed with chronic myelocytic leukemia 2 months ago. He is 6 feet tall and weighs 176 lb. He is currently being treated with busulfan (155.1), 4 mg po daily, and allopurinol (155.2), 300 mg po daily. His busulfan (155.1) dose is monitored and adjusted based on weekly white blood cell counts. TL's physician wants to add a second antineoplastic agent to TL's regimen and has decided to use 6-mercaptopurine (155.3). The initial average daily oral dose of 6-mercaptopurine (155.3) is 2.5 mg/kg. TL enters your pharmacy with a prescription that reads, "purinethiol 50 mg tabs, 2 tabs bid."

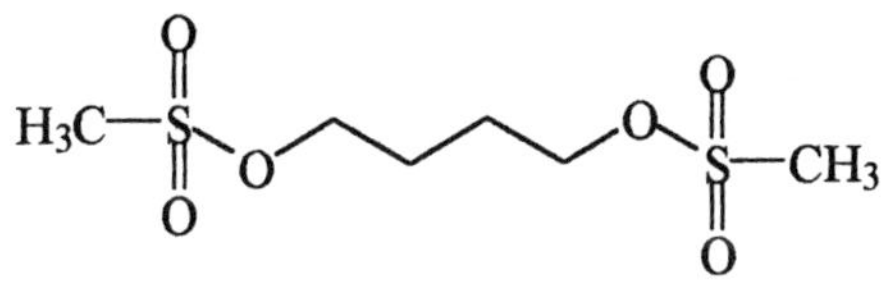

155.1

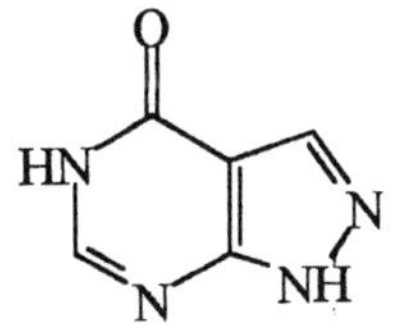

155.2

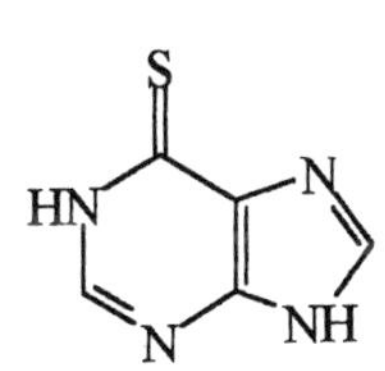

155.3

1. **When TL presents the prescription to you, he says, "I'm confused. I suppose I should have questioned my doctor about this, but he seemed to be very busy. As I was leaving, my doctor gave me this prescription and told me that he thought that remission of my leukemia was possible if I took this additional drug. He said that having two drugs working against the leukemia is better than just using one. I thought I was already taking two drugs for my leukemia, or is that not correct?" You tell him that he indeed is mistaken and that only one of the two original medications is for the leukemia. He then asks, "Which one is for the leukemia, and why am I taking the other drug?" How do you respond?**

2. **Should you dispense the prescription as written, or should you call TL's physician? If you call TL's physician, what would you tell him?**

Case 156

Submitted by Marc W. Harrold

WY has testicular cancer and is being treated with a combination therapy of ifosfamide (156.1) and mesna (156.2). The ifosfamide (156.1) is given as a single daily dose for 5 consecutive days. It is administered IV using a dosage of 1.2 gm/m^2. The mesna (156.2) is administered three times a day over the same 5-day period. It is administered at the time of ifosfamide (156.1) administration as well as 4 and 8 hours later. Each of the three mesna (156.2) doses are given as IV bolus injections at a dosage that is 20% of the ifosfamide (156.1) dosage. Thus, the total daily mesna (156.2) dose is 60% of the daily ifosfamide (156.1) dose.

It is 6:00 PM, and you are working the evening shift at the local hospital. George, who has been working the 9:30 to 6:00 shift in the IV room, is about to leave. Before he leaves, he tells you that he was able to prepare all the new IV orders from the 4:30 run, but that he wasn't able to prepare WY's 9:00 PM mesna (156.2) dose. He further informs you that there is already an open ampule of mesna (156.2) in the laminar flow hood, and that there should be enough mesna (156.2) left in the ampule for the 9:00 PM dose.

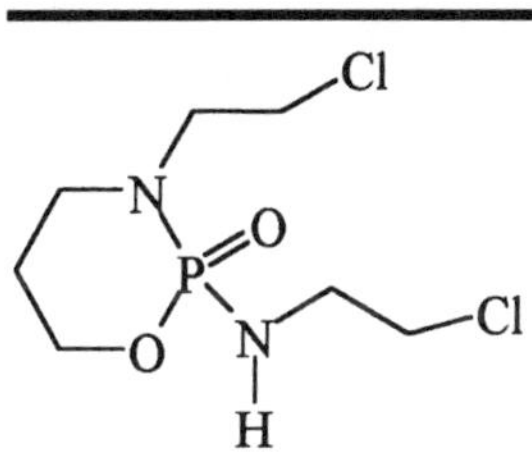

156.1

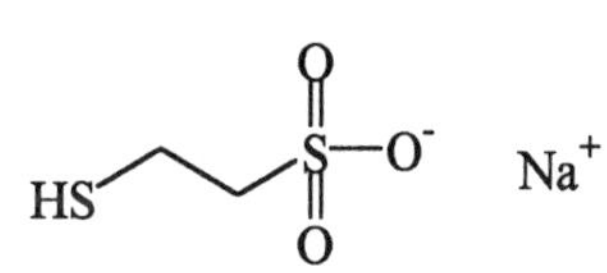

156.2

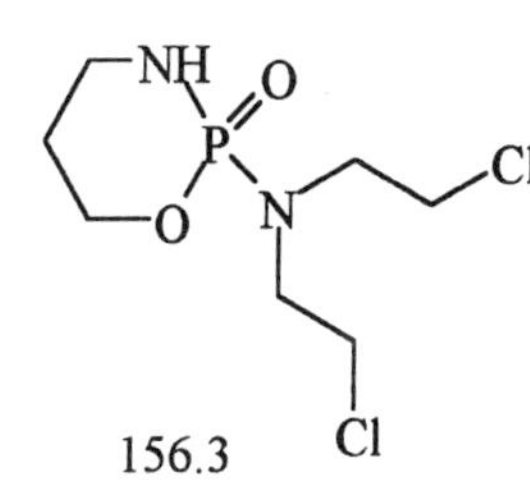

156.3

1. Why is mesna (156.2) used in combination with ifosfamide (156.1)?

2. **Cyclophosphamide (156.3) is a positional isomer of ifosfamide (156.1) and has an identical mechanism of action, yet it is not usually used in combination with mesna (156.2). Why?**

3. **At 8:00 PM you go into your pharmacy's IV room to prepare WY's 9:00 PM dose. You find unopened 2-mL, 4-mL, and 10-mL ampules of mesna (156.2), each containing a concentration of 100 mg/mL. In addition, there is an open 10-mL ampule in the IV hood containing approximately 5.5 mL. The only directions George left for you state that the mesna (156.2) must be diluted to a final concentration of 20 mg/mL using 5% D_5W. Using a m^2 value of 1.75 for WY, what dose of mesna (156.2) should WY receive and how should it be prepared?**

Case **157**

Submitted by Matthias C. Lu and George Udeani

A 45-year-old woman with ovarian cancer is treated every 21 days with the following chemotherapy regimen:

cyclophosphamide (157.1)	1000 mg/m^2 IV
cisplatin (157.2)	50 mg/m^2 IV

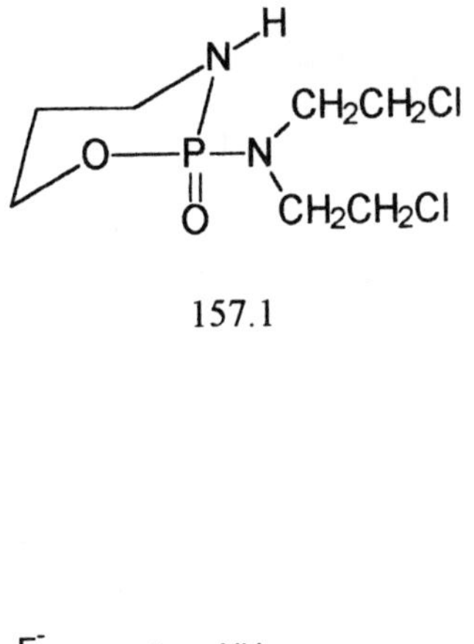

157.1

157.2

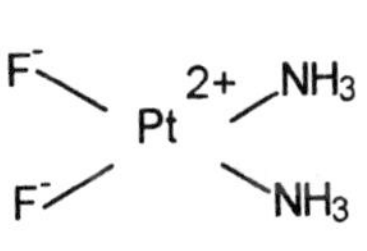

157.3

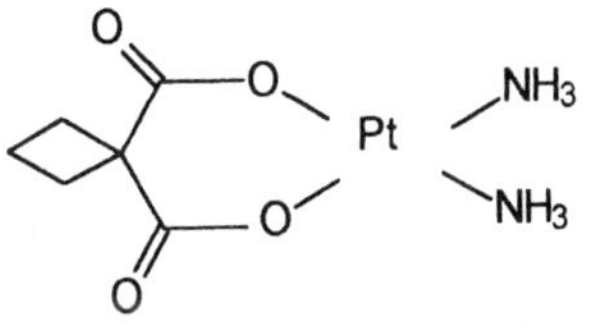

157.4

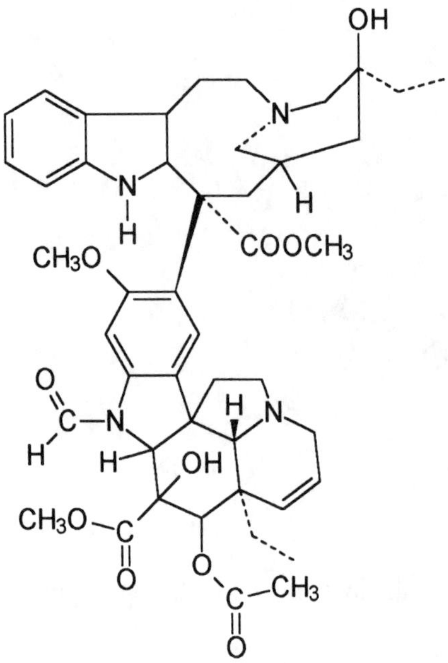

157.5

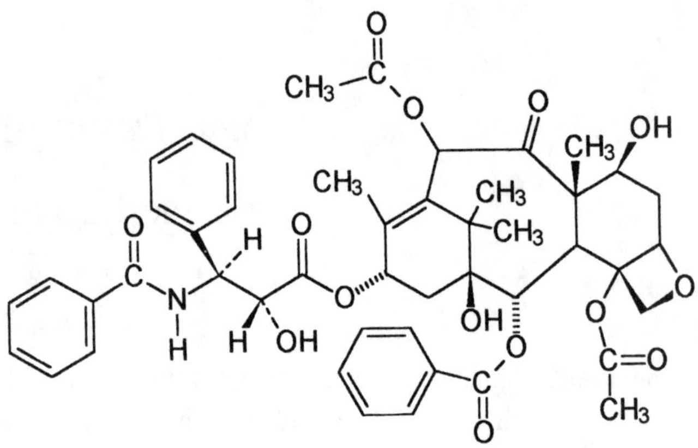

157.6

1. **What are the mechanisms of antitumor action for cyclophosphamide (157.1) and cisplatin (157.2)? Explain why transplatin is inactive as an antitumor agent even though it binds to DNA in vitro.**

2. **After the third course of chemotherapy, the patient develops renal complications. Her oncologist decides to use another related compound in place of cisplatin (157.2). Which of the drugs (157.3 or 157.4) should be used, and why?**

3. **The patient's tumor progresses despite completion of the previously described chemotherapy regimen. Her physician decides to initiate taxol (157.6) therapy, 135 mg/m² every 21 days. What is the mechanism of action of taxol (157.6)? How does it differ from using vincristine (157.5) as an anticancer agent?**

Case 158

Submitted by Victoria F. Roche

IR, a 65-year-old woman new to your community, has a history of mild congestive heart failure currently being controlled with quinapril. She has recently been diagnosed with breast cancer, and she is about to start a course of intravenous therapy at an outpatient clinic. One of the antitumor agents in her regimen is the anthracycline molecule shown as structure 158.1. IR is taking a holistic approach to her treatment, and she has purchased several OTC vitamin supplements, one of which contains ferrous sulfate (structure 158.2).

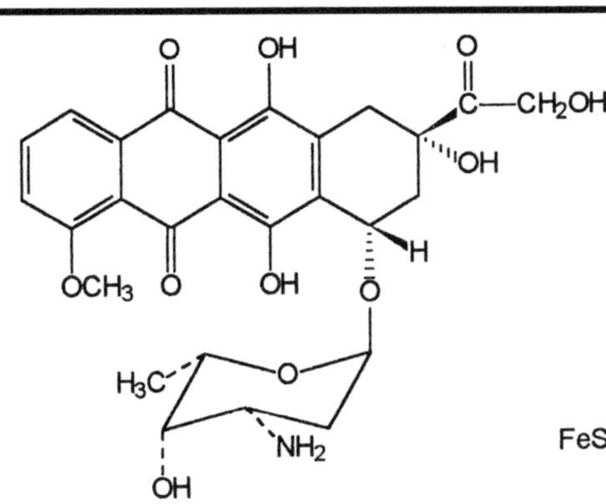

158.1

$FeSO_4$

158.2

1. What are the mechanisms by which anthracyclines, like structure 158.1, exert their antineoplastic activity?

2. What problems are inherent in this clinical situation?

3. Can you suggest an alternative anthracycline-based antitumor agent that would be less cardiotoxic?

Case 159

Submitted by S. William Zito and Victoria F. Roche

You are an archeologist in the year 2169. Your early training was in pharmacy, with a particular interest in medicinal chemistry. You are currently excavating a site in Omaha, Nebraska that is believed to have been the location of a college of pharmacy. The dig has progressed 1000 feet down, and you appear to be in what are the remains of the basement of a building when you discover a well-preserved laboratory research notebook. Paging through the book, you see the following:

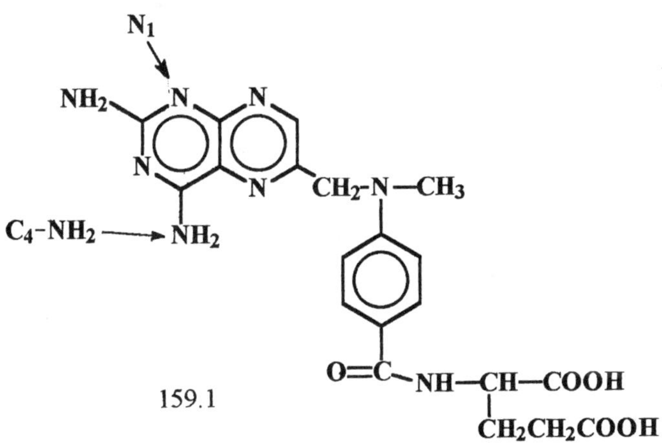

159.1

Spacers

$HOOCCH_2CH_2COOH$

159.2

$NH_2CH_2CH_2NH_2$

159.3

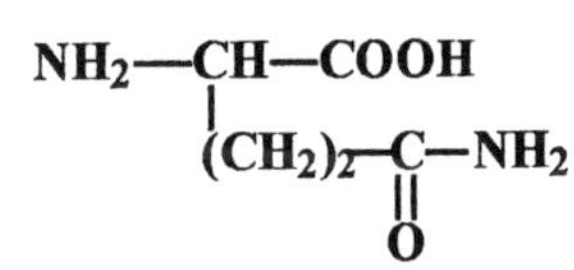

159.4

ANTIBODY

159.5

You quickly realize that this was the idea of a medicinal chemist to construct a site-directed immunoconjugate using methotrexate (159.1) and a variety of spacers (159.2–159.4).

1. What is the biochemical basis for the antineoplastic activity of methotrexate (159.1)?

2. **What groups on methotrexate (159.1) did this chemist intend to link to the spacers (159.2–159.4)? Why?**

3. **How will the *drug-spacer* be linked to the *antibody* (159.5)? Is using ester or amide links important? Why? Are all the spacers appropriate?**

4. **How many immunoconjugates was this immunochemist able to make?**

Case 160

Submitted by S. William Zito

You are a pharmacist in a hospital setting, and you will be participating in the administration of some new experimental granulocyte colony-stimulating factor (G-CSF) derivatives (160.1–160.3). The derivatives have all been made to the amino terminus of the 175 amino acid protein. G-CSF is indicated to decrease the incidence of infection in patients with non-myeloid malignancies receiving myelosuppressive anticancer drugs. The physician requests the derivative with the longest in vivo stability.

H-Leu-Gly-Gly-[G-CSF]-OH	**H-Tyr-Gly-Ala-Phe-[G-CSF]-OH**	**H-Met-Ala-Tyr-[G-CSF]-OH**
160.1	160.2	160.3

1. **Using your knowledge of the factors that contribute to the in vivo stability of protein drugs, which of the three derivatives (160.1–160.3) would you recommend? Your answer should include a discussion of the biochemical basis for the in vivo stability of peptides as well as the importance of amino acid sequence and type.**

Case 161

Submitted by Robert A. Wiley

DS is a 45-year-old woman admitted to the hospital 11/16 with a diagnosis of metastatic recurrence of breast adenocarcinoma (ACa).

History of Present Illness—Breast ACa diagnosed 4 years earlier; radical mastectomy performed. Multiple chest wall recurrences of ACa during the 4 years, treated with radiotherapy, tamoxifen (161.4), and fluoxymesterone (161.3). At the present time, there is a recurrence of ACa in the left neck.

Present Examination—On admission, BP was 140/80, pulse 100.

Plan—Bilateral adrenalectomy. Performed 11/26. Drugs administered as on accompanying chart. On 12/4, there was a sudden fall in blood pressure to 80/60 (pulse 130), and the patient was very lethargic.

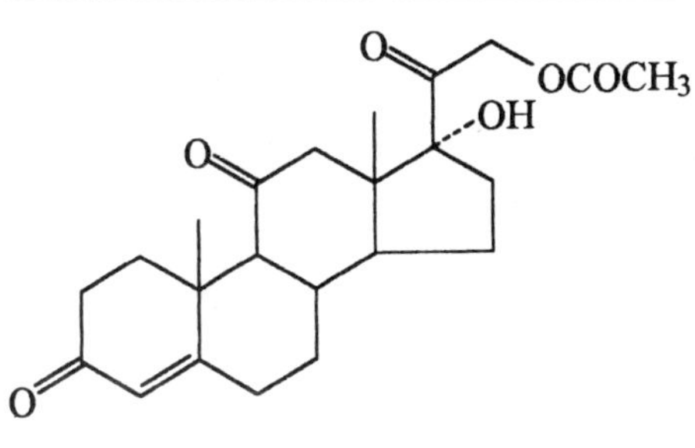

161.1

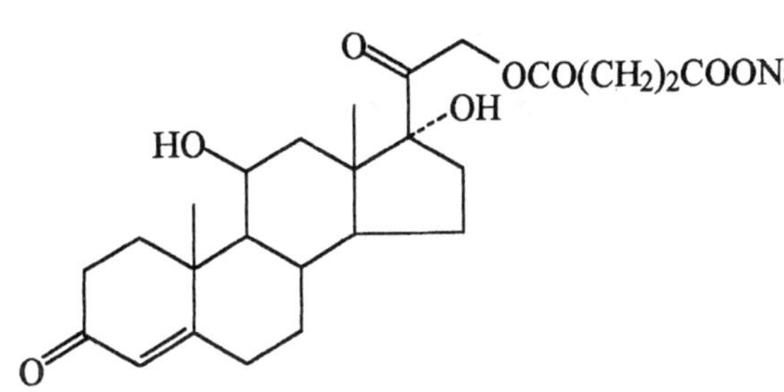

161.2

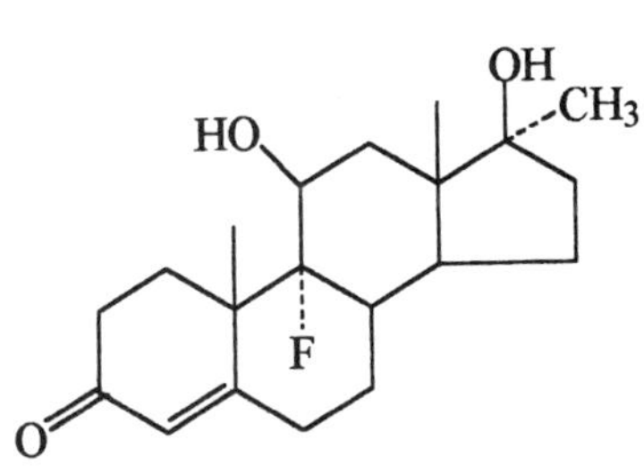

161.3

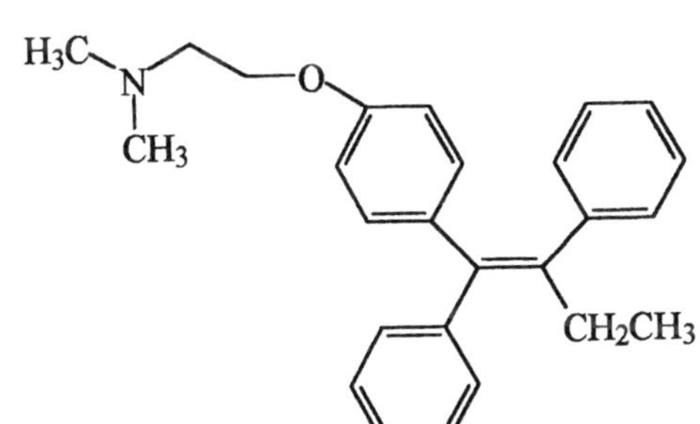

161.4

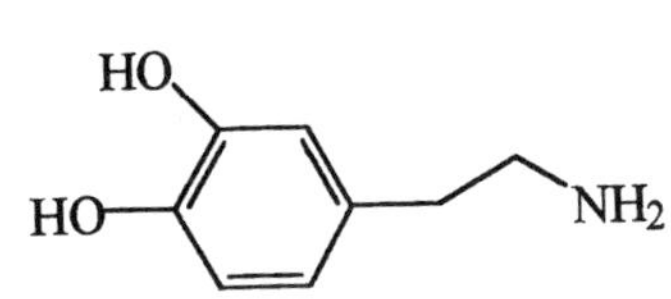

161.5

Case 161: Patient Chart

Date (*=surgery)	11/25	26 *	27	28	29	30	12/1	2	3	4	5	6	7	8	9	10	11	12	13	14
Mg citrate 8 oz	x																			
Dalmane 30 mg hs		x																		
aminophylline 200 mg iv			x																	
Compazine 10 mg po	x		x																	
M.S. 5-10 mg q 4-6 h			x	x	x	x														
heparin 5000 sc q 12 h								x	x	x	x	x	x							
cimetidine 300 mg q 6 h iv				x	x	x	x	x	x	x	x	x	x							
insulin NPH 15-20 u /d	x	x	x	x	x	x	x	x	x	x	x	x	x	x	x	x	x	x	x	
cefoxiten 1 Gm q 6 h									x	x	x	x	x	x	x	x	x			
dopamine drip										x	x									
Solu-cortef 100 mg q 8 h			x																	
Solu-cortef 100 mg q 12 h				x	x															
Solu-cortef 50 mg q 12 h						x	x													
cortisone ac im 75/25								x	x	x										
Solu-cortef 50 mg ivp										x										
Solu-cortef 100 mg q 12 h										x	x	x	50	25						
cortisone ac po 50/25 mg												x	x	x						
cortisone ac po 25/12.5 mg															x	x	x			
pulse	100									130	124									
blood pressure	140/80									80/60	108/94									

1. **What are the roles of tamoxifen (161.4) and fluoxymesterone (161.3) in the treatment of ACa of the breast and recurrences thereof?**

2. **What is the rationale behind use of bilateral adrenalectomy to treat metastatic ACa of the breast?**

3. **What happened to cause the patient to go into shock on 12/4? What treatment would you advocate to overcome this problem?**

4. **The patient was saved by use of the drug therapy shown on the chart for the few days following the hypotensive crisis. Why was this treatment effective? What is the function of dopamine (161.5)?**

Appendix A. Drug List

Appendix B. Drug Interactions

acebutolol/isoproterenol	pharmacologic	108
acetaminophen/n-acetylcysteine	chemical	37
acetohydroxamic acid/ferrous sulfate	chemical	140
acetylsalicyclic acid/warfarin potassium	pharmacokinetic	35
aluminum aspirin/ammonium chloride	chemical	2
aluminum hydroxide antacid/tetracycline	chemical	134
aluminum hydroxide-magnesium hydroxide/tetracycline	chemical	135
amobarbital sodium/ammonium chloride	chemical	5
astemizole/erythromycin	pharmacokinetic	30
atenolol/imipramine	pharmacologic	63
atenolol/isocarboxazid	pharmacologic	63
atenolol/viloxazine	pharmacologic	63
bismuth salicylates/tetracycline	chemical	227
bismuth subsalicylate/doxycycline	chemical	138
bupivacaine hydrochloride/cetylpyridinium bromide	chemical	1
calcium/ciprofloxacin	chemical	139
captopril/hydrochlorothiazide-triamterene	pharmacologic	120
cephalosporin/warfarin	pharmacologic	146
chloral hydrate/ethanol	pharmacologic	56
chlorprothixene hydrochloride/sodium bicarbonate	chemical	7
cholestyramine/warfarin	pharmacokinetic	105
cimetidine/diazepam	pharmacokinetic	24
cimetidine/diazepam	pharmacokinetic	49
ciprofloxacin/theophylline	pharmacokinetic	141
co-trimoxazole/procaine	chemical	23
co-trimoxazole/tetracaine	chemical	23
deprenyl/carbidopa	pharmacokinetic	70
diphenhydramine/amitriptyline	pharmacologic	64
diphenhydramine/desipramine	pharmacologic	64
diphenhydramine/fluoxetine	pharmacologic	64
diphenhydramine/phenelzine	pharmacologic	64
doxepin hydrochloride/narcolepsy	pharmacologic	75
doxorubicin/ferrous sulfate	chemical	158
droxidopa/deprenyl	pharmacokinetic	70
enalapril/triamterene-hydrochlorothiazide	pharmacologic	94